Physical Processes in Lakes

Physical Processes in Lakes

Editor

Lars Bengtsson

MDPI • Basel • Beijing • Wuhan • Barcelona • Belgrade • Manchester • Tokyo • Cluj • Tianjin

Editor
Lars Bengtsson
Water Resources Engineering
Lund University
Lund
Sweden

Editorial Office
MDPI
St. Alban-Anlage 66
4052 Basel, Switzerland

This is a reprint of articles from the Special Issue published online in the open access journal *Water* (ISSN 2073-4441) (available at: www.mdpi.com/journal/water/special_issues/physical_processes_lakes).

For citation purposes, cite each article independently as indicated on the article page online and as indicated below:

LastName, A.A.; LastName, B.B.; LastName, C.C. Article Title. *Journal Name* **Year**, *Volume Number*, Page Range.

ISBN 978-3-0365-3698-9 (Hbk)
ISBN 978-3-0365-3697-2 (PDF)

Cover image courtesy of Lars Bengtsson

© 2022 by the authors. Articles in this book are Open Access and distributed under the Creative Commons Attribution (CC BY) license, which allows users to download, copy and build upon published articles, as long as the author and publisher are properly credited, which ensures maximum dissemination and a wider impact of our publications.
The book as a whole is distributed by MDPI under the terms and conditions of the Creative Commons license CC BY-NC-ND.

Contents

About the Editor vii

Preface to "Physical Processes in Lakes" ix

Sofya Guseva, Peter Casper, Torsten Sachs, Uwe Spank and Andreas Lorke
Energy Flux Paths in Lakes and Reservoirs
Reprinted from: *Water* **2021**, *13*, 3270, doi:10.3390/w13223270 1

Andrew Folkard
The Multi-Scale Layering-Structure of Thermal Microscale Profiles
Reprinted from: *Water* **2021**, *13*, 3042, doi:10.3390/w13213042 23

David Birt, Danielle Wain, Emily Slavin, Jun Zang, Robert Luckwell and Lee D. Bryant
Stratification in a Reservoir Mixed by Bubble Plumes under Future Climate Scenarios
Reprinted from: *Water* **2021**, *13*, 2467, doi:10.3390/w13182467 45

Galina Zdorovennova, Nikolay Palshin, Sergey Golosov, Tatiana Efremova, Boris Belashev and Sergey Bogdanov et al.
Dissolved Oxygen in a Shallow Ice-Covered Lake in Winter: Effect of Changes in Light, Thermal and Ice Regimes
Reprinted from: *Water* **2021**, *13*, 2435, doi:10.3390/w13172435 73

Sergey Bogdanov, Roman Zdorovennov, Nikolay Palshin and Galina Zdorovennova
Deriving Six Components of Reynolds Stress Tensor from Single-ADCP Data
Reprinted from: *Water* **2021**, *13*, 2389, doi:10.3390/w13172389 93

Bushra Tasnim, Xing Fang, Joel S. Hayworth and Di Tian
Simulating Nutrients and Phytoplankton Dynamics in Lakes: Model Development and Applications
Reprinted from: *Water* **2021**, *13*, 2088, doi:10.3390/w13152088 105

Bushra Tasnim, Jalil A. Jamily, Xing Fang, Yangen Zhou and Joel S. Hayworth
Simulating Diurnal Variations of Water Temperature and Dissolved Oxygen in Shallow Minnesota Lakes
Reprinted from: *Water* **2021**, *13*, 1980, doi:10.3390/w13141980 139

Bertram Boehrer, Sylvia Jordan, Peifang Leng, Carolin Waldemer, Cornelis Schwenk and Michael Hupfer et al.
Gas Pressure Dynamics in Small and Mid-Size Lakes
Reprinted from: *Water* **2021**, *13*, 1824, doi:10.3390/w13131824 165

Xiamei Man, Chengwang Lei, Cayelan C. Carey and John C. Little
Relative Performance of 1-D Versus 3-D Hydrodynamic, Water-Quality Models for Predicting Water Temperature and Oxygen in a Shallow, Eutrophic, Managed Reservoir
Reprinted from: *Water* **2021**, *13*, 88, doi:10.3390/w13010088 185

About the Editor

Lars Bengtsson

The editor, Lars Bengtsson, is senior professor at the Department of Water Resources Engineering. He has been full professor, head and research responsible at the Department of Water Research at LuleåTechnical University, at the Department of Hydrology, Uppsala University, and at Water Resources Engineering, Lund University. He has been guest professor at the University of Florida, at McMaster University, Canada, and at Vituki Research Institute, Budapest, Hungary. Professor Bengtsson has supervised more than 40 students to their doctoral degree. He has published about 200 international contributions most of them being single or first author. The contributions cover many fields within hydrology and environmental engineering. His special interest is cold climate hydrology, physical processes in lakes and green urban hydrology.

Preface to "Physical Processes in Lakes"

There are nine million lakes larger than two ha in the world, and many more small ponds. Humans use the lakes for many purposes such as water supply and as receiving waters, for fishing, and for recreation. Lakes are beautiful parts of the landscape, but may be most important to constitute the environment for many habitats.

The numerous climatological studies of today are related to large-scale processes. In environmental studies, the physical processes of lakes are seldom considered in detail, treating the lake only as a point. However, biological and chemical processes in lakes relate to physical processes. They are keys for the status of a lake. Physical processes in lakes include mixing and circulation within the lake, thermal processes, light penetration, sedimentation, ice processes, oxygen and gas dynamics, exchange processes between water-atmosphere and sediment. The physical processes are different in tropical, temperate and arctic climate and during different parts of the year.

When lakes are used for water supply or as receiving waters, these technical systems must be adapted to lake physics considering how the water is mixed and the thermal structure within the lake.

Oxygen is important for the life in lakes. Several contributions of this Special Issue relate to dissolved gases. Dissolved gases may form bubbles leading to exchange of climate-relevant gases between water and the atmosphere. A highly theoretical approach is compared with observations. During ice covered conditions, the dissolved oxygen is consumed mainly close to the sediments, which is discussed in one paper. Usually processes within a lake are discussed on a daily or longer time scale. However, wind and solar radiation varies through the day. The importance of using shorter time step, when simulating dissolved oxygen and temperature in the vertical is shown in another contribution. One paper shows how nutrients and phytoplankton dynamics relate to dissolved oxygen and temperature. Again, in a study of oxygen conditions, it is shown how bubble plumes aerate the water and how the water temperature profile may change slightly in a warmer climate. Depending on the purpose of modelling, the used models may be more or less advanced. In a contribution, a comparison between 1D and 3D models is performed for a shallow lake.

Energy from the wind produces currents, waves and turbulent mixing. There are three contributions relating to this, all based on field measurements. In the first paper, the momentum flux from atmosphere to water is measured. Energy contributions to waves dominate. Energy dissipation is shown to vary over the day. Additionally, in a study in an ice-covered lake, the dissipation rate was measured and the Reynold stresses determined. In the third contribution relating to energy, thermal microstructure profiling was analyzed in a novel way showing how turbulent diffusivity could be determined.

I hope that this issue to some extent may contribute to increased understanding of physical processes and encourage studies on lake physics.

Lars Bengtsson
Editor

Article

Energy Flux Paths in Lakes and Reservoirs

Sofya Guseva [1,*], **Peter Casper** [2], **Torsten Sachs** [3], **Uwe Spank** [4] **and Andreas Lorke** [1]

[1] Institute for Environmental Sciences, University of Koblenz-Landau, 76829 Landau, Germany; lorke@uni-landau.de
[2] Department of Experimental Limnology, Leibniz-Institute of Freshwater Ecology and Inland Fisheries, 12587 Berlin, Germany; pc@igb-berlin.de
[3] GFZ German Research Centre for Geosciences, 14473 Potsdam, Germany; torsten.sachs@gfz-potsdam.de
[4] Institute of Hydrology and Meteorology, Chair of Meteorology, Technische Universität Dresden, 01069 Dresden, Germany; Uwe.Spank@tu-dresden.de
* Correspondence: guseva@uni-landau.de

Abstract: Mechanical energy in lakes is present in various types of water motion, including turbulent flows, surface and internal waves. The major source of kinetic energy is wind forcing at the water surface. Although a small portion of the vertical wind energy flux in the atmosphere is transferred to water, it is crucial for physical, biogeochemical and ecological processes in lentic ecosystems. To examine energy fluxes and energy content in surface and internal waves, we analyze extensive datasets of air- and water-side measurements collected at two small water bodies (<10 km^2). For the first time we use directly measured atmospheric momentum fluxes. The estimated energy fluxes and content agree well with results reported for larger lakes, suggesting that the energetics governing water motions in enclosed basins is similar, independent of basin size. The largest fraction of wind energy flux is transferred to surface waves and increases strongly nonlinearly for wind speeds exceeding 3 m s^{-1}. The energy content is largest in basin-scale and high-frequency internal waves but shows seasonal variability and varies among aquatic systems. At one of the study sites, energy dissipation rates varied diurnally, suggesting biogenic turbulence, which appears to be a widespread phenomenon in lakes and reservoirs.

Keywords: energy fluxes; energy content; lakes; reservoirs; internal waves; surface waves; biogenic turbulence

Citation: Guseva, S.; Casper, P.; Sachs, T.; Spank, U.; Lorke, A. Energy Flux Paths in Lakes and Reservoirs. *Water* **2021**, *13*, 3270. https://doi.org/10.3390/w13223270

Academic Editor: Lars Bengtsson

Received: 6 August 2021
Accepted: 10 November 2021
Published: 18 November 2021

Publisher's Note: MDPI stays neutral with regard to jurisdictional claims in published maps and institutional affiliations.

Copyright: © 2021 by the authors. Licensee MDPI, Basel, Switzerland. This article is an open access article distributed under the terms and conditions of the Creative Commons Attribution (CC BY) license (https://creativecommons.org/licenses/by/4.0/).

1. Introduction

The spatial distribution and temporal dynamics of mechanical energy play vital roles in the physical, biogeochemical and ecological functioning of lentic ecosystems. At the water surface, wind-generated turbulence regulates the vertical distribution of heat that is exchanged with the atmosphere, affects thermal stratification [1] and controls gas exchange with the atmosphere [2], which can be enhanced by surface waves [3]. Vertical turbulent mixing in the surface layer controls the exposure of planktonic organisms to light, therewith regulating primary production and community composition of phytoplankton [4–6]. Wind-induced upwelling [7,8], as well as internal waves [9], affect phytoplankton and water quality by transporting nutrients from the stratified hypolimnion to the surface layer. The state of mixing of water bodies can result in the persistence of harmful cyanobacteria in the thermocline throughout summer stratification [10]. At the bottom of lakes and reservoirs, boundary layer turbulence controls the oxygen flux into the sediments [11] and therewith the rate of carbon burial and methane production [12], as well as the internal loading of the lake with nutrients [13].

The major source of mechanical energy in lentic systems is wind, which exerts a shear force at the water surface. However, only a small fraction of the vertical wind energy flux in the atmospheric boundary layer is transferred to water motions (~1.9% according to [14] and ~22% including surface waves in [15]). Water motions are distributed over a wide range of temporal and spatial scales ranging from low to high frequencies of the energy spectrum [15]. Recently, energy transfer from wind to water has been found to be more efficient when the lake is thermally stratified, resulting in an enhancement of mean kinetic energy throughout the water column during seasonal stratification [16,17]. The majority of the wind energy flux is dissipated in the atmospheric and water surface boundary layers. A considerable part of the non-dissipated wind energy (~20%) is fed to the surface wave field [15]. In [18], however, this percentage is lower—1.5–3.5%, although both estimates were derived from the same set of observations conducted in a Swiss lake. Approximately 1% of the energy input is stored in large-scale currents such as basin-scale internal waves [15]. Shear instabilities lead to the degeneration of large-scale internal waves into propagating high-frequency internal waves. They appear at frequencies being some fraction of the maximum buoyancy frequency, which is related to the strength of vertical density stratification [19–21]. It has been estimated that about 90% of the energy in the internal waves is dissipated within the bottom boundary layer [14,15].

Current knowledge about the partitioning and distribution of energy fluxes in lakes and reservoirs described above is mainly based on observations from the same lake [16,17], or information compiled from asynchronously conducted measurements in different systems [15]. The generality of current figures, their transferability to water bodies of different size and depth, and their temporal dynamics remain largely unexplored. Moreover, all existing estimates are based on bulk parameterization of wind energy fluxes, as observations lack direct measurements of atmospheric fluxes [14–18]. The role of surface waves in the energy budget appears to be constraint by observations, which are restricted to a single study in a large lake [15,18].

To address these gaps in research, we analyze the most relevant components of energy fluxes and energy content in various types of water motions in response to wind energy fluxes in two small (<10 km^2) water bodies. The study sites, a lake and a reservoir, differ in surface area by one order of magnitude, but have comparable water depth (~10 m). Given the difference in surface area and the fact that the reservoir experiences water level variations, we expect the hydrodynamic processes in these two water bodies to be different. The selected lake is considered representative of a large number of small lakes, belonging to the most abundant size class which contributes 54% of the global lake surface area [22,23]. To overcome shortcomings of previous studies, we used direct measurements of momentum fluxes in the atmosphere above the water surface for the estimation of the wind energy flux into the lakes. We investigate the wind speed and fetch dependence on the surface wave characteristics based on measurements covering nearly a complete annual cycle. The predominant modes of basin-scale internal waves and the presence of high-frequency internal waves are identified and examined. The data are used to complement and to re-examine mean energy budgets of small lentic systems, their temporal dynamics, and their variation with water body size.

2. Materials and Methods

2.1. Study Sites and Measurements

Measurements were conducted at a small reservoir (Bautzen Reservoir, surface area: 5.33 km^2, volume: 39.2×10^6 m^3, maximum depth: 12.2 m) and a small lake (Lake Dagow, surface area: 0.3 km^2, volume: 1.2×10^6 m^3, maximum depth: 9.5 m)—both situated in Germany. Bautzen Reservoir is a part of the dammed river Spree in southeastern Germany, with a mean water residence time of 164 days [24]. It can be classified as a small storage-type reservoir [25,26] with additional purposes of flood control and leisure activities. Besides, the reservoir is used to regulate the water supply for wetlands and power stations located downstream of the river. The outlet tower located near the dam regulates water discharge through the bottom of the reservoir. Major water withdrawal in summer is associated with a gradual decrease of water level [27]. The reservoir is often not persistently stratified throughout the summer due to a lack of shelter against strong winds and experiences several full mixing events [24,28]. Lake Dagow is a glacial lake in the Lake Stechlin area in northern Germany. It is a small eutrophic lake with a water residence time of 5 years [29]. The lake develops persistent density stratification every year leading to anoxic conditions in its hypolimnion [29]. Lake Dagow was impacted by wastewater, duck and carp farming in the 1960–1970s, with restoration activities in the 1980s.

At both water bodies, a similar set of instruments was installed for a period resolving the seasonal dynamics of stratification and mixing in Bautzen Reservoir (3 April until 3 December in 2018) and the transition from stratified to mixed conditions during the autumn overturn in Lake Dagow (11 September until 25 November in 2017). Meteorological measurements, including radiation fluxes and eddy covariance (EC) measurements of vertical momentum fluxes, were conducted from floating platforms (Points A and E, Figure 1). The platform was 3 × 3 m in size in Bautzen Reservoir and about 2.5 × 5 m in size in (Lake Dagow). Both platforms were attached with steel chains (Bautzen Reservoir) or guy wires (Lake Dagow) to four concrete anchors at the bottom. Vertical profiles of flow velocity in the water column, including turbulent velocity fluctuations, were measured by acoustic Doppler current profilers (ADCP), which were mounted downward-facing at the platforms (see Table 1). In Bautzen reservoir, the profiling range of the platform-mounted ADCP did not cover the entire water column and an additional ADCP was deployed at the bottom at ~10 m distance from the platform. The bottom-mounted instrument did not resolve turbulent velocity fluctuations and provided mean flow velocity profiles only. In Lake Dagow, there were three sequential ADCP deployments with time gaps in between. The ADCP was installed at the bottom during the second and third deployments. Thermistor chains were deployed at the platform in Bautzen Reservoir and in the middle of Lake Dagow to observe vertical temperature stratification in water. Wave recorders (high-frequency pressure loggers) were placed near the shore at both locations and an additional wave recorder was installed at the platform in Bautzen Reservoir. The measurement campaign lasted from 3 April until 3 December in 2018 in Bautzen Reservoir and from 11 September until 25 November in 2017 in Lake Dagow.

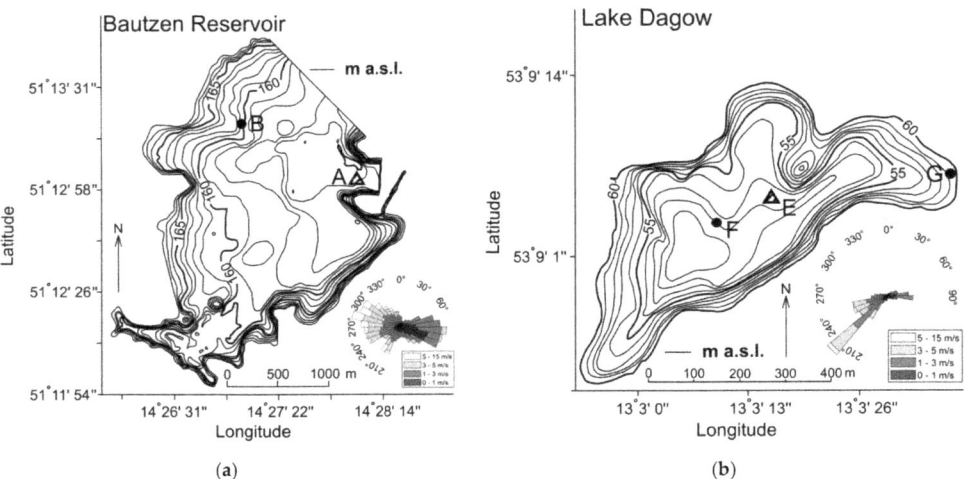

Figure 1. Bathymetric maps of the study sites: (**a**) Bautzen Reservoir; (**b**) Lake Dagow. The plots were created based on the published maps in [24,29]. Black lines show isolines with equal elevation in meters above sea level (m a.s.l.). Small panels on the right show wind roses with wind directions and windspeed. The locations of the instruments are indicated by triangles and circles labeled with capital letters. Points A and E (triangles) mark the locations of the platforms for micrometeorological measurements and the acoustic Doppler current profiler (ADCP). The ADCPs were deployed at the bottom, at a distance of approximately 10 m from the platforms. Points A and F mark the locations of the thermistor chains. A, B and G mark the location of the surface wave observations (pressure sensors).

Table 1. Instrumentation and resolution for the water-side and atmospheric measurements conducted in Bautzen Reservoir and Lake Dagow.

Type of Measurements	Water Body	Instrument	Resolution	Location on the Map (Figure 1)
Flow velocity	Bautzen Reservoir	(a) ADCP RDI Workhorse 600 kHz (range: 1.4–10 m); (b) Workhorse 1200 kHz (range: 0.8–4.7 m)	(a) 10 min with 200 pings with 0.25 m bin size; (b) 1 s with 0.1 m bin size	(a) Bottom deployment (facing upward) ~10 m from southern corner of the platform; (b) platform deployment (facing down, southwest corner)
	Lake Dagow	ADCP RDI Workhorse 600 kHz (3 deployments, range: (1) 0.5–6.8 m; (2–3) 0.8–7.1 m)	5 s with 12 pings with 0.1 m bin size	3 deployments: (1) platform (facing down, west corner), point E; (2–3) ~6–7 m from northern corner of the platform (facing upward)
Water temperature	Bautzen Reservoir	(a) Thermocouples (type T, Copper/Constantan)	10 min averages from measurements in 30 s intervals	Platform, point A
	Lake Dagow	RBR solo	10 s	Point F
Wave measurements	Bautzen Reservoir Lake Dagow	RBR duet	10 min with 512 measurements of 16 Hz	Platform, point A; shore, point B Shore, point G
Wind speed	Bautzen Reservoir	Campbell Scientific, CSAT3 (1.8 m)	20 Hz, as well as 10 min and 30 min averages	Platform, point A
	Lake Dagow	Gill Instruments HS-50 (1.97 m)	20 Hz	Platform, point E
Radiation	Bautzen Reservoir	Kipp and Zonen, CNR1	10 and 30 min averages from measurements in 30 s intervals	Platform, point A
	Lake Dagow	Kipp and Zonen, CNR4	measured at 1 Hz, logged at 1 min averages	Platform, point E

Detailed information about instrumentation and resolution is provided in Table 1. As described in the following sections, the collected data were analyzed to characterize energy fluxes from wind to water and the energy content in different types of water motion. Meteorological data were screened based on plausibility limits, logged information about maintenance work at the measurement station, and by detection of errors and outliers. ADCP velocity data were filtered using a threshold for signal correlation (>70 (-)) and were despiked following [30,31]. All final parameters were estimated with a temporal resolution of 30 min.

2.2. Energy Content

2.2.1. Internal Waves

Standing, basin-scale internal waves (internal seiches) are characterized by oscillations of the vertical density structure that appears due to wind forcing acting on the stratified lake. Waves are typically present in the form of one or several energetic modes depending on the density layer structure [32]. In our study, we identified the major modes of internal waves using the "Internal Wave Analyzer" software (IWA) [33] and selected wave "events" when visual evidence of their presence was observed. Visual evidence appeared in vertical displacements of isotherms and as a pronounced peak in the power spectral density estimated for variations in isothermal depths and for the selected velocity components. Similarly, events were identified for propagating (high frequency) internal waves, which were mainly present in the spectra of the vertical velocity component and displacement of isotherms during the stratified season. The high-frequency limit of the wave band in spectra is limited by the maximum buoyancy frequency N_{max} (Hz) [19,34]:

$$N_{\max} = \max\left[\left(-\frac{g}{\rho_{w0}}\frac{\partial \rho_w}{\partial z}\right)^{\frac{1}{2}}\right] \quad (1)$$

where g (m s^{-2}) is gravitational acceleration, ρ_w (kg m^{-3}) is water density, ρ_{w0} (kg m^{-3}) is the mean water density, z (m) is the height above the bottom with positive direction upwards. The conversion from temperature to density was done based on the freshwater equation of state following [35].

The energy content in a linear internal wave field is equally partitioned into potential energy and kinetic energy [36]. An appropriate approach for estimation of the energy is calculation of the locally available potential energy (APE (J m^{-3})) from temporal variations of water density observed at a single mooring location [37,38]:

$$APE = \int_{z-\zeta}^{z} g[\rho_w(z) -]\rho_{w0}(z\prime)]dz\prime \quad (2)$$

where ζ (m) is the vertical displacement of a fluid particle and z' is the integration variable. We estimated potential energy at 30 min resolution (or 1 min for high-frequency internal waves) by using different time intervals for estimating the mean (background) density stratification. As a rule of thumb, we considered 10 periods of the observed wave, either the basin-scale or high-frequency internal wave, for calculating mean density profiles. In addition, for basin-scale internal waves we estimated kinetic energy from the spectra of the three velocity components within the frequency band corresponding to the range of the wave periods provided by IWA. For high-frequency internal waves a fixed frequency range from 1×10^{-3} to 6.1×10^{-3} Hz was selected.

For comparison with surface energy fluxes, we integrated the volumetric energy content over depth. Both integrated potential and kinetic energies (as well as dissipation rate described in Section 2.4), were normalized by depth-dependent surface area, i.e., for a given quantity X, integration over the water column was computed as:

$$\frac{1}{A_{\text{surf}}} \int_0^H X(z)\, A(z)\, dz \quad (3)$$

where A_{surf} is the surface area of the water body, H is the water depth, A is the depth-dependent cross-sectional area.

Following [14], we used the depth of the thermocline as the upper limit of integration for APE (APE in J m^{-2}), which was estimated using the "Lake Analyzer" software [39].

2.2.2. Surface Waves

Significant wave height H_{sig} (m) and energy content in surface waves E_{wave} (J m^{-2}) were calculated from pressure fluctuations measured by the wave recorders. The calculation was carried out following standard procedures based on linear wave theory [40] by using the "Ruskin" software provided by the manufacturer [41]. The calculations take into account the attenuation of wave-induced pressure fluctuations at the sampling depth of the sensor. Significant wave height is defined as the average height of the highest one third of the waves during each sampling interval. Mean wave energy was calculated from the variance of water surface elevation. Note that for Bautzen Reservoir we used only those wave measurements that corresponded to the acceptable wind directions (195–355°) to avoid the possible sheltering effect of the measurement platform (the sensor was deployed at the south-western corner).

2.2.3. Schmidt Stability

The Schmidt stability Sc (J m^{-2}) describes the integrated potential energy in vertical density stratification of the entire basin. It is equivalent to the work required for vertical mixing, i.e., the energy required to move the vertical coordinate of the center of mass of all water in the basin to the corresponding center of volume z_v:

$$Sc = \frac{g}{A_{\text{surf}}} \int_0^H (z - z_V) \rho_w(z) A(z) dz \quad (4)$$

$$z_V = \frac{1}{V} \int_0^H A(z) z dz, \quad V = \int_0^H A(z) dz \quad (5)$$

where V is the volume of the basin.

2.3. Energy Fluxes
2.3.1. Wind Energy Flux and Rate of Working

The vertical energy flux at a standard height of 10 m above the water surface P_{10} (W m^{-2}) is equivalent to the vertical shear stress multiplied by the horizontal wind velocity [42]:

$$P_{10} = \tau_s U_{10} = \rho_a u_{*a}^2 U_{10} \quad (6)$$

where $\tau_s = \rho_a u_{*a}^2$ (kg m^{-1} s^{-1}) is the shear stress, u_{*a} (m s^{-1}) is the atmospheric friction velocity and ρ_a (kg m^{-3}) is air density. In our analysis, we estimated u_{*a} from measurements of turbulent velocity fluctuations in the atmospheric boundary layer using the eddy-covariance (EC) method [43]. The mean wind speed measured at a height of 1.8 m (Bautzen Reservoir) and 1.97 m (Lake Dagow) above the water surface was corrected to a standard height of 10 m (U_{10} (m s^{-1})) by considering atmospheric stability [44,45]. We calculated the fraction of the wind energy that is transferred to the water as the total rate of working RW (W m^{-2}) following [16]:

$$(\tau_x, \tau_y) = C_{\text{DN10}} \rho_a U_{10}(U, V) \quad (7)$$

$$RW = \tau_x u + \tau_y v \quad (8)$$

where the wind (U, V) and flow velocity (u, v) components were rotated along the wind direction averaged over 12 h since the shape of two water bodies does not have any preferred elongated direction. The drag coefficient C_{DN10} was first determined as $C_{\text{D10}} = u_*^2/U_{10}^2$ (-) and then corrected to its neutral counterpart $C_{\text{DN10}} = C_{\text{D10}}\left(1 + C_{\text{D10}}^{1/2} \kappa^{-1} \psi(10/L)\right)^{-2}$, where $\kappa = 0.4$ (-) is the von Kármán constant, L is the Monin-Obukhov length scale and ψ are the stability functions described in [44]. Note that, we used the first acceptable measurement of flow velocity below the surface: for Bautzen Reservoir, it corresponds to ~1 m depth, for Lake Dagow ~0.6 m (first deployment), ~1.3 m (second and third deployment).

2.3.2. Surface Wave Energy Flux

The horizontal wave energy flux per unit length of the wave crest of surface waves (P_{wave} (W m^{-1})) was calculated as the product of the wave energy and the wave group velocity. The group velocity is a function of wave period, which we assign to the period corresponding to the maximum in the wave spectrum. The estimation of the wave energy flux proceeds in the same way as in [46]. To compare P_{wave} with the wind energy flux P_{10}, we considered the ratio $P_{\text{wave}}/(P_{10} \times F)$ 100 (%), where F (m) is the wind fetch at the wave measurement location. The wind fetch is interpolated from distances obtained from the map corresponding to the standard grid of wind direction. Note that, as in Section 2.2.2, we disregarded data with unacceptable wind directions for Bautzen Reservoir.

2.3.3. Surface Heat Flux and Buoyancy Flux

The net surface energy flux in form of heat and radiation H_{net} (W m^{-2}) is expressed as the sum of net shortwave radiation Q_{SW}, longwave radiation Q_{LW}, and latent and sensible heat fluxes H_L, H_S (W m^{-2}). Latent and sensible heat fluxes were calculated following the standard EC methodology using the "Eddy Pro Version 6.2.1" software (LI-COR, Inc., Lincoln, NE, USA).

The surface buoyancy flux J_{BO} (W m^{-2}) was calculated as:

$$J_{BO} = z_V \frac{g\alpha}{c_p} H_{net} \qquad (9)$$

where α (K^{-1}) is the temperature-dependent thermal expansion coefficient of water and c_p (J kg^{-1} K^{-1}) is the specific heat capacity of water.

2.3.4. Energy Flux to Basin-Scale Internal Waves

For the estimation of the fraction of the wind energy input attributed to basin-scale internal waves, we manually selected isolated episodes with solitary wind event and corresponding enhancement of the available potential energy in internal waves. This fraction (%) was calculated as a ratio of APE averaged over one cycle of the wave right after the wind event to the wind energy flux integrated over the time for the respective wind event.

2.4. Dissipation Rates

Dissipation rates of turbulent kinetic energy ε (W kg^{-1}) were estimated following two methods: inertial subrange fitting (ISF) [47] and second-order structure function (SF) [48]. Both methods have been widely applied and validated for obtaining dissipation rates from velocity data measured by ADCP [49–52]. Under the assumption of isotropic turbulence, the theoretical power spectrum of turbulent velocity fluctuations S (m^3 s^{-2} rad^{-1}) is expressed as:

$$S(k) = C_1 \alpha_K \varepsilon^{2/3} k^{-5/3} \qquad (10)$$

where $\alpha_K = 1.5$ (-) is the universal Kolmogorov constant, k (rad m^{-1}) is the spatial wavenumber and C_1 (-) is the isotropy constant, which depends on the direction of the velocity component $18/55 \leq C_1 \leq 4/3 \times 18/55$. We used a constant value of $C_1 = 18/55$ as we used beam-averaged velocity spectra from the ADCP, which measures along-beam velocity fluctuations without directional information [53]. Power spectra estimated from measurements were fitted to Eq. 10 to estimate the dissipation rates. The upper wavenumber limit for the fit was found as a breakpoint where the power spectral density became smaller than the level of noise. The noise level was determined as the logarithmically averaged high-frequency end of the spectra at frequencies higher than 0.2 Hz. To find the lower frequency limit for inertial subrange fitting, we used the optimization procedure described in [54]. We used three criteria for quality assurance for calculated dissipation rates: validity of Taylor's frozen turbulence hypothesis, coefficient of determination of the fit (for both—see [47]) and the length of observed inertial subrange (set to a minimum of 10/8 of decade). The application of these quality criteria led to significant reduction in dissipation rate estimated using ISF (~70% of the data were removed).

Due to the presence of surface waves in velocity spectra, we could not apply ISF for the entire period of measurements. We manually selected velocity spectra where no sur-face wave peak was observed. For periods when no inertial subrange could be observed in the spectra, we applied the SF method, which can be corrected for the case when surface waves are present [55]:

$$D(z,r) = C_2 \varepsilon^{2/3} r^{2/3} + C_3 \left(r^{2/3} \right)^3 + N_m \qquad (11)$$

where $D(z,r)$ (m² s⁻²) is the mean squared velocity difference at two locations separated by the distance r (m), C_2 = 2.1 (-) is a constant, C_3 (-) is a coefficient describing wave orbital motion and N_m (m² s⁻²) is the measurement noise. C_3, $C_2\,\varepsilon^{2/3}$ and N_m were determined using least square fits of measured along-beam velocity fluctuations to Equation (11). We used fixed numbers of ADCP bins for the fitting. First, we applied the procedure with 5 bins (for the purpose of the calculations the number of bins should be odd). We noticed that noise could be negative in cases when the theoretical structure function was not long enough to reach its "plateau". Therefore, we used the procedure with seven bins and replaced the values of dissipation rates from the previous step for cases when the noise was negative. We disregarded fits, if either N_m, $C_2\,\varepsilon^{2/3}$, the difference between the first point of the structure function and the noise, or the difference between the second and first point of the structure function were negative. The application of these criteria led to ~51% and ~30% reduction of the dissipation rate estimates for Bautzen Reservoir and Lake Dagow, respectively.

The dissipation rates obtained from both ISF and SF methods agreed reasonably well (see Figure S1, Supplementary Material). However, the scatter at low dissipation rates increases towards the bottom and the structure function estimates were a factor of 2 to 3 lower than the ISF estimates for dissipation rates exceeding 1×10^{-8} W kg⁻¹. The final dissipation rates combined both estimates using ISF and SF techniques considering the ISF calculations as default value and gap-filling with estimates from SF.

The ISF method could be applied for the full depth range of the velocity measurements. Application of the SF method results in dissipation profiles that lack several bins at the beginning and at the end of the profiling range due to the calculation procedure but could be applied in the presence of surface waves. Thus, we combined the advantages of both methods. However, this procedure was applied and validated for Bautzen Reservoir data, while for Lake Dagow only the SF method was used. That was because there were only few periods during which the velocity spectra were not affected by surface waves during the first deployment and application of the ISF method was not possible for most of the time.

Logarithmic velocity profiles in the bottom boundary layer (BBL) were not resolved by our ADCP measurements at both locations due to the limited profiling range. Visual observation of the flow velocity profiles revealed that the BBL extended up to ~2 m distance above the bottom. Dissipation rates in the BBL were calculated using the flow velocity u_m (m s⁻¹) at the measurement depth (ADCP bin) closest to the bottom using the law of the wall [56]:

$$\varepsilon = \frac{u_{*w}^3}{\kappa z} \qquad (12)$$

where $u_{*w} = (C_{Db} u_m^2)^{\frac{1}{2}}$ (m s⁻¹) is the bottom friction velocity. Following [56], the bottom drag coefficient C_{Db} was corrected for the distance from the bottom at which the flow speed was measured, using a standard value of 1.5×10^{-3} at 1 m height. Resulting dissipation rate profiles were integrated over depth as in Equation (3) and multiplied by density of the water ρ_w to obtain areal estimates of depth-integrated dissipation rates (in W m⁻²).

3. Results

3.1. Overview of the Measurements

The measurements include both stratified and mixed conditions throughout 243 days in Bautzen Reservoir and the transition from seasonal summer stratification to mixed conditions during the autumn overturn (76 days) in Lake Dagow (Figure 2). In Bautzen Reservoir, the temperature stratification during summer was occasionally disrupted by winds exceeding 7 m s⁻¹. These mixing events are consistent with observations at this reservoir reported in previous studies [24,28,57]. The maximum temperature at the water surface was 29.2 °C (4 August) in Bautzen Reservoir and 18 °C (11 September) in Lake Dagow. The maximum temperature difference between surface and bottom was 15.2 °C (10 June) and 8 °C (12 September) in Bautzen Reservoir and Lake Dagow, respectively.

In Lake Dagow, the thermocline was located close to the bottom and the thickness of the hypolimnion was only ~0.7 m. Average Schmidt stability was 4.3 and 41.8 J m^{-2} and maximum values were 21 J m^{-2} (17 September) and 178 J m^{-2} (01 June) in Lake Dagow and Bautzen Reservoir, respectively. The heat and buoyancy fluxes varied between -155 and 1113 W m^{-2} (-5.1×10^{-4} and 4.2×10^{-3} W m^{-2}) in Bautzen Reservoir and between -130 and 763 W m^{-2} (-2.6×10^{-4} and 3.1×10^{-3} W m^{-2}) in Lake Dagow (buoyancy flux in parenthesis, Table S1, Figure S8, Supplementary Material).

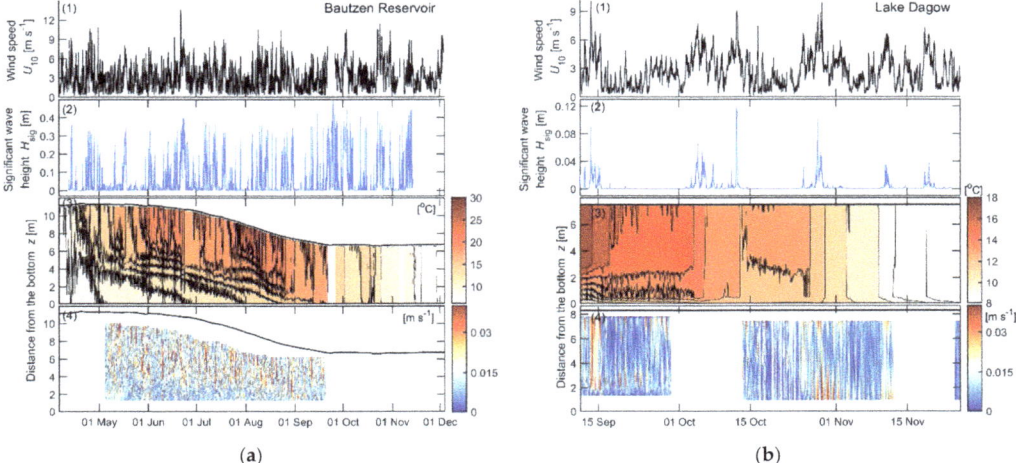

Figure 2. Overview of wind forcing and hydrodynamic conditions in (**a**) Bautzen Reservoir and (**b**) Lake Dagow. From the top to the bottom: (1) wind speed corrected to a height of 10 m; (2) significant wave height; (3) temperature profile (color denotes temperature, lines show isothermal depths); (4) flow velocity profiles (velocity magnitude). The black lines mark the location of the water surface. All data are shown at 30 min resolution.

Wind forcing in both systems was of the same order of magnitude: the wind speed at 10 m height was 3.0 ± 1.9 and 2.7 ± 1.7 m s^{-1} (here and further, $\pm$ denotes standard deviation) with maximum values of 13.7 and 10.1 m s^{-1} in Bautzen Reservoir and Lake Dagow, respectively. West-northwestern (280–300°) and south-western (220–240°) wind directions were predominant for Bautzen Reservoir and Lake Dagow, respectively. The water level continuously declined throughout the study period from 11.2 m to 6.8 m at the platform location in Bautzen Reservoir, while in Lake Dagow it remained constant (~8.3 m at the ADCP location). Water discharge at the inflow and at the outlet tower varied between 0.6 and 3.9 m^3 s^{-1}, with mean values of 1.2 and 1.9 m^3 s^{-1}, respectively (Figure S2, Supplementary Material). In both water bodies, flow velocities were relatively small for most of the time (~0.01–0.02 m s^{-1}). The maximum flow speed was 0.1 m s^{-1} in Bautzen Reservoir and 0.07 m s^{-1} in Lake Dagow. The mean significant wave height H_{sig} was $(3.9 \pm 9.6) \times 10^{-3}$ m and $(1.9 \pm 2.7) \times 10^{-2}$ m at shore sampling locations (Point B in Bautzen Reservoir, Point G in Lake Dagow, see Figure 1) and $(7.4 \pm 9.9) \times 10^{-2}$ m at the platform in Bautzen Reservoir. The maximum value of significant wave height was 0.1 and 0.2 m for the shore sampling locations in Lake Dagow and Bautzen Reservoir, respectively, and 0.5 m at the platform in Bautzen Reservoir.

3.2. Wind Energy Flux and Rate of Working

The ratio of the wind energy flux at 10 m height (P_{10}) to the rate of working associated with shear stress in the surface layer of the water column (RW) provides an estimate of the efficiency of the energy transfer from wind to water. We analyzed the ratio separately for mixed and for stratified conditions, but we did not find significant differences between both conditions (Figure 3). We used a fixed, although arbitrary, threshold for Schmidt stability (Sc = 5 J m^{-2}) to distinguish between mixed and stratified conditions. Distributions of the ratio of RW to P_{10} for periods when the Schmidt stability was smaller or larger than 5 J m^{-2} were in close agreement (Figure S3, Supplementary Material). The median values of the ratio were 1.8×10^{-3} and 1.6×10^{-3} in Bautzen Reservoir, and 1.7×10^{-3} and 0.7×10^{-3} in Lake Dagow for non-stratified and stratified conditions, respectively. We applied linear regressions of RW as a function of P_{10} considering P_{10} less than 2 W m^{-2}, as most of the data belonged to this interval. Different values of the threshold in Schmidt stability to distinguish stratified and mixed conditions did not result in significant changes in slopes. However, we noticed that the slope of the regression was sensitive to the inclusion of few high-magnitude values. The slope coefficient for all data, which describes the efficiency of energy transfer is equal to $(1.3 \pm 0.1) \times 10^{-3}$ and $(2.61 \pm 0.05) \times 10^{-3}$ ($\pm$ here denotes standard error for the slope) for Bautzen Reservoir and Lake Dagow, respectively. Data from Bautzen Reservoir were additionally filtered based on the same wind directions as for surface waves (Section 2.2.2) to avoid potential sheltering by the measurement platform. These values were comparable to the mean efficiency of 1.3×10^{-3} estimated under mixed conditions in a lake by [16].

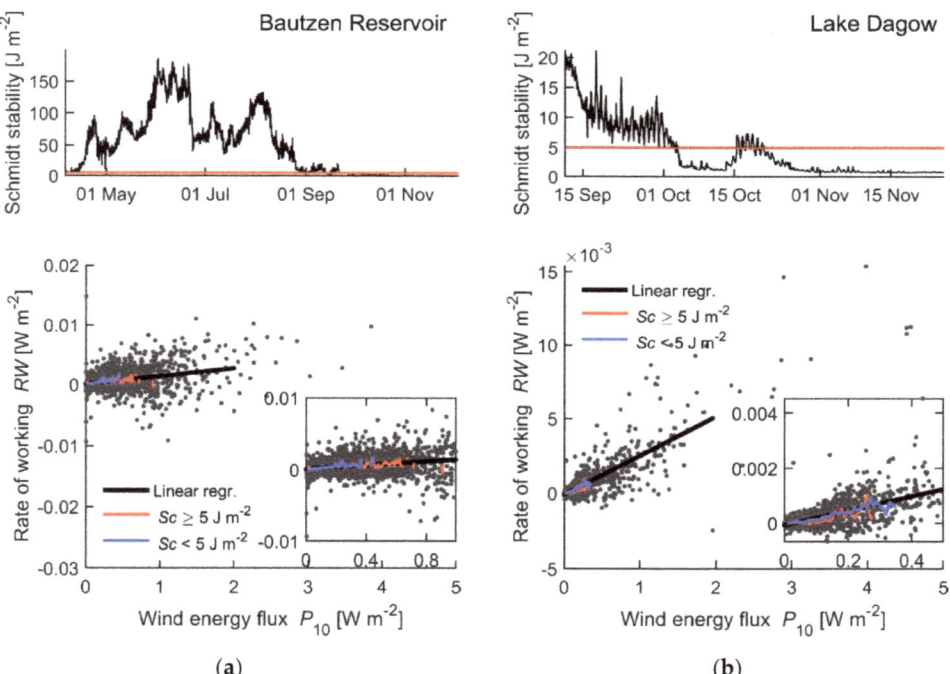

Figure 3. Potential energy in stratification and efficiency of energy transfer from wind to water in (**a**) Bautzen Reservoir and (**b**) in Lake Dagow. From top to bottom: (1) the black line shows the time series of Schmidt stability (Sc). The horizontal red lines mark the threshold value (5 J m^{-2}) to separate mixed and stratified conditions. (2) Relationship between rate of working RW and wind energy input P_{10}. Gray dots show all data. Red and blue lines represent bin-averages for two selected cases: $Sc \geq 5$ and $Sc < 5$ J m^{-2}—indicating stratified and mixed conditions, respectively. A minimum of 5 data points was considered for the bin-averaging. The black line shows a linear regression for all data with $P_{10} < 2$ W m^{-2}. Inset graphs in the lower panels show a detailed view of the data at small energy fluxes. The slope coefficient, i.e., the efficiency of energy transfer from wind to water is equal to $(1.3 \pm 0.1) \times 10^{-3}$ and $(2.61 \pm 0.05) \times 10^{-3}$ for Bautzen Reservoir and Lake Dagow, respectively.

3.3. Surface Waves

We examined the relationship between significant wave height and wind speed at 10 m height (Figure 4a) and compared our wave measurements with the only—to the best of our knowledge—other existing dataset of significant wave height in relation to wind speed in a lake. These data were measured in a large lake in Switzerland over a two-week period [18]. Although the significant wave height reached higher values in [18] (~0.5 m) than in our measurements (~0.3 m, at the platform location in Bautzen Reservoir), the wind-speed dependence is remarkably consistent in both datasets. Wave heights were relatively small ($<1 \times 10^{-2}$ m) with weak dependence on wind speed for winds below ~3 m s^{-1}, while wave heights strongly increased for wind speeds exceeding 3–4 m s^{-1}. Significant wave heights measured at the shore locations were generally smaller in amplitude and showed a weaker increase with increasing wind compared to open-water measurements (Figure 4a). This could be related to the interference with waves reflected from the shore and shallow water depth at the sampling locations.

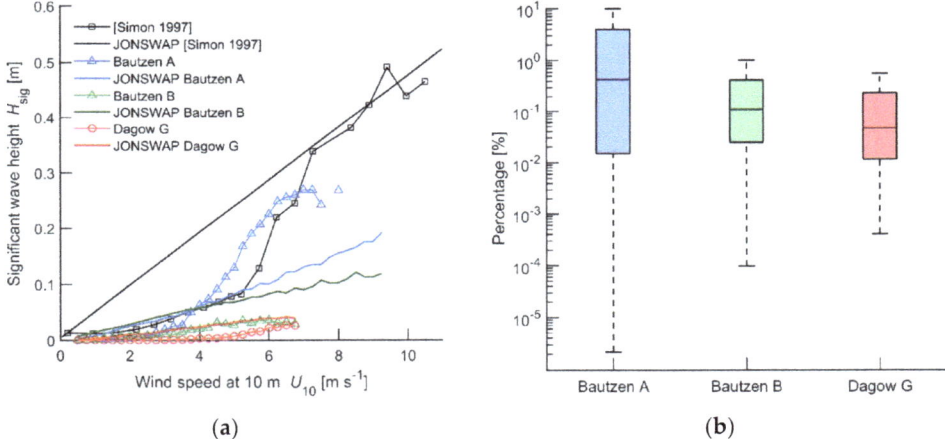

Figure 4. (**a**) Significant wave H_{sig} as a function of wind speed U_{10}. Lines with markers represent measurements: black color shows data from [18]. Blue, green and red colors represent the bin-averaged data from the present study: Bautzen Reservoir A (platform), B (shore) and Dagow Lake G (shore). Solid lines show a commonly applied empirical parameterization of the significant wave height based on the wind speed and fetch length (JONSWAP). The parameterization was applied to the observed wind speed and direction with a resolution of 30 min. (**b**) Boxplots showing the percentages of the ratio of wave energy flux per unit length of wave crest (P_{wave}) to the fetch-integrated wind energy flux (approximated as P_{10} multiplied by the fetch length) for three measurement locations. The central horizontal line in boxes indicates the median; the bottom and top edges of the boxes denote the 25th and 75th percentiles; the whiskers extend to the largest data points which are not considered outliers.

We applied an empirical approach for the prediction of significant wave height based on the wave frequency, wind speed and fetch length, which was originally derived from measurements in the North Sea but is also commonly used in lake models (the Joint Sea Wave Project or JONSWAP, [58]). The parameterization overestimated the significant wave height up to wind speeds of approximately 4 m s^{-1} and underestimated wave heights for higher wind speeds (see blue lines, Figure 4a). This supported the earlier finding of [18]; however, their "breakpoint" was at around 6 m s^{-1} and the JONSWAP predictions agreed well with the measurements at higher wind speed. The authors of [18] note that the possible reasons for underestimation at low wind speeds could either be insufficient sensor accuracy to resolve the small amplitude of waves, or a failure to compute the wind fetch correctly as the variability in wind direction at low wind speeds is very high. The wind fetch indeed varied considerably due to the variations in wind direction; however, smoothing of the calculated fetch prior to applying the wave model did not significantly reduce the predicted values at low wind speeds. The JONSWAP model overestimated the significant wave height for both measurements at shore locations.

Energy content in surface waves varied between 1.3×10^{-4} and 9.1×10^2 J m^{-2} with a log-averaged value of 0.3 J m^{-2} for the measurements at the platform in Bautzen Reservoir. In Lake Dagow, it varied between 1.6×10^{-4} and 1.1×10^1 J m^{-2} with a log-averaged value of 1.5×10^{-3} J m^{-2}. Wave energy measured at the shore in Bautzen Reservoir (1.3×10^{-4}–2.1×10^1 J m^{-2} with log-averaged value 2×10^{-2} J m^{-2}) was comparable in magnitude to the shore measurements in Lake Dagow. Wave energy showed strong dependence on the wind speed exceeding 3 m s^{-1} with a power-law exponent of ~8–9 (Figure S4, Supplementary Material).

The median values of the fraction of the wind energy input attributed to the surface waves (see Section 2.3.2) varied between 0.05% and 0.4% between sampling locations (Figure 4b). Mean values varied between 0.5% and 4%. The fraction strongly increased for wind speeds exceeding 3 m s^{-1}.

3.4. Internal Waves

Basin-scale and high-frequency internal waves were observed in both water bodies. The prevailing mode of basin-scale waves was the second vertical mode/first horizontal mode (V2H1), with a period of approximately 8–9 and 12 h and maximum amplitude of 1–2 m and 0.5–1 m in Bautzen Reservoir and Dagow Lake, respectively (Figure 5). Note, that we also observed waves with a period of approximately 21 h in both water bodies; however, they occurred only once during a short period of time. Although a clear periodic structure of the waves was present below the thermocline (Figure 5a), close to the surface it was often masked by wind-driven flow and mixed layer dynamics. Unlike basin-scale internal waves, high-frequency internal waves occurred not only during summer stratification but also during autumn and spring mixing. The period of the major wave peak in the power spectra of velocity and isothermal depths varied throughout the measurements from 5–6 min during the stratified season up to 15–20 min during mixing in spring and autumn. High-frequency internal waves appeared in the frequency range 0.02–0.2 N, which was lower than values reported for larger lakes (0.1–0.7 N) [59–62].

Energy content in basin-scale and high-frequency internal waves is comprised of available potential energy (APE) and kinetic energy (see Section 2.2.1). We analyzed its seasonal dynamics, using the long-term measurement from Bautzen Reservoir (Figure 6a). During the deepening of the upper mixed layer in spring, the average APE during internal wave events (for the definition of "event" see Section 2.2.1) had a maximum of 6.7 J m^{-2}, while it remained nearly constant below 1 J m^{-2} during most of the time. Its average value for 13 analyzed internal wave events was 1.2 ± 1.0 J m^{-2}. Note that the periods of the internal waves in the two last events (October and November, see Figure 6a) were not supported by the predictions of the Internal Wave Analyzer; however, we visually observed internal waves as a peak in the velocity spectrum. The average APE in high-frequency internal waves evaluated for 210 events reached its maximum value of 0.45 J m^{-2} in summer during the strongest stratification (end of June–beginning of September). The average value over the entire measurement period was 0.06 ± 0.05 J m^{-2}.

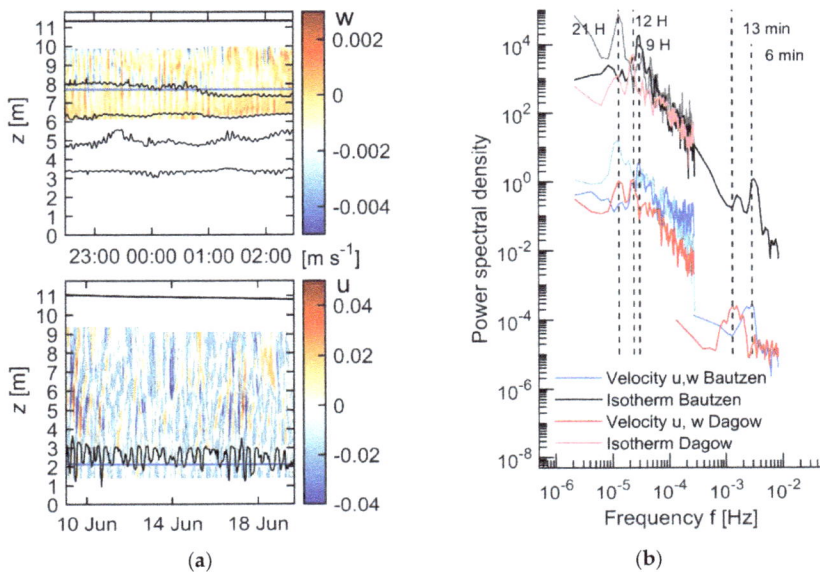

Figure 5. (**a**) Sample data from Bautzen Reservoir illustrating high-frequency internal waves with a period of 6 min (upper panel) and basin-scale internal waves with the period of 9 h (lower panel). The upper and lower panels show vertical (w) and horizontal (u) velocity components, respectively. Black lines show temperature isotherms; blue lines show the distance above the bed for which the velocity power spectra are shown in panel (**b**). (**b**) Power spectral density estimates for flow velocity components (m^2 s^{-1}) and isotherms (m^2 s). Blue, light blue and red solid lines show velocity spectra; gray, black and light red show isotherm spectra for Bautzen Reservoir and Lake Dagow, respectively. Internal waves are associated with distinct spectral peaks and vertical dashed lines denote major internal wave periods.

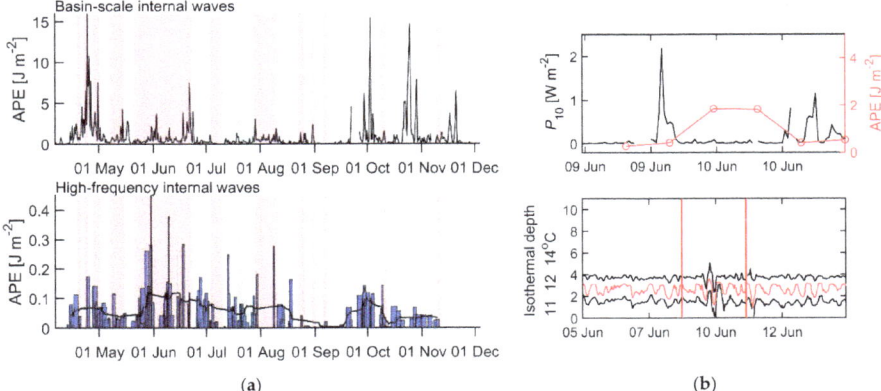

Figure 6. (**a**) Average available potential energy (APE) in basin-scale internal waves (upper panel) and in high-frequency internal waves (lower panel) throughout the measurement period in Bautzen Reservoir. The red-shaded areas correspond to the presence of the basin-scale waves with 8–9 h period, the green-shaded area shows the presence of a wave with 21 h period. The width of the bars is not to scale but is proportional to the event duration (i.e., narrow bars indicate shorter events where waves are present, and wider bars indicate longer events). The black line in the lower panel (**a**) shows a moving average of the APE. (**b**) Sample data demonstrating the transfer of wind energy to basin-scale internal waves. The upper panel shows the time series of P_{10} and APE averaged over one wave cycle. After the wind event stopped, APE grows first and then decays after three wave cycles. The lower panel shows isothermal depths with the event duration marked by red vertical lines.

The mean kinetic energy during the selected basin-scale internal wave events was on average a factor of four smaller than the APE. This inconsistency with linear wave theory can be explained by the location of measurements relative to the lake center. In standing waves, the kinetic energy is higher near the center, whereas the potential energy is higher along the edges. However, for high-frequency internal waves, the average kinetic energy for all selected events was 0.05 J m^{-2}, which was comparable to the APE (note that the difference between them is greater during April to August, probably because velocity measurements cover only the upper part of the water column). In Lake Dagow, we identified only one event with basin-scale internal wave activity and the corresponding APE of 0.05 J m^{-2} was two orders of magnitude smaller than in Bautzen Reservoir because the thermocline was close to the bottom. Following [14], we used the double value of APE as a measure of the total energy in internal waves.

The fraction of the wind energy flux, which is transferred to basin-scale internal waves was estimated based on nine selected episodes with solitary wind events in Bautzen Reservoir and three in Lake Dagow (see Section 2.3.4). One example is presented in Figure 6b, which demonstrates how waves are energized by a wind event. The average percentage of wave energy and integrated wind energy flux amounted up to 0.1% with an average value of 0.04 ± 0.03%. In Lake Dagow, the energy flux to internal waves was 0.002%, which is one order of magnitude lower than the average value in Bautzen.

3.5. Dissipation Rate in Surface and Bottom Boundary Layers

The log-averaged dissipation rates of turbulent kinetic energy were of the same order of magnitude (~10^{-8} W kg^{-1}) in Bautzen Reservoir and Lake Dagow (Figure 7). They tended to increase towards the water surface, while remaining nearly constant in the middle of the water column. Estimates of dissipation rate in the bottom boundary layer (Equation (12)) were on average one (Bautzen Reservoir) and two (Lake Dagow) orders of magnitude smaller than dissipation rates in the interior and the surface layer.

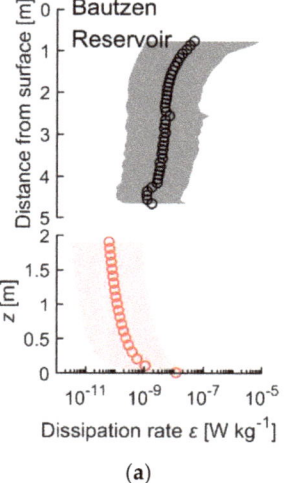
(a)

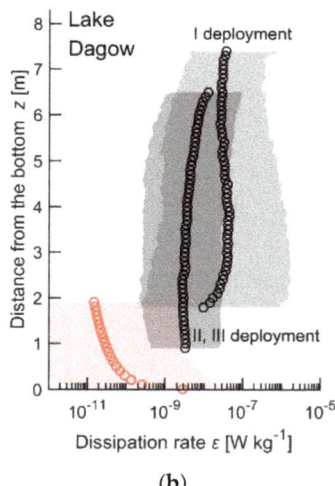
(b)

Figure 7. Vertical profiles of log-averaged energy dissipation rates (ε) for all available data (gray and red shaded areas show the 5th to 95th range of data): (**a**) combined data with dissipation rates calculated using the structure function and inertial subrange fitting methods (black circles, the platform-mounted ADCP facing downward) and using bottom boundary layer approach (see Section 2.4, red circles, the ADCP deployed at the bottom) in Bautzen Reservoir. The vertical axis is split into two subaxes with identical scaling: The lower axis corresponds to the distance from the bottom, the upper—to the distance from the surface. Thus, we avoid averaging over the entire water column because the water level change was significant throughout the measurements. Data are based on measurements in Bautzen Reservoir. (**b**) Dissipation rate calculated using the structure function method and the BBL approach based on measurements in Lake Dagow (separately for the first and for the following two ADCP deployment periods).

In Lake Dagow, the BBL approach may not be appropriate as there was a possible influence of biogenic activity, which may contribute additional turbulence or interfere with the analysis method. We observed a strong diurnal variation in the dissipation rate (primarily during the first deployment; see Figures S5 and S8, Supplementary Material). However, in contrast to similar observations [17,63], we observed high acoustic backscatter during the night (see Figure S5b, Supplementary Material) and an increase in dissipation rates and vertical velocity occurred during the day (Figure S5b, Supplementary Material). It is important to note that high dissipation rates were present in the entire water column and not just close to the bottom where elevated acoustic backscatter was observed. We also observed a "trace" of high dissipation rates starting from midnight when migrating zooplankton species begin to swim towards the bottom. We suggest that strong acoustic backscatter during the night and high dissipation rate and vertical flow velocity patterns during the day (see average profiles for days and nights in Figure S5, Supplementary Material) are related to diurnal migrations of different types of organisms. The pronounced diurnal pattern of dissipation rates and backscatter strength became less obvious in the second and third ADCP deployments suggesting the reduction of the number of species in autumn.

Depth-integrated dissipation rates were typically three orders of magnitude smaller than the wind energy flux (Figures S6–S8, Supplementary Material). The dissipation rates integrated over the BBL were on average two orders of magnitude lower than the dissipation rates integrated over the interior and surface boundary layer (Figure S7, Supplementary Material). The share of the wind energy input that was dissipated in turbulence increased with wind speed. In Bautzen Reservoir, highest values of depth-integrated dissipation rates were comparable in magnitude to the corresponding wind energy flux (Figure S6a, Supplementary Material). We observed a strong relationship between the integrated dissipation rate and wind energy flux with a power-law exponent of ~2.6 for wind speeds exceeding ~3 m s^{-1} in Bautzen Reservoir (Figure S7). We did not find this relationship in the data from Lake Dagow, where the wind energy flux reached comparable magnitude, but integrated dissipation rates remained below the high values observed in Bautzen Reservoir. However, rare events with high integrated energy dissipation rates did not contribute to mean conditions. The average ratio of total integrated dissipation rate and wind energy flux was 0.23% and 0.5% for Bautzen Reservoir and Lake Dagow, respectively. Similarly, the mean values of depth-integrated dissipation rates were a factor of two lower in Bautzen (1.7×10^{-5} W m^{-2}) compared to Lake Dagow (3.4×10^{-5} W m^{-2}).

4. Discussion

4.1. Overall Energy Budget

Based on simultaneous measurements of energy fluxes in the atmospheric boundary layer and along the water column in two small water bodies, we compiled energy budgets in terms of energy fluxes and energy content in different types of water motions (Figure 8, Table S1, Supplementary Material). Despite having different origin and differing in size by one order of magnitude, the reservoir and the lake feature similar hydrodynamic processes and energy flux paths. Most of the vertical energy flux at 10 m above the water surface is dissipated in the atmospheric boundary layer (~95% of the wind energy flux) and is not transferred to the water body. The remaining fraction is distributed into various types of water motions.

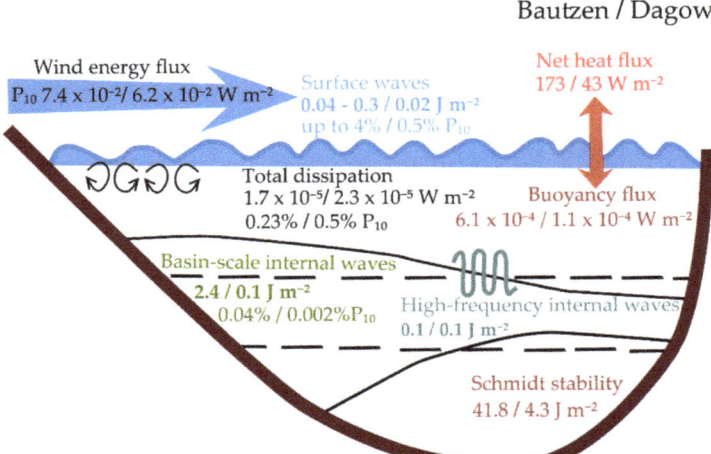

Figure 8. Scheme showing mean energy fluxes in (W m^{-2}) and mean energy content (J m^{-2}) for the two studied water bodies. Numbers written before/after a slash correspond to the values obtained from Bautzen Reservoir and Lake Dagow measurements, respectively. Solid and dashed lines illustrate the motion and equilibrium positions of vertical layers of water with the same density.

The largest fraction of wind energy flux (0.5–4%) is transferred to surface waves. Energy dissipation in turbulent flows accounted for 0.2–0.5% of the wind energy flux, while one order of magnitude less energy is transferred to basin-scale internal waves (0.002–0.04%). The energy content is largest in basin-scale internal waves in Bautzen Reservoir, where it is about one order of magnitude higher than the energy content in surface waves, as well as in high-frequency internal waves. Lake Dagow, the energy content is of comparable magnitude in all type of waves. Generally, the energy content in water motions is small compared to the potential energy in thermal stratification (Schmidt stability), which differed on average by one order of magnitude between both water bodies. The average wind energy flux in the atmosphere exceeded the average buoyancy flux by two orders of magnitude in both water bodies (Figure 8).

Some important components of the energy budget are lacking, including turbulent kinetic energy and kinetic energy in large-scale water motions such as water currents (except for waves). As discussed in the following sections, our results revealed a number of important findings related to specific energy flux paths, as well as to the potential generalization of the derived energy budget to other lakes and reservoirs.

4.2. Energy Transfer Efficiency

The estimated efficiency of energy transfer from wind to water derived from wind energy flux and rate of working in the surface layer was in close agreement for both water bodies, despite their difference in surface area. The estimated rate of wind working can be considered as an independent estimate of the energy transfer from wind to water. It can be obtained by summing up the relevant components of the energy budget (Table S1). Taking the mean shear and flow velocity at the water surface into account, the RW should be equal to the sum of all energy fluxes, except for surface waves, which are not resolved in the estimation of the RW. The sum of the components was dominated by depth-integrated energy dissipation rates (0.23–0.5% of P_{10}), which were slightly higher but of the same order of magnitude as RW (0.14–0.27% of P_{10}). Generally, this agreement supports the magnitudes of the energy fluxes compiled in Figure 8.

In contrast to earlier studies on Lake Windermere [16,17], there was no evidence of intensification of the energy transfer from the wind to the water during the stratified season compared to the period of lake mixing. The efficiency of wind energy input for both water bodies was found to be within the values reported for non-stratified conditions [14]. However, we did find a strong, non-linear increase of the efficiency with increasing wind speed in Bautzen reservoir. This fact was in line with our finding that estimates of a mean transfer efficiency are sensitive to including the few largest observations. We observed a close agreement between the mean transfer efficiencies estimated in the present study for two water bodies differing by an order of magnitude in surface area and those reported in [14] for Lake Windermere South Basin, which is slightly larger in size than Bautzen Reservoir but considerably deeper. This agreement suggests that the values represent a robust order-of-magnitude estimate for aquatic systems of small to medium size.

4.3. Surface and Internal Waves

The fraction of the wind energy flux attributed to surface waves was up to 4% on average and within the range reported by [13,16] based on measurements in a large (214 km^2) lake in Switzerland. However, this fraction strongly increased with the wind speed exceeding 3 m s^{-1} and depended on the sampling location. We demonstrated that the JONSWAP model for the estimation of the significant wave height may not be an appropriate approach for estimating significant wave heights in smaller lakes and reservoirs as it significantly overestimated wave height at low wind speed. At high wind speed, we found an extremely strong increase of wave energy with a power-law coefficient of ~8–9 with wave height exceeding the JONSWAP predictions. More wave observations in different lakes and reservoirs and further detailed investigation of the relationship between the wave characteristics and wind speed are needed to improve predictions of wave height and wave energy fluxes in lakes and reservoirs.

The energy content in basin-scale internal waves was on average lower than values reported for a larger and deeper alpine lake in [14] (0.1–2.4 J m^{-2} versus 22 ± 3 J m^{-2}) and slightly higher than the values in [15] (10^{-2}–1 J m^{-2}). We assume that this difference can be related to the strength of stratification and lake depth. The alpine lakes studied in [14] and [15] showed persistent and large-amplitude internal seiching, which occurred rather sporadically and with smaller amplitudes in Bautzen and Dagow. In addition, the energy content in basin-scale internal waves varied with season and was on average five-fold higher in spring than for the remaining sampling period in Bautzen Reservoir. This can be explained by the deepening of the thermocline and the way of calculation of the energy content with the thermocline depth being the upper limit for vertical integration. Also, lake bathymetry can affect the seasonal variation of the internal waves [64]. The energy flux to the basin-scale internal waves can be up to 0.1% of the wind energy flux but is on average two orders of magnitude smaller than that reported for the alpine lake (0.04% versus 1% in [13]). Energy content in high-frequency internal waves was on average one order of magnitude smaller than in basin-scale internal waves (Bautzen Reservoir) and comparable with basin-scale internal waves in Lake Dagow. During the stratified season, high-frequency waves can contain on average twice as much energy than during the remaining period.

4.4. Energy Dissipation Rates

Average energy dissipation rates were of the same order of magnitude in both water bodies (~5 × 10^{-9} W kg^{-1}). More energy was dissipated with increasing wind energy flux. Although in Bautzen Reservoir at high wind speeds almost all of the wind energy flux was dissipated, this was not observed at Lake Dagow, which may be related to the smaller size of the lake and the sheltering effect of the surrounding forest. On average, a similar percentage of the wind energy flux was dissipated in both water bodies. However, the dissipation rates estimated for the bottom boundary layer were on average one (Bautzen Reservoir) and two (Lake Dagow) orders of magnitude smaller than those calculated for the remaining water column. This can potentially be explained by the fact that flow velocities were generally very low, and the boundary layer may not be observed within the ADCP profiling range, making an underestimation of the dissipation rates possible.

The other possible explanation for such a difference in the dissipation rates in Lake Dagow is biogenic activity. We observed high acoustic backscatter in the upper part of the water column during the night and at larger depth during daytime, suggesting diurnal vertical migration of zooplankton [65]. Recent studies found higher dissipation rates at depths of high acoustic backscatter [17,63], suggesting a contribution of migrating organisms to energy dissipation. In contrast to these observations, we observed enhanced dissipation rates at depths and at times of low acoustic backscatter. The difference in dissipation rates between day and night can be up to two orders of magnitude. As pointed out in [65], the acoustic backscatter strength is affected by both the abundance and the acoustic properties of the scatterers and can be dominated by organisms containing gas bubbles. Other organisms, such as small fish, may follow opposing migration patterns, which remain hidden in the volume backscatter strengths [66]. Aggregations of small swimming fish with densities of 5–8 m^{-3} have been shown to enhance dissipation rates by one order of magnitude [67]. The role of biogenic turbulence in marine in inland waters has been widely discussed and analyzed in the past decade (see reviews in [68,69]). The main conclusion was that although small swimmers may generate additional flow and energy dissipation, they are unlikely to contribute to vertical mixing. Small swimmers generate flow disturbances at the scale of some multiple body length [70]. The dissipation rate estimates from ADCP are limited by the relatively large size of the sampling volume (bin size) and are theoretically based on turbulent energy transfer from large to small scales. These limitations may challenge the measurement of energy dissipation rates with ADCP in the presence of small swimmers. The increasing reporting of diurnal patterns in energy dissipation rates in relation to acoustic backscatter in recent studies calls for careful validation of these estimates using alternative methods for estimating dissipation rates, such as microstructure profiling and particle image velocimetry.

4.5. Study Limitations

In addition to the methodological limitations and uncertainties mentioned above, a number of limitations of our analysis should be noted. We made single-point measurements, and internal waves were not spatially resolved. Additional measurements along the direction of the internal wave may allow to estimate the potential and kinetic energy in wave motions more precisely. To assess the possible effect of lake bathymetry and morphology on internal wave structure and seasonal variability, spatially resolved flow measurements can be combined with numerical models in future studies.

In addition, we did not analyze the potential impact of the water level variation in Bautzen Reservoir on the energy fluxes. Reduction of the water level in a medium-sized reservoir shortens the stratification period [71,72]. This would affect the energy content in the basin-scale internal waves, at least in the way we calculated it. Possible interference of the internal waves with the lakebed may cause more turbulence in the bottom boundary layer. However, it is difficult to distinguish between the effect of water level variation and the seasonal change in the lake thermal structure without additional modeling.

Furthermore, we assumed that the wind speed is horizontally homogeneous over the lake for all sampling locations. This may not be applicable and additional, spatially resolving meteorological measurements should be included in future studies. The uncertainties in the estimation of dissipation rates could be related to the selection of the constants (in inertial subrange fitting and structure function methods). For example, [52,73] showed that these constants depend on the distance from the boundary and the use of the "canonical" constants may lead to significant errors. In addition, it remains unclear how flow and energy dissipation generated by small swimming organisms affect measurements and bulk parameterizations of energy dissipation rates.

5. Conclusions

For the first time, we related observations of energy content and fluxes in different types of water motion to simultaneously measured energy fluxes in the atmospheric boundary layer. Although the two studied water bodies were expected to be governed by contrasting hydrodynamic processes and conditions, we did not find any significant differences in energy fluxes. The observed and estimated energy fluxes and energy content agree well with results reported for larger water bodies, suggesting that the energetics governing water motions in enclosed basins is similar, independent of basin size.

Only a small fraction (<5%) of the vertical wind energy flux in the atmosphere is transferred to water motions. By disregarding surface waves, the efficiency of energy transfer does not differ strongly between various water bodies of different sizes. The transfer efficiency increases with increasing wind energy flux, but we could not observe significant differences of the energy efficiency under stratified and mixed conditions, as it has been reported for a deeper lake.

Our measurements highlight the importance of surface waves, which receive the largest share of the wind energy flux into the water and have mostly been neglected in previous studies. The wave energy flux increases strongly nonlinearly with increasing wind speed for wind speed exceeding 3 m s^{-1}. Existing parametrizations of wave height as a function of wind speed and fetch length fail to reproduce observed wave amplitudes in small water bodies.

The largest energy content was observed in basin-scale internal waves, which was found to be within the range reported for larger lakes. However, the energy fluxes and energy content in internal waves seem to vary strongly among lakes having different size and depth. Internal waves appear to be more important in mean energy budgets in larger and deeper lakes.

Dissipation rates of turbulent kinetic energy show similar structure and dynamics and are of comparable magnitude in water bodies of different size. Similar to surface waves, depth-integrated dissipation rates increase strongly nonlinearly if the wind energy flux exceeds a threshold value, which corresponds to a wind speed of 3 m s^{-1}. We observed a pronounced diurnal pattern in dissipation rates at one of our study sites, which is most likely related to vertically migrating organisms. The reliability of commonly applied measurement and analysis procedures for estimating energy dissipation rates in the presence of swimming organisms needs to be confirmed in future studies.

Supplementary Materials: The following are available online at https://www.mdpi.com/article/10.3390/w13223270/s1, Figure S1: Dissipation rate estimated using structure function method (SF) versus dissipation rate calculated using inertial subrange fitting method (ISF) at ~1.5 m depth for measurements in Bautzen Reservoir (gray dots). Figure S2: Discharge of the inflow (red line) and outflow (at the outlet tower) of Bautzen reservoir. The data was provided by the the Landestalsperren-Verwaltung Sachsen (LTV). Figure S3: Probability distribution of the ratio of rate of working (RW) to wind energy flux (P10) selected for two cases: Schmidt stability (Sc) < 5 J m-2 (area shown in gray, corresponding to non-stratified conditions) and Sc $\geq$ 5 J m-2 (area shown in red, corresponding to stratified conditions) for (a) Bautzen Reservoir. The median values are $1.8 \times 10-3$ and $1.6 \times 10-3$, the average values ($\pm$standard deviation) are $(2 \pm 4) \cdot 10-3$ and $(0.6 \pm 21.6) \cdot 10-2$, for non-stratified and stratified conditions, respectively. (b) Lake Dagow. Figure S4: Surface wave energy versus wind speed at 10 m height at the platform location in Bautzen Reservoir (gray dots). Wave energy shows strong dependence on wind speed exceeding 2–3 m s-1. The black line shows bin-average data, the red line represents a power-law relationship with an exponent of nine. The latter was obtained from a linear regression of log-transformed data. Figure S5: (a) Dissipation rates of turbulent kinetic energy averaged over night and over daytime during the first ADCP deployment in Lake Dagow. (b) Acoustic backscatter strength recorded by the ADCP (upper panel), vertical flow velocity (middle panel), dissipation rate (lower panel). Figure S6: Depth-integrated dissipation rate (including surface and bottom boundary layers and interior of the water bodies) versus the vertical wind energy flux above the water surface in (a) Bautzen Reservoir; (b) Lake Dagow. Figure S7: Dissipation rate integrated over the bottom boundary layer (the thickness of 2 m, light gray dots) and over the rest of the water column where the ADCP measurements are available (dark gray dots) using data from (a) Bautzen Reservoir; (b) Lake Dagow. The gray solid line represents a 1:1 relationship. Figure S8: Temporal dynamics of wind energy flux (black line, upper panel), dissipation rates integrated over the water depth (red dots, upper panel) and buoyancy flux (lower panel) for data measured in (a) Bautzen Reservoir; (b) Lake Dagow. Note the pronounced diurnal pattern in integrated energy dissipation rates in Lake Dagow during the first ADCP deployment (cf. Figure S4). Table S1: Energy content and energy fluxes.

Author Contributions: Conceptualization by A.L., S.G. and U.S.; methodology by U.S., P.C., T.S. and A.L.; formal analysis by S.G. and A.L.; writing—original draft preparation by S.G., U.S., P.C., T.S, A.L; data curation by U.S., S.G. and T.S.; visualization by S.G; supervision by A.L. All authors have read and agreed to the published version of the manuscript.

Funding: S.G. and A.L. were supported by the German Research Foundation (Deutsche Forschungsgemeinschaft, DFG) under the grant LO1150/12-1. Eddy Covariance measurements at Lake Dagow (T.S.) used infrastructure of the Terrestrial Environmental Observatories Network (TERENO) and were supported by the Helmholtz Young Investigators Grant (VH-NG-821) of the Helmholtz Association of German Research Centers. U. S. was participated in project "Greenhouse Gas Emissions from Reservoirs: Mechanisms and Quantification (TREibhausGAsemissionen von TAsperren—TREGATA)" which was funded by the DFG and was listed under the project number 288267759.

Institutional Review Board Statement: Not applicable.

Informed Consent Statement: Not applicable.

Data Availability Statement: The water-side measurements from Bautzen Reservoir and Lake Dagow are openly available in Zenodo at 10.5281/zenodo.5159088. The meteorological data from Lake Dagow presented in this study is openly available in FLUXNET at https://doi.org/10.18140/FLX/1669633 (accessed 20 August 2021) [74]. The meteorological and temperature data from Bautzen Reservoir are available on request from U. S. The data are not publicly available due to ongoing research. The data of discharge at the outlet tower (Bautzen Reservoir) is property of the Landestalsperren-Verwaltung Sachsen (LTV).

Acknowledgments: We strongly appreciate the help and support with instrumentation and maintenance from Christoph Bors, Gonzalo Santaolalla, Jens Nejstgaard, Tim Walles, Christian Wille, Philipp Keller, Matthias Koschorreck, Christian Bernhofer, Heiko Prasse, Uwe Eichelmann, Markus Hehn, and Martin Wieprecht. We thank Christian Wille for providing materials for the study.

Conflicts of Interest: The authors declare no conflict of interest.

References

1. Imberger, J. Flux paths in a stratified lake: A review. In *Physical Processes in Lakes and Oceans*; American Geophysical Union: District of Columbia, WA, USA, 1998; pp. 1–17.
2. Jähne, B.; Haußecker, H. Air-Water Gas Exchange. *Annu. Rev. Fluid Mech.* **1998**, *30*, 443–468. [CrossRef]
3. Perolo, P.; Fernández Castro, B.; Escoffier, N.; Lambert, T.; Bouffard, D.; Perga, M.-E. Accounting for Surface Waves Improves Gas Flux Estimation at High Wind Speed in a Large Lake. In *Dynamics of the Earth System*; Interactions: Franklin, MA, USA, 2021.
4. MacIntyre, S. Vertical Mixing in a Shallow, Eutrophic Lake: Possible Consequences for the Light Climate of Phytoplankton. *Limnol. Oceanogr.* **1993**, *38*, 798–817. [CrossRef]
5. Huisman, J.; Sharples, J.; Stroom, J.M.; Visser, P.M.; Kardinaal, W.E.A.; Verspagen, J.M.H.; Sommeijer, B. Changes in Turbulent Mixing Shift Competition for Light between Phytoplankton Species. *Ecology* **2004**, *85*, 2960–2970. [CrossRef]
6. Peeters, F.; Straile, D.; Lorke, A.; Ollinger, D. Turbulent Mixing and Phytoplankton Spring Bloom Development in a Deep Lake. *Limnol. Oceanogr.* **2007**, *52*, 286–298. [CrossRef]
7. Corman, J.R.; McIntyre, P.B.; Kuboja, B.; Mbemba, W.; Fink, D.; Wheeler, C.W.; Gans, C.; Michel, E.; Flecker, A.S. Upwelling Couples Chemical and Biological Dynamics across the Littoral and Pelagic Zones of Lake Tanganyika, East Africa. *Limnol. Oceanogr.* **2010**, *55*, 214–224. [CrossRef]
8. Bocaniov, S.A.; Schiff, S.L.; Smith, R.E.H. Plankton Metabolism and Physical Forcing in a Productive Embayment of a Large Oligotrophic Lake: Insights from Stable Oxygen Isotopes: Plankton Metabolism and Physical Forcing. *Freshw. Biol.* **2012**, *57*, 481–496. [CrossRef]
9. MacIntyre, S.; Jellison, R. Nutrient Fluxes from Upwelling and Enhanced Turbulence at the Top of the Pycnocline in Mono Lake, California. *Hydrobiologia* **2001**, *466*, 13–29. [CrossRef]
10. Sepúlveda Steiner, O. *Mixing Processes and Their Ecological Implications: From Vertical to Lateral Variability in Stratified Lakes*; EPFL: Lausanne, Switzerland, 2020.

11. Lorke, A.; Müller, B.; Maerki, M.; Wüest, A. Breathing Sediments: The Control of Diffusive Transport across the Sediment-Water Interface by Periodic Boundary-Layer Turbulence. *Limnol. Oceanogr.* **2003**, *48*, 2077–2085. [CrossRef]
12. Sobek, S.; Durisch-Kaiser, E.; Zurbrügg, R.; Wongfun, N.; Wessels, M.; Pasche, N.; Wehrli, B. Organic Carbon Burial Efficiency in Lake Sediments Controlled by Oxygen Exposure Time and Sediment Source. *Limnol. Oceanogr.* **2009**, *54*, 2243–2254. [CrossRef]
13. Søndergaard, M.; Jensen, J.P.; Jeppesen, E. Role of Sediment and Internal Loading of Phosphorus in Shallow Lakes. *Hydrobiologia* **2003**, *506–509*, 135–145. [CrossRef]
14. Wüest, A.; Piepke, G.; Van Senden, D.C. Turbulent Kinetic Energy Balance as a Tool for Estimating Vertical Diffusivity in Wind-Forced Stratified Waters. *Limnol. Oceanogr.* **2000**, *45*, 1388–1400. [CrossRef]
15. Imboden, D.M. The Motion of Lake Waters. In *The Lakes Handbook, Volume 1*; O'Sullivan, P.E., Reynolds, C.S., Eds.; Blackwell Science Ltd.: Malden, MA, USA, 2003; pp. 115–152. ISBN 978-0-470-99927-1.
16. Woolway, R.I.; Simpson, J.H. Energy Input and Dissipation in a Temperate Lake during the Spring Transition. *Ocean Dyn.* **2017**, *67*, 959–971. [CrossRef]
17. Simpson, J.H.; Woolway, R.I.; Scannell, B.; Austin, M.J.; Powell, B.; Maberly, S.C. The Annual Cycle of Energy Input, Modal Excitation and Physical Plus Biogenic Turbulent Dissipation in a Temperate Lake. *Water Res.* **2021**, *57*, e2020WR029441. [CrossRef]
18. Simon, A. Turbulent Mixing in the Surface Boundary Layer of Lakes. Ph.D. Thesis, Swiss Federal Institute of Technology, Zürich, Switzerland, 1997.
19. Heyna, B.; Groen, P. On Short-Period Internal Gravity Waves. *Physica* **1958**, *24*, 383–389. [CrossRef]
20. Boegman, L.; Ivey, G.N.; Imberger, J. The Energetics of Large-Scale Internal Wave Degeneration in Lakes. *J. Fluid Mech.* **2005**, *531*, 159–180. [CrossRef]
21. Preusse, M.; Peeters, F.; Lorke, A. Internal Waves and the Generation of Turbulence in the Thermocline of a Large Lake. *Limnol. Oceanogr.* **2010**, *55*, 2353–2365. [CrossRef]
22. Downing, J.A.; Prairie, Y.T.; Cole, J.J.; Duarte, C.M.; Tranvik, L.J.; Striegl, R.G.; McDowell, W.H.; Kortelainen, P.; Caraco, N.F.; Melack, J.M.; et al. The Global Abundance and Size Distribution of Lakes, Ponds, and Impoundments. *Limnol. Oceanogr.* **2006**, *51*, 2388–2397. [CrossRef]
23. Choulga, M.; Kourzeneva, E.; Zakharova, E.; Doganovsky, A. Estimation of the Mean Depth of Boreal Lakes for Use in Numerical Weather Prediction and Climate Modelling. *Tellus A Dyn. Meteorol. Oceanogr.* **2014**, *66*, 21295. [CrossRef]
24. Rinke, K.; Hübner, I.; Petzoldt, T.; Rolinski, S.; König-Rinke, M.; Post, J.; Lorke, A.; Benndorf, J. How Internal Waves Influence the Vertical Distribution of Zooplankton. *Freshw. Biol.* **2007**, *52*, 137–144. [CrossRef]
25. Poff, N.L.; Hart, D.D. How Dams Vary and Why It Matters for the Emerging Science of Dam Removal. *BioScience* **2002**, *52*, 659. [CrossRef]
26. Tundisi, J.G. *Limnology*; CRC Press: Boca Raton, FL, USA, 2017; ISBN 978-1-138-07204-6.
27. Kerimoglu, O.; Rinke, K. Stratification Dynamics in a Shallow Reservoir under Different Hydro-Meteorological Scenarios and Operational Strategies. *Water Resour. Res.* **2013**, *49*, 7518–7527. [CrossRef]
28. Wagner, A.H.; Janssen, M.; Kahl, U.; Mehner, T.; Benndorf, J. Initiation of the Midsummer Decline of Daphnia as Related to Predation, Non-Consumptive Mortality and Recruitment: A Balance. *Arch. Hydrobiol.* **2004**, *160*, 1–23. [CrossRef]
29. Casper, S.J. *Lake Stechlin: A Temperate Oligotrophic Lake*; Springer: Berlin/Heidelberg, Germany, 2012; Volume 58.
30. Goring, D.G.; Nikora, V.I. Despiking Acoustic Doppler Velocimeter Data. *J. Hydraul. Eng.* **2002**, *128*, 117–126. [CrossRef]
31. Wahl, T.L. Discussion of "Despiking Acoustic Doppler Velocimeter Data" by Derek G. Goring and Vladimir I. Nikora. *J. Hydraul. Eng.* **2003**, *129*, 484–487. [CrossRef]
32. Münnich, M.; Wüest, A.; Imboden, D.M. Observations of the Second Vertical Mode of the Internal Seiche in an Alpine Lake. *Limnol. Oceanogr.* **1992**, *37*, 1705–1719. [CrossRef]
33. De Carvalho Bueno, R.; Bleninger, T.; Lorke, A. Internal Wave Analyzer for Thermally Stratified Lakes. *Environ. Model. Softw.* **2021**, *136*, 104950. [CrossRef]
34. Antenucci, J.P.; Imberger, J. On Internal Waves near the High-Frequency Limit in an Enclosed Basin. *J. Geophys. Res.* **2001**, *106*, 22465–22474. [CrossRef]
35. Chen, C.-T.A.; Millero, F.J. Thermodynamic Properties for Natural Waters Covering Only the Limnological Range. *Limnol. Oceanogr.* **1986**, *31*, 657–662. [CrossRef]
36. Kundu, P.K.; Cohen, I.M.; Dowling, D.R. Gravity Waves. In *Fluid Mechanics*; Elsevier: Amsterdam, The Netherlands, 2012; pp. 253–307. ISBN 978-0-12-382100-3.
37. Holliday, D.; Mcintyre, M.E. On Potential Energy Density in an Incompressible, Stratified Fluid. *J. Fluid Mech.* **1981**, *107*, 221. [CrossRef]
38. Kang, D.; Fringer, O. On the Calculation of Available Potential Energy in Internal Wave Fields. *J. Phys. Oceanogr.* **2010**, *40*, 2539–2545. [CrossRef]
39. Read, J.S.; Hamilton, D.P.; Jones, I.D.; Muraoka, K.; Winslow, L.A.; Kroiss, R.; Wu, C.H.; Gaiser, E. Derivation of Lake Mixing and Stratification Indices from High-Resolution Lake Buoy Data. *Environ. Model. Softw.* **2011**, *26*, 1325–1336. [CrossRef]
40. RBR Ltd. Wave Parameters. Available online: https://docs.rbr-global.com/support/ruskin/ruskin-features/waves/wave-parameters (accessed on 25 July 2021).
41. Ruskin Software. Available online: https://rbr-global.com/products/software (accessed on 25 July 2021).
42. Imboden, D.M.; Wüest, A. Mixing Mechanisms in Lakes. In *Physics and Chemistry of Lakes*; Lerman, A., Imboden, D.M., Gat, J.R., Eds.; Springer: Berlin/Heidelberg, Germany, 1995; pp. 83–138. ISBN 978-3-642-85134-6.
43. Foken, T.; Nappo, C.J. *Micrometeorology*; Springer: Berlin/Heidelberg, Germany, 2008; ISBN 978-3-540-74665-2.

44. Paulson, C.A. The Mathematical Representation of Wind Speed and Temperature Profiles in the Unstable Atmospheric Surface Layer. *J. Appl. Meteorol. Climatol.* **1970**, *9*, 857–861. [CrossRef]
45. Large, W.G.; Pond, S. Open Ocean Momentum Flux Measurements in Moderate to Strong Winds. *J. Phys. Oceanogr.* **1981**, *11*, 324–336. [CrossRef]
46. Hofmann, H.; Lorke, A.; Peeters, F. The Relative Importance of Wind and Ship Waves in the Littoral Zone of a Large Lake. *Limnol. Oceanogr.* **2008**, *53*, 368–380. [CrossRef]
47. Bluteau, C.E.; Jones, N.L.; Ivey, G.N. Estimating Turbulent Kinetic Energy Dissipation Using the Inertial Subrange Method in Environmental Flows: TKE Dissipation in Environmental Flows. *Limnol. Oceanogr. Methods* **2011**, *9*, 302–321. [CrossRef]
48. Wiles, P.J.; Rippeth, T.P.; Simpson, J.H.; Hendricks, P.J. A Novel Technique for Measuring the Rate of Turbulent Dissipation in the Marine Environment. *Geophys. Res. Lett.* **2006**, *33*, L21608. [CrossRef]
49. Guerra, M.; Thomson, J. Turbulence Measurements from Five-Beam Acoustic Doppler Current Profilers. *J. Atmos. Ocean. Technol.* **2017**, *34*, 1267–1284. [CrossRef]
50. McMillan, J.M.; Hay, A.E. Spectral and Structure Function Estimates of Turbulence Dissipation Rates in a High-Flow Tidal Channel Using Broadband ADCPs. *J. Atmos. Ocean. Technol.* **2017**, *34*, 5–20. [CrossRef]
51. Lorke, A. Boundary Mixing in the Thermocline of a Large Lake. *J. Geophys. Res.* **2007**, *112*, C09019. [CrossRef]
52. Jabbari, A.; Rouhi, A.; Boegman, L. Evaluation of the Structure Function Method to Compute Turbulent Dissipation within Boundary Layers Using Numerical Simulations. *J. Geophys. Res. Oceans* **2016**, *121*, 5888–5897. [CrossRef]
53. Lorke, A.; Wüest, A. Application of Coherent ADCP for Turbulence Measurements in the Bottom Boundary Layer. *J. Atmos. Ocean. Technol.* **2005**, *22*, 1821–1828. [CrossRef]
54. Guseva, S.; Aurela, M.; Cortes, A.; Kivi, R.; Lotsari, E.S.; Macintyre, S.; Mammarella, I.; Ojala, A.; Stepanenko, V.M.; Uotila, P.; et al. Variable Physical Drivers of Near-Surface Turbulence in a Regulated River. *Water Resour. Res.* **2021**, *57*. [CrossRef]
55. Scannell, B.D.; Rippeth, T.P.; Simpson, J.H.; Polton, J.A.; Hopkins, J.E. Correcting Surface Wave Bias in Structure Function Estimates of Turbulent Kinetic Energy Dissipation Rate. *J. Atmos. Ocean. Technol.* **2017**, *34*, 2257–2273. [CrossRef]
56. Lorke, A.; MacIntyre, S. The Benthic Boundary Layer (in Rivers, Lakes, and Reservoirs). In *Encyclopedia of Inland Waters*; Elsevier: Amsterdam, The Netherlands, 2009; pp. 505–514. ISBN 978-0-12-370626-3.
57. Benndorf, J.; Kranich, J.; Mehner, T.; Wagner, A. Temperature Impact on the Midsummer Decline of *Daphnia galeata*: An Analysis of Long-Term Data from the Biomanipulated Bautzen Reservoir (Germany): Temperature Impact on Midsummer Decline. *Freshw. Biol.* **2001**, *46*, 199–211. [CrossRef]
58. Hasselmann, K.F.; Barnett, T.; Bouws, E.; Carlson, H.; Cartwright, D.; Enke, K.; Ewing, J.; Gienapp, H.; Hasselmann, D.; Meerburg, A.; et al. Measurements of Wind-Wave Growth and Swell Decay during the Joint North Sea Wave Project (JONSWAP). *Ergänzungsheft zur Dtsch. Hydrogr. Z. Reihe A* **1973**, *12*, 1–95. Available online: http://hdl.handle.net/21.11116/0000-0007-DD3C-E (accessed on 14 October 2021).
59. Mortimer, C.H.; McNaught, D.C.; Stewart, K.M. Short Internal Waves near Their High-Frequency Limit in Central Lake Michigan. In Proceedings of the 11th Annual Conference on Great Lakes Research, Milwaukee, WI, USA, 18–20 April 1968; pp. 454–469.
60. Thorpe, S.A. High-Frequency Internal Waves in Lake Geneva. *Phil. Trans. R. Soc. Lond. A* **1996**, *354*, 237–257. [CrossRef]
61. Stevens, C.L. Internal Waves in a Small Reservoir. *J. Geophys. Res.* **1999**, *104*, 15777–15788. [CrossRef]
62. Gímez-Giraldo, A.; Imberger, J.; Antenucci, J.P.; Yeates, P.S. Wind-Shear-Generated High-Frequency Internal Waves as Precursors to Mixing in a Stratified Lake. *Limnol. Oceanogr.* **2008**, *53*, 354–367. [CrossRef]
63. Ishikawa, M.; Bleninger, T.; Lorke, A. Hydrodynamics and Mixing Mechanisms in a Subtropical Reservoir. *Inland Waters* **2021**, *11*, 286–301. [CrossRef]
64. Fricker, P.D.; Nepf, H.M. Bathymetry, Stratification, and Internal Seiche Structure. *J. Geophys. Res.* **2000**, *105*, 14237–14251. [CrossRef]
65. Lorke, A.; McGinnis, D.F.; Spaak, P.; Wuest, A. Acoustic Observations of Zooplankton in Lakes Using a Doppler Current Profiler. *Freshw. Biol.* **2004**, *49*, 1280–1292. [CrossRef]
66. Lorke, A.; Weber, A.; Hofmann, H.; Peeters, F. Opposing Diel Migration of Fish and Zooplankton in the Littoral Zone of a Large Lake. *Hydrobiologia* **2008**, *600*, 139–146. [CrossRef]
67. Lorke, A.; Probst, W.N. In Situ Measurements of Turbulence in Fish Shoals. *Limnol. Oceanogr.* **2010**, *55*, 354–364. [CrossRef]
68. Kunze, E. Biologically Generated Mixing in the Ocean. *Annu. Rev. Mar. Sci.* **2019**, *11*, 215–226. [CrossRef]
69. Simoncelli, S.; Thackeray, S.J.; Wain, D.J. Can Small Zooplankton Mix Lakes? *Limnol. Oceanogr.* **2017**, *2*, 167–176. [CrossRef]
70. Wickramarathna, L.N.; Noss, C.; Lorke, A. Hydrodynamic Trails Produced by Daphnia: Size and Energetics. *PLoS ONE* **2014**, *9*, e92383. [CrossRef] [PubMed]
71. Nowlin, W.H.; Davies, J.-M.; Nordin, R.N.; Mazumder, A. Effects of Water Level Fluctuation and Short-Term Climate Variation on Thermal and Stratification Regimes of a British Columbia Reservoir and Lake. *Lake Reserv. Manag.* **2004**, *20*, 91–109. [CrossRef]
72. Bonnet, M.-P.; Poulin, M.; Devaux, J. Numerical Modeling of Thermal Stratification in a Lake Reservoir. Methodology and Case Study. *Aquat. Sci.* **2000**, *62*, 105–124. [CrossRef]
73. Jabbari, A.; Boegman, L.; Valipour, R.; Wain, D.; Bouffard, D. Dissipation of Turbulent Kinetic Energy in the Oscillating Bottom Boundary Layer of a Large Shallow Lake. *J. Atmos. Ocean. Technol.* **2020**, *37*, 517–531. [CrossRef]
74. Sachs, T.; Wille, C. FLUXNET-CH4 DE-Dgw Dagowsee, Dataset, 2015–2018. Available online: https://doi.org/10.18140/FLX/1669633 (accessed on 20 August 2021).

Article

The Multi-Scale Layering-Structure of Thermal Microscale Profiles

Andrew Folkard

Lancaster Environment Centre, Lancaster University, Lancaster LA1 4YQ, UK; a.folkard@lancaster.ac.uk

Abstract: Thermal microstructure profiling is an established technique for investigating turbulent mixing and stratification in lakes and oceans. However, it provides only quasi-instantaneous, 1-D snapshots. Other approaches to measuring these phenomena exist, but each has logistic and/or quality weaknesses. Hence, turbulent mixing and stratification processes remain greatly under-sampled. This paper contributes to addressing this problem by presenting a novel analysis of thermal microstructure profiles, focusing on their multi-scale stratification structure. Profiles taken in two small lakes using a Self-Contained Automated Micro-Profiler (SCAMP) were analysed. For each profile, buoyancy frequency (N), Thorpe scales (L_T), and the coefficient of vertical turbulent diffusivity (K_Z) were determined. To characterize the multi-scale stratification, profiles of d^2T/dz^2 at a spectrum of scales were calculated and the number of turning points in them counted. Plotting these counts against the scale gave pseudo-spectra, which were characterized by the index D of their power law regression lines. Scale-dependent correlations of D with N, L_T and K_Z were found, and suggest that this approach may be useful for providing alternative estimates of the efficiency of turbulent mixing and measures of longer-term averages of K_Z than current methods provide. Testing these potential uses will require comparison of field measurements of D with time-integrated K_Z values and numerical simulations.

Keywords: fractal; lakes; mixing; multi-scale; stratification; turbulence

1. Introduction

Stratification and mixing of lakes and oceans govern vertical fluxes of dissolved and particulate matter—including nutrients, pollutants and planktonic biota—and therefore are of great importance for understanding chemical and biological aspects of surface waterbodies [1]. They are also fundamentally important processes for the global heat energy budget [2]. Their actions lead to lakes and oceans having vertical density profiles that can be divided into distinct layers. At the simplest, macroscale level, these include, in oceans (lakes), a surface mixed layer (epilimnion), thermocline (metalimnion) and deeper, generally more quiescent layers (hypolimnion). This macroscale, three-layer structure is the classic example of a relatively strongly-stratified layer being found between two relatively well-mixed layers. The development of high spatial-resolution microstructure profilers in the last decades of the 20th century led to the discovery that increasingly subtle forms of this layering structure occurred at ever finer scales (Figure 1). Notwithstanding its subtlety, this finer-scale layering is significant because even small changes in density can have significant effects on vertical fluxes of plankton and dissolved and particulate materials [3].

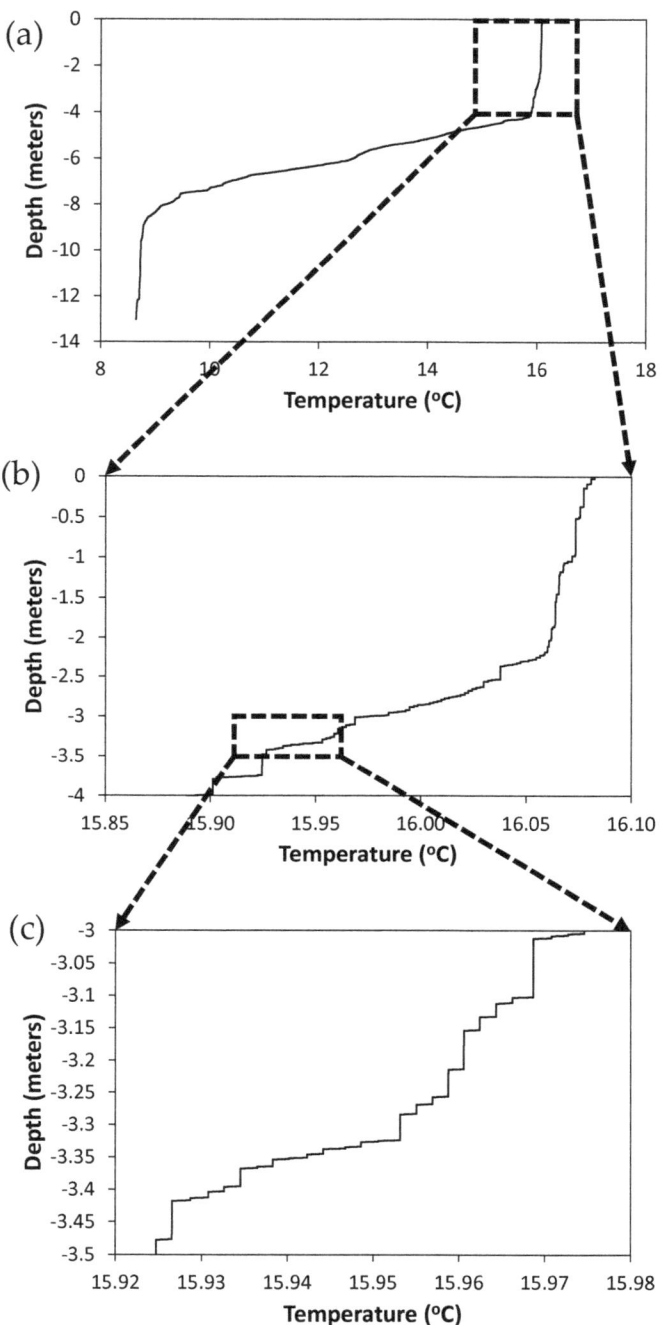

Figure 1. Example of the multi-scale layering structure in thermal microstructure profiles (profile recorded in Blelham Tarn at 11:00 on 11 August 2008): (**a**) full-depth profile; (**b**) zoomed into top 4 m; (**c**) zoomed into 3–3.5 m below surface. See Figure 2 and associated text for explanation of the "chunkiness" in the fine structure shown here.

A wide variety of microstructure profilers have been developed which have revealed fluctuations in temperature, velocity shear, conductivity, chlorophyll, turbidity and chemical concentrations at the scale of the smallest turbulent eddies: centimeter- or millimeter-scale [4,5]. Specifically, thermal microstructure profiling has been an established technique for investigating mixing and stratification in lakes and oceans for many years [2,6], and continues to be widely used [7–9]. In freshwater lakes—the object of the study reported here—the use of thermal profilers is particularly useful, because in the absence of salinity variations, density (and therefore buoyancy and the transport and mixing it induces) is primarily determined by temperature alone. Data from these thermal microstructure profilers reveal not only the layered stratification structure of the water column in great detail, they also capture snapshots of the turbulent stirring of the profile, which disturb it from a monotonic state. Length scales of the turbulent overturns causing this stirring are most commonly quantified as Thorpe scales L_T. These are calculated from Thorpe displacements δ_T (differences in height of temperature records between raw, unsorted and monotonically-sorted profiles) as $L_T = <\delta_T^2>^{1/2}$ i.e., the root mean square value [10]. The presence of these overturns is used to identify actively mixing layers, while the stratification structure of the profile can be used to identify already mixed layers (see [11] for discussion of the importance of this distinction). The edges of active turbulent layers are often defined as the point where δ_T falls to zero [12,13]; hereinafter this is referred to as the "$\delta_T = 0$" method of mixing layer identification. The statistics of stratification and mixing in these layers, especially the surface mixed layer, are important parameters for a wide variety of ocean and lake phenomena, including biological productivity, air-water exchange processes, and long-term climate change [14,15].

Thermal microstructure is also widely used to derive values of the coefficient of vertical turbulent diffusivity, K_Z. These may then be used to quantify the rate of vertical turbulent fluxes [16], which are essential parameters in models of lake heat budgets, nutrient cycling, plankton population dynamics and many other aspects of surface waterbodies. When deriving K_Z from thermal microscale profiles, it is most commonly calculated using a method commonly known as the "Batchelor method", after its creator [17]. This entails transforming the temperature profile into a temperature gradient spectrum and then obtaining the value of the rate of turbulent kinetic energy dissipation (ε) by fitting the theoretical Batchelor spectrum [17] to that temperature gradient spectrum at high wavenumbers [18]. K_Z is then calculated as

$$K_Z = \Gamma \varepsilon / N^2 \qquad (1)$$

Here, N is the buoyancy frequency, $N = (g(\partial \rho / \partial z)/\rho_0)^{1/2}$, and the parameter Γ is defined as

$$\Gamma = R_f / (1 + R_f) \qquad (2)$$

where R_f is the flux Richardson number, which measures the efficiency of the mixing process. In physical terms, this can be thought of as the proportion of turbulent kinetic energy that is converted into irreversible changes in the potential energy of the profile, rather than being dissipated down the turbulent energy cascade. In this sense, it can be written as $Rf = b/(b + \varepsilon)$ where b represents buoyancy flux (i.e., changes in potential energy) and ε is turbulent dissipation. Values of vertical diffusivity calculated using the Batchelor method have also been compared with those calculated assuming a direct relationship between K_Z and the Thorpe scale L_T. Using the parameterization $K_Z = 2\nu(\varepsilon/\nu N^2)^{1/2}$ for the energetic turbulent regime $\varepsilon/\nu N^2 > 100$, this relationship can be written as [19,20]:

$$K_Z = 1.6 \nu^{1/2} L_T N^{1/2} \qquad (3)$$

This method is referred to hereinafter as the "Thorpe scale method".

A significant problem with microstructure measurements, and therefore the mixing and stratification parameters derived from them, remains that they reflect quasi-instantaneous, 1-D snapshots of the turbulence field at specific locations. Other approaches to measuring K_Z, such as eddy correlation or tracer diffusion techniques [21,22] provide temporal averages of the vertical flux effects of the turbulence, but are much more time-consuming and logistically difficult to set up, and still only provide data on a very small spatial and temporal scale compared to that required to understand the global ocean-scale, or even whole-lake scale, effects of turbulent mixing. This is because of the great intermittency in time and space of turbulent activity, which means that sampling of turbulent activity and thus its effective parameterisation for use in predictive models is highly problematic. To date, the rate and density of sampling of these turbulent events is far too low to give us a clear picture of their distribution and global characteristics [23]. This is also true for lakes, where most information of vertical mixing in lake thermoclines is based on laboratory measurements and simulations [22], as it is for oceans.

As a result, the distribution (in space and time) of turbulent mixing and stratification processes, and their characterization in terms of parameters such as L_T and K_Z continue to be widely-studied [24–27]. A novel approach to this problem has been proposed by [28], who adopted a statistical physics perspective and characterised the efficiency (increase in potential energy to total energy input ratio) as a distinction between changes in the coarse-grained buoyancy profile, which represents the irreversible increase in potential energy, to the remaining energy, which is lost to fine scale fluctuations of velocity and buoyancy. They found that the variation of mixing efficiency with the Richardson number strongly depended on the background buoyancy profile, and that the mixing efficient has a maximum value of 0.25, which agreed well with predictions based on the more usual kinematics-based approach. This is consistent with measurements from the field [29] and laboratory [30].

This perspective suggests that considering the scale spectrum of layers in the microscale temperature profile might provide a way of understanding how both the currently active turbulence, and that which created the multi-scale layering of the temperature profile prior to its recording, might provide insights regarding the parameterisation of the turbulent mixing process that is required for it to be robustly incorporated into lake and ocean models. Therefore, given that this aspect of microstructure profiles does not appear to have been considered in the literature previously, this paper aims to:

- investigate the layered structure of microscale temperature profiles;
- identify its essential properties and determine whether any of them are universal;
- determine whether they vary consistently in relation to other parameters; assess whether (and how) they can be interpreted as a diagnostic tool for understanding turbulence mixing processes and their consequences better.

2. Materials and Methods

2.1. Site Description and Data Collection

To explore the multi-scale layering structure concept described above, thermal microstructure profiles taken in two lakes, Esthwaite Water and Blelham Tarn, during the summer stratified period of 2008 using a Self-Contained Automated Micro-Profiler (SCAMP, Precision Measurement Engineering Inc., San Diego, CA, USA) were used. This data does not have any characteristics that distinguish it from microstructure profile data taken in any other waterbodies, it was chosen only because it was readily available. Both lakes are small, glacially-scoured and lie within the catchment of Windermere in the Lake District of Northwest England. The larger of the two, Esthwaite Water (54.36° N, 2.99° W), has a surface area of 0.96 km^2, a total volume of 6.7×10^6 m^3 and a mean depth of 6.9 m [31]. Blelham Tarn (54.40° N, 2.98° W) has a surface area of 0.1 km^2 and a mean depth of 6.8 m [32]. Thus, they are of similar depth, but Esthwaite Water has a surface area approximately ten times that of Blelham Tarn.

SCAMP was operated in upward-looking mode (see, for example, [33]). It was deployed from a boat with weights and a baffle attached, which caused it to drop through the water column at an angle of approximately 45° to the vertical. When it reached a user-defined depth, a pressure sensor caused a screw to turn, which released the weights. This made the instrument positively buoyant, so that it rose vertically through water thus undisturbed by its descent, at a speed of approximately 0.1 ms^{-1}, recording temperature and depth (pressure) at 100 Hz (thus generating data points with approximately 1 mm spatial resolution). Once it reached the surface, the weights were recovered and re-attached, the pressure sensor re-set and the next profile initiated.

The analysed profiles were recorded on eight days in Esthwaite Water (22 May, 16 June, 21 July, 31 July, 4 August, 1 September, 2 October and 3 October) and four days in Blelham Tarn (9 June, 23 June, 28 July and 11 August). All of these days fell within the summer stratified period for the lakes, which runs from onset in March or April to turnover in October. The profiles were all recorded during the daytime, between approximately 9:30 a.m. and 5:00 p.m., the time for each day being dependent on logistical arrangements. On each day, a group of six profiles were recorded, separated by between five and ten min, thus covering a total of 30 to 60 min in total. All measurements were taken at the deepest point in each lake, the profiles having maximum depths of approximately 14 m.

2.2. Data Processing

Each profile was converted from its raw form into ASCII format files using processing software provided by the manufacturer. The profiles were then truncated at the bottom, where data recording had begun before the SCAMP's upward travel had started, and at the top, where recording had continued after it had breached the surface. The truncated profiles thus began at the deepest point recorded in the raw profiles, and ended where the (pre-calibrated) pressure measurements indicated zero depth.

As the mm-resolution of the raw profiles was only approximate, the truncated profiles were then interpolated to exactly 1 mm-depth resolution using an inverse distance-weighted mean of all recorded temperature data within 5 mm of each point on the 1 mm-resolution scale. To remove noise from the data, and thus prevent false identification of turbulent overturns, the de-noising method of [34] was then applied to each profile, following [19]. In the original version of this method, the noise threshold is defined in terms of the density. Since the data used here was temperature data, the density threshold needed to be converted to a temperature threshold. However, in standard practice, density is calculated from temperature using a fifth order polynomial [35] and inversion of fifth order polynomials is intractable. To convert the previously-used density noise threshold to a temperature noise threshold, therefore, a linear fit of the density-temperature relationship in the range 8–22 °C (where all of our temperature data lay) was performed, and the gradient of this was used to convert the density threshold of [34] (5×10^{-4} kgm^{-3}) to the temperature threshold of 3.5×10^{-3} °C, which was used in this cleaning process. The results of both the interpolation and noise removal processes are illustrated in Figure 2. At the relatively fine scale indicated in this figure, the "cleaned" (noise removed) profile may appear chunky, and possibly artificial, at first sight. However, this is simply the result of cleaning the profile to appropriate spatial and temperature resolutions of 1 mm and 3.5×10^{-3} °C, respectively, following [19,34]. The profiles thus cleaned are those which are analysed as described below.

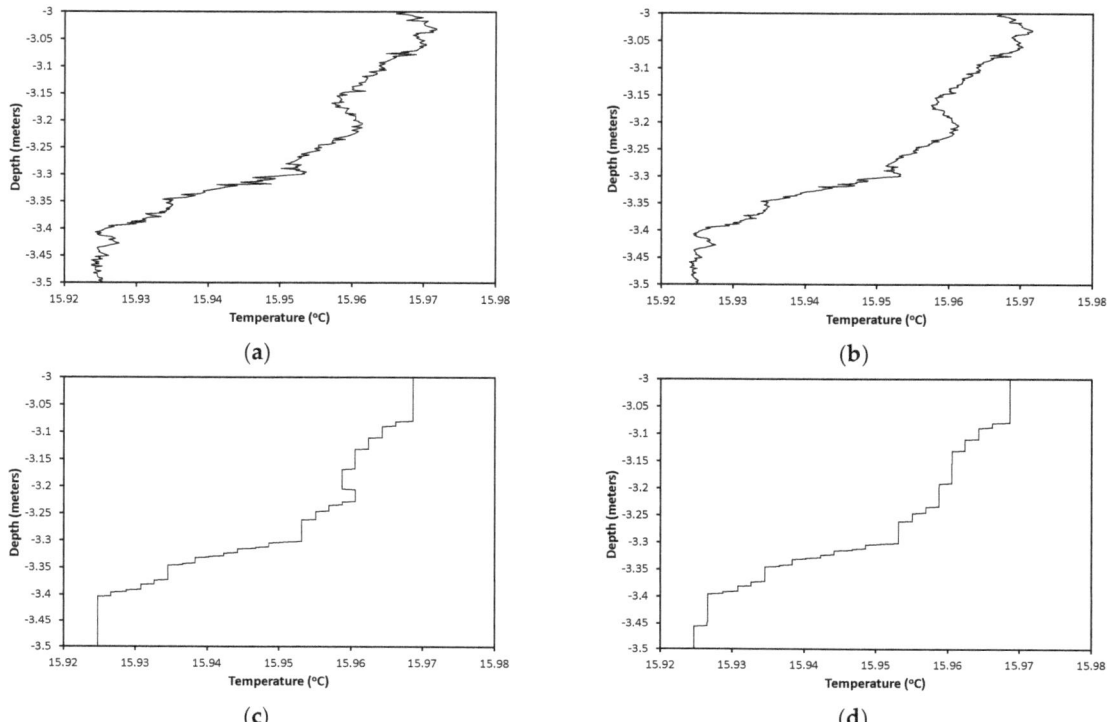

Figure 2. Examples of profile segments illustrating the effects of interpolation to 1 mm vertical spatial resolution and temperature resolution of 3.5×10^{-3} °C: (**a**) raw profile; (**b**) spatially interpolated profile; (**c**) spatially and temperature interpolated profile; (**d**) Thorpe-ordered version of (**c**). Profile recorded in Blelham Tarn at 11:00 on 11 August 2008).

2.3. Identification of Mixing Layers and Thorpe Scales

The Thorpe scale at each depth in each mm-resolution profile was calculated from a 0.5-m depth window centered on that depth (truncated by the start or end of the profile for points within 0.25 m of the top or bottom of the profile). These were converted to centered Thorpe displacements for the purpose of illustration using the method of [36].

To identify and delineate actively mixing patches within the profiles, a slightly different approach to the $\delta_T = 0$ method was adopted. Each temperature-depth profile was sorted by temperature (as one would to calculate Thorpe displacements), then cumulatively summed via the (re-sorted) depths of this profile, and at each depth this sum was compared with the cumulative sum of the unsorted (i.e., monotonically-ordered) depth profile. The point where these two sums are first equal (working from the top downwards) is the point above which the mixing is entirely 'self-contained': all of the points in this region in the sorted profile are also within it in the unsorted profile, and no points from outside it have been moved into it by the sorting process. This is defined as the uppermost mixing patch. Since the sums at this point are equal, the process re-sets and the next point down at which the sums again become equal marks the bottom of the next mixing patch down. In theory, this is a more robust method of patch identification than the $\delta_T = 0$ method, since it avoids the possibility of a point within the sorted profile being (coincidentally) at the same point in the unsorted profile (which would give $\delta_T = 0$ at that depth) but being surrounded by points that have been mixed upwards and downwards across it (i.e., being within a mixing patch, not at its edge). In practice, however, this method segmented the profiles in a manner that was indistinguishable by visual inspection from the segmentation done using the $\delta_T = 0$ method. Nevertheless, it was used hereinafter and is presented as a very simple, and potentially more robust, method of profile segmentation into distinct mixing patches.

2.4. Calculation of K_Z

The Batchelor method of calculating K_Z was applied to each of the six profiles from each date, using software incorporating this method provided by SCAMP's manufacturer, PME Inc. Although there has been much discussion of the variability of R_f and Γ in the literature [13,37], the standard approach of assigning a value of $\Gamma = 0.2$ was taken. Mean and standard deviation values of K_Z were calculated for each 0.5-m depth bin from the surface downwards and plotted against the center point of each bin (0.25 m, 0.75 m etc.). The Thorpe scale method was then used to calculate K_Z at every depth (i.e., at millimeter resolution) in the profile using 0.5-m centered windows as described above for L_T, calculating the dynamic viscosity, ν, as a function of temperature. Mean and standard deviation values of these Thorpe-scale derived K_Z values were then calculated at the same depths used for the Batchelor method (i.e., 0.25 m, 0.75 m etc.).

2.5. Layer Structure Analysis

Analysis of the multi-scale layering of the Thorpe-ordered profiles was carried out using best-fit straight lines ("rulers") of lengths from 3 mm to the full profile length (Figure 3). For the 3 mm case, a straight line was first fitted to the first, second and third points in the profile, and its gradient (dT/dz) assigned to the center point of this set (i.e., the depth of the second point). This was repeated for the second, third and fourth points, assigning the gradient to the depth of the third point, and so on down to the bottom of the profile, the last gradient value being assigned to the penultimate depth point in the profile. The second derivative (d^2T/dz^2) profile was then calculated from this set of dT/dz values, using the same 3-mm spatial scale. This process was then repeated using a 5-mm spatial scale, and so on, up to the largest odd number less than the full length of the profile (only odd numbers were used so that the center point of each fit line corresponded to a specific depth in the profile). Thus, a full set of d^2T/dz^2 profiles at a spectrum of different spatial scales was obtained.

At depths where d^2T/dz^2 peaks, the temperature profile is changing most rapidly from a low-gradient (well-mixed) section to a high-gradient (highly-stratified) section; where there are troughs in d^2T/dz^2, the temperature profile is changing most rapidly the other way. Thus, these points identify "shoulders" in the temperature profile that distinguish relatively-mixed layers from relatively-stratified ones (Figure 4). The total number of these shoulders was calculated for each ruler-length, and a pseudo-spectrum (ruler-length, or scale S vs. number of shoulders or layers N_L) ws then plotted for each two-meter depth bin from the surface down to 14 m. These pseudo-spectra were generally closely fitted by power law regression lines, implying a fractal structure. Therefore, the fractal dimension, D, was calculated for each profile such that $N_L = aS^{-D}$. As well as the fractal dimension of the full spectrum, separate values of D for the fine, intermediate and coarse-scale sections of the spectrum were calculated. Correlations of all of these forms of D with N, L_T and K_Z were investigated using Pearson's product-moment correlation coefficients and their associated statistical significance (*p*-values).

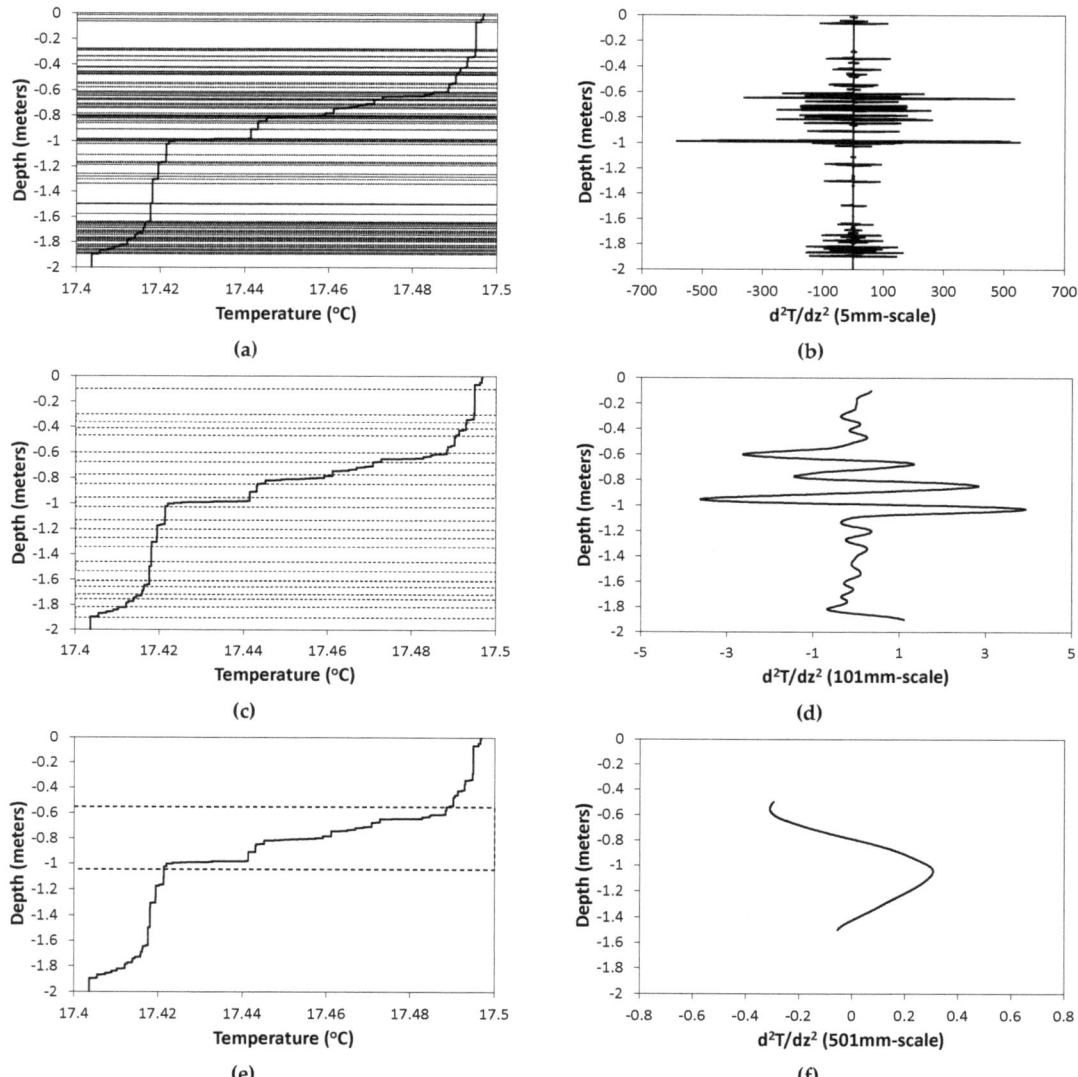

Figure 3. Illustration of the process of calculating the number of layers identified at different spatial scales, by counting turning points (peaks and troughs) in the d^2T/dz^2 profile: (**a**) location of layer edges ("shoulders") at 5 mm scale; (**b**) d^2T/dz^2 at 5 mm scale (**c**) location of layer edges at 101 mm scale; (**d**) d^2T/dz^2 at 101 mm scale; (**e**) location of shoulders at 501 mm scale; (**f**) d^2T/dz^2 at 501 mm scale.

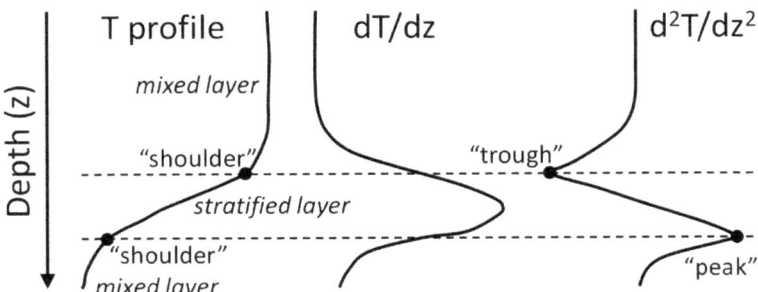

Figure 4. Sketch showing how troughs and peaks in the d^2T/dz^2 profile identify boundaries between alternately (relatively) strongly- and weakly-stratified layers in the temperature profile.

3. Results

3.1. Mixing and Stratification Structure

SCAMP temperature profiles from each of the twelve data collection days are presented in the left-hand panels of Figure 5 (Esthwaite Water) and Figure 6 (Blelham Tarn). These are cleaned, individual, unsorted profiles (rather than averages of each set of six). The segmentation of the profiles into separate mixing layers also provide a clear visual illustration of the proportion of the profile that is actively mixing at the time of each profile, and the locations of the mixing and non-mixing regions. This is also indicated by the centered Thorpe displacements, showing the coincidence of the mixing layers identified by the method used here and the $\delta_T = 0$ method. The seasonal timescale transition through a maximally-stratified state (e.g., 4 August) to a deepening surface mixed layer as overturn approaches (e.g., 3 October) can be seen clearly, especially in Esthwaite Water (Figure 5). Diurnal timescale changes can also be seen clearly by comparing the 2 and 3 October profiles in Figure 5. On 2 October, the profiles were taken in the afternoon, and the surface layers shows some stratification and a relatively quiescent state in terms of Thorpe displacements. On 3 October, the profiles were taken in the morning, the surface layers is much more completely mixed (presumably due to overnight convection) and much more actively stirring, as indicated by the much larger Thorpe displacements.

3.2. Comparison of K_Z Values

Plots of K_Z values for 0.5-m depth bins calculated using both the Batchelor and Thorpe scale methods are shown in the right-hand panels of Figure 5 (Esthwaite Water) and Figure 6 (Blelham Tarn). In general, the plots have the expected structure for a stratified lake, with relatively high K_Z values in the surface mixing layer and at the bottom of the profile in the near-bed layer, and lower values in the "quiet interior" at mid-depths [6]. In many places, the K_Z-values from the two methods agree very well. This agreement appears stronger in the surface layer, particularly later in the year in Esthwaite Water, and stronger in general in Blelham Tarn than in Esthwaite Water. In many other places, however, the agreement is poor. This is particularly the case at depth (and particularly in Esthwaite Water), where the Batchelor method values are generally larger than the Thorpe-scale method values, by at least an order of magnitude (and in some places several orders of magnitude) and are also as large or larger than the surface layer values.

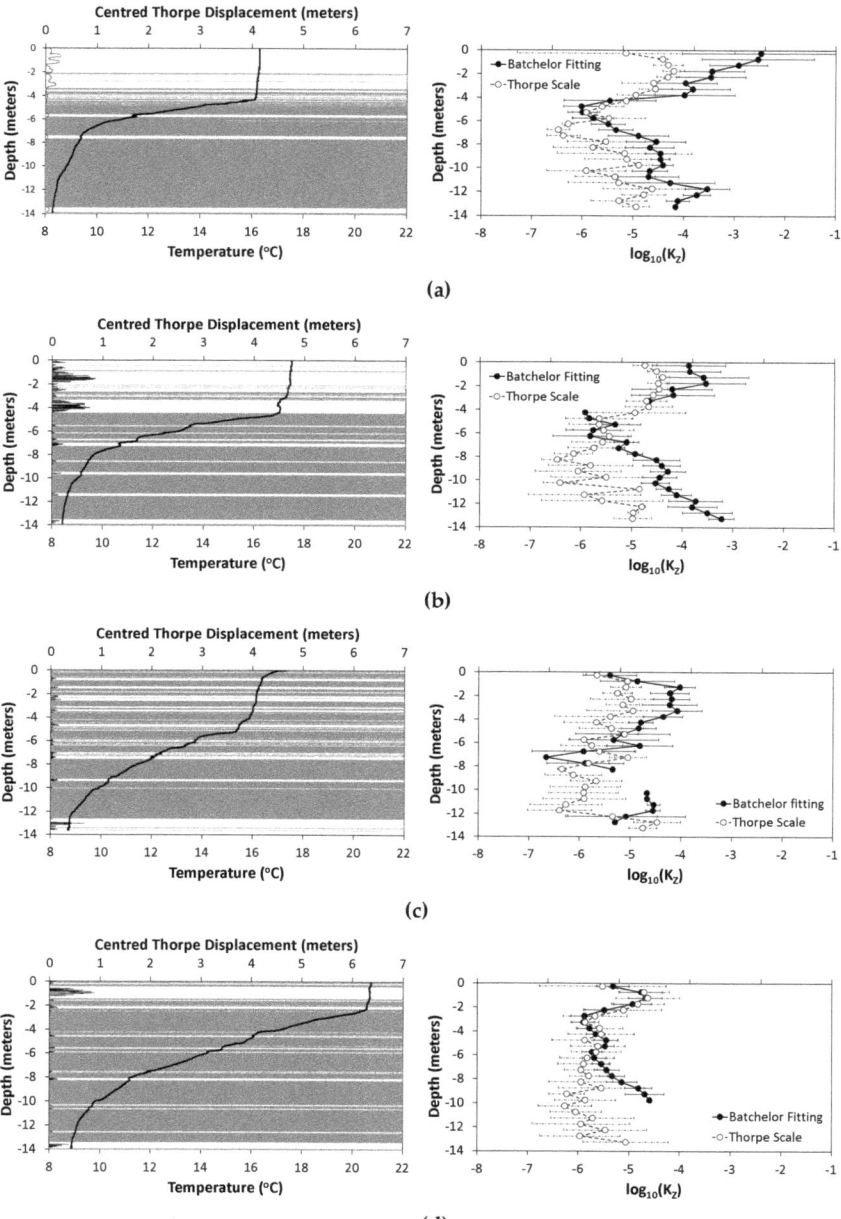

Figure 5. Cont.

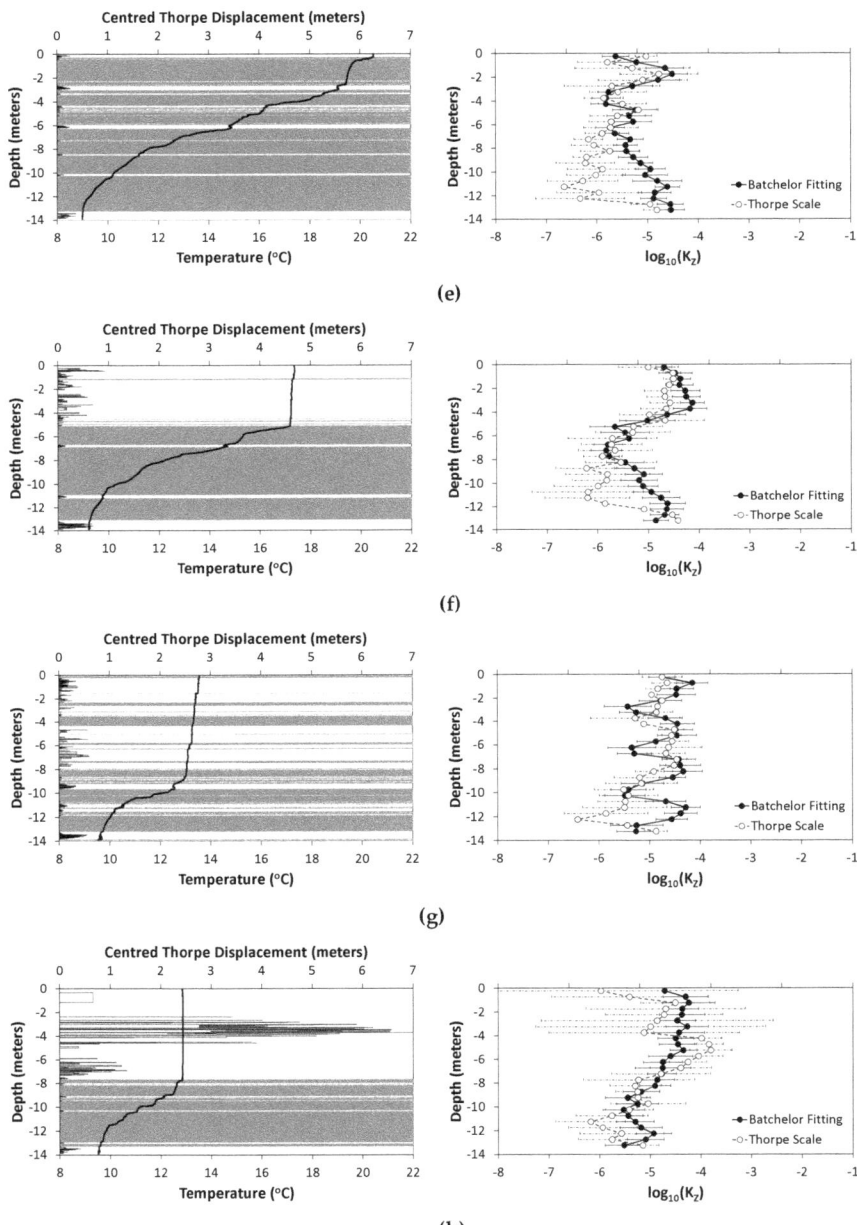

Figure 5. Plots showing (in left hand panels) temperature profile (thick black line), mixing layer segmentation (horizontal gray lines) and centered Thorpe displacements (black line at left hand side); and (in right hand panels) values of the coefficient of turbulent diffusivity (K_Z) using the two methods described in the text, for all sampling dates for Esthwaite Water. Profiles are individual examples from the set of six profiles taken on each date, recorded at (**a**) 16:21 on 22 May; (**b**) 12:33 on 16 June; (**c**) 12:15 on 21 July; (**d**) 11:25 on 31 July. (**e**) 12:07 on 4 August; (**f**) 12:38 on 1 September; (**g**) 16:05 on 2 October; (**h**) 09:56 on 3 October.

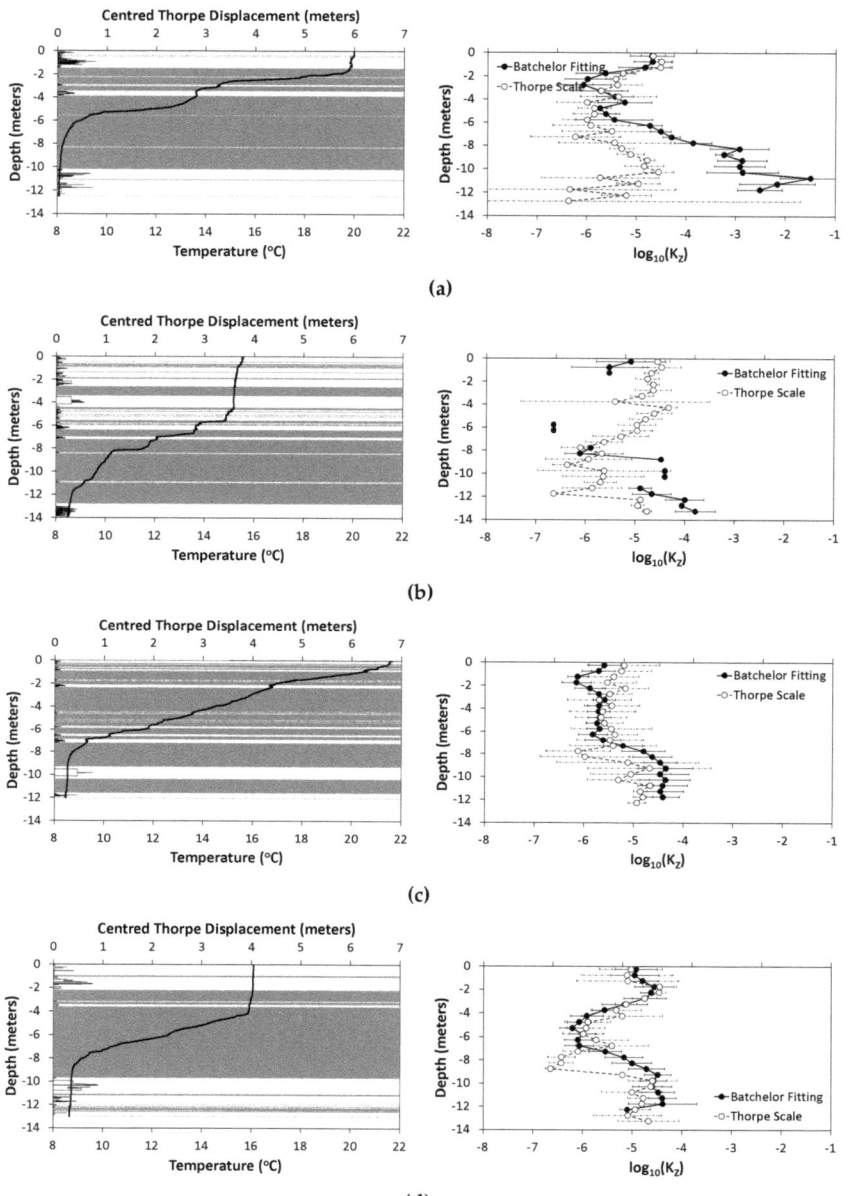

Figure 6. As for Figure 5, for Blelham Tarn. Profiles recorded at (**a**) 13:33 on 9 June; (**b**) 12:41 on 23 June; (**c**) 11:07 on 28 July; (**d**) 11:00 on 11 August.

3.3. Layer Structure Pseudo-Spectra

Figure 7a,b show two contrasting examples of layer structure pseudo-spectra, with power law lines fitted, taken from individual depth bins of the same profile (the one taken at 12:28 on 16 June). The equations of the best-fit power law lines are shown on these plots: the power index, which, as explained above, is denoted D (c.f. fractal dimension) is used to quantify the gradient of the dataset. Figure 7c shows the pseudo-spectra from all depth bins of this example profile together. Figure 7a illustrates a fractal structure across all scales (i.e., the gradient is much the same across the whole range of data values), whereas in Figure 7b the fitted power law line digresses strongly from the data, especially at smaller scales. The shapes of these pseudo-spectra are interpreted as follows: where they are steeper, the number of extras layers identified as the scale is reduced is higher, i.e., there is a lot of layering-structure at that particular scale; where they are flatter, there is little layering structure. Thus, in Figure 7b, the flatness of the data at smaller scales (say, <20 mm) implies that there are very few layers smaller than 20 mm in size, i.e., there is very little fine-scale structure in the profile.

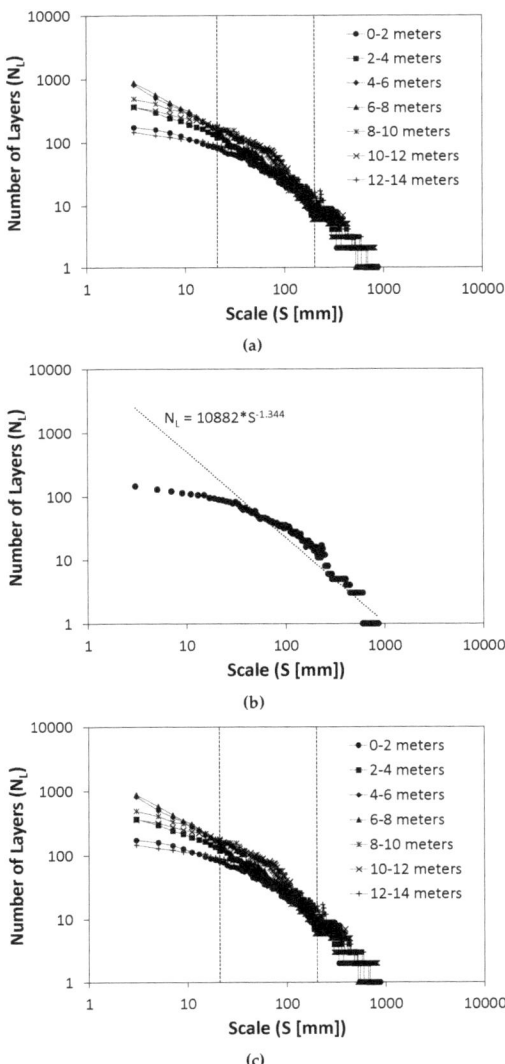

Figure 7. Examples of pseudo-spectra of the layering structure, all for the profile recorded at 12:28 on 16 June 2008 in Esthwaite Water. (**a**) depth bin from 4 to 6 m; (**b**) depth bin from 12 to 14 m; (**c**) all 2 m-depth bins together.

Following this interpretative approach, the gradient of the best fit power law line to the whole data set in each plot can be said to give a quantitative indication of the overall level of layering structure in the temperature profile. To gain insights into the amount of layering structure at different spatial scales, each spectrum is divided into three spatial scales, which are referred to hereinafter as "fine", "intermediate" and "coarse", and best-fit power law lines are calculated for each one. This division is based on a subjective judgment of where the breaks in the gradient of the spectra tend to fall, derived from visual inspection of spectra from all depth bins of all profiles. Thus, the fine-scale is defined as <20 mm, the intermediate scale as >20 mm but <200 mm, and the coarse scale as >200 mm. These divisions are shown in Figure 7c. The gradients for each of these scales are denoted D_F, D_I and D_C, respectively.

3.4. Variation of D with Depth, between Lakes and over Time

To explore the variation of the fractal dimension of the profiles, their variations with depth, between lakes and over seasonal and diel timescales are first considered. Differences in the mean values of D in each 2-m depth bin between the two lakes are shown in Figure 8. The mean values and standard deviations in this plot have been calculated from data from all profiles, thus seasonal variations are both hidden in, and serve to smear out, the standard deviation bars. The value of D peaks deeper in Esthwaite Water than in Blelham Tarn, and is smaller at the bottom of Blelham Tarn than in Esthwaite Water. But overall, both lakes show a similar pattern of lower values at the surface and bottom, and higher values at mid-depths. This suggests an association between less layering structure (lower values of D) and more weakly-stratified and mixed (or mixing) parts of the profiles.

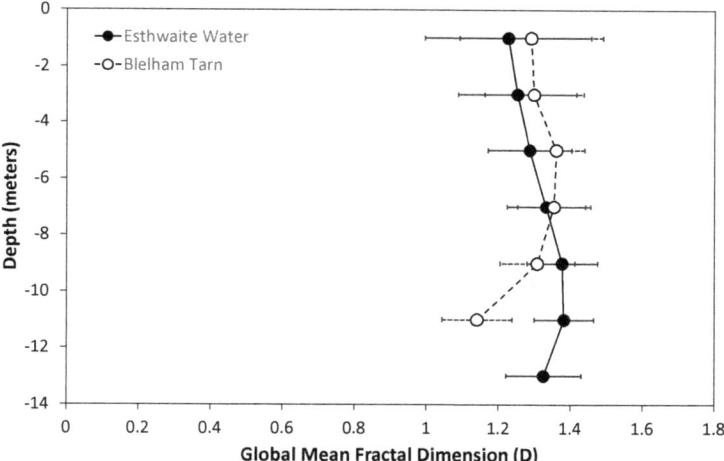

Figure 8. Mean (± one standard deviation) values of D in each 2 m-depth bin, averaged over all profiles from all dates in each lake.

The variation in profiles of D over seasonal and diel timescales is illustrated in Figure 9. Again, the very weakly-stratified, strongly mixing upper part of the 3 October profile is characterised by smaller values of D, indicating less layering structure, while the values of D for 2 October are higher, and those for 4 August are higher still.

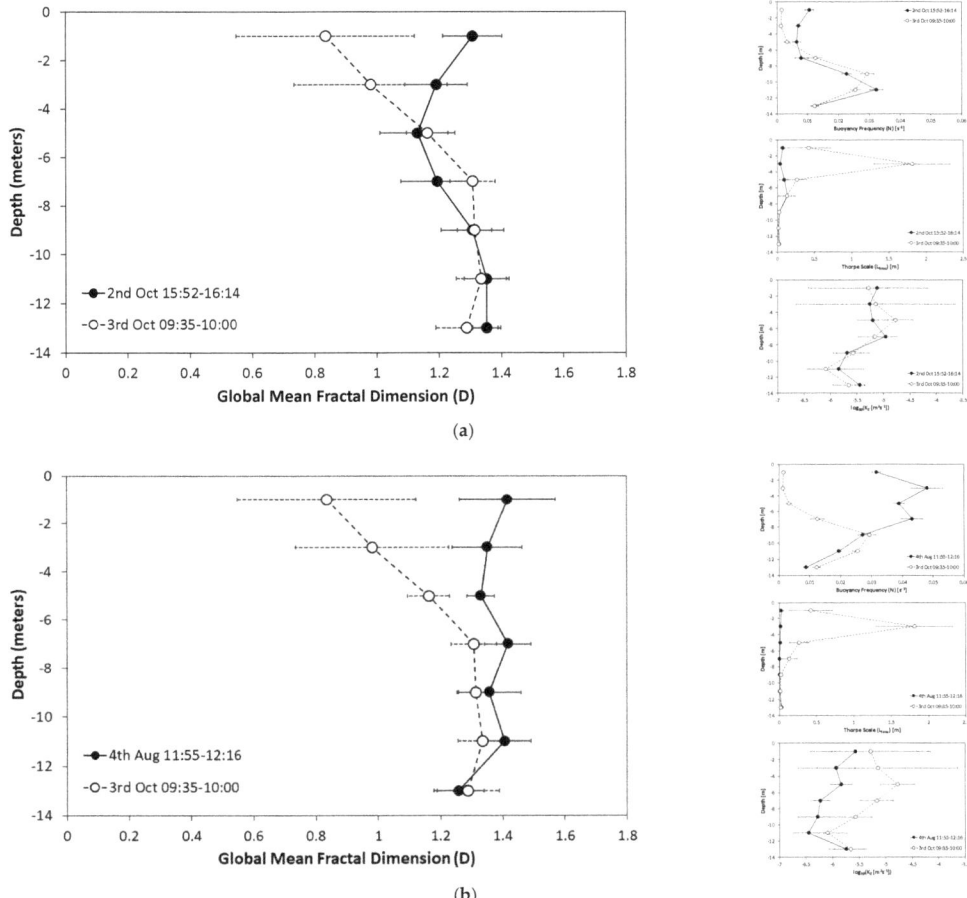

Figure 9. Comparisons of mean (± one standard deviation) values of D in each 2 m-depth bin (left hand panels) for profiles separated by (**a**) daily variations and (**b**) seasonal variations. Right hand panels show buoyancy frequency (top), root mean square Thorpe scale (middle) and $\log_{10}(K_Z)$ (bottom) for each date, for comparison.

3.5. Variation of D with N, L_T and K_Z

To explore further the way in which D varies, it and its scale-specific forms D_F, D_I and D_C were plotted against the three standard parameters which quantify stratification and mixing—the buoyancy frequency N, the Thorpe scale L_T and the coefficient of vertical turbulent diffusivity K_Z (using the Thorpe-scale derived version of the last parameter). The relationship between D and N (Figure 10) is not well-fitted by any standard form of regression line but does have very clear limits. There are no cases where D < 1.15 and N > 0.02 s^{-1}, and no cases where N > 0.01 s^{-1} and D < 1. For D > 1.15, there appears to be no relationship between D and N. The maximum D is approximately 1.55 for all N > 0.01 s^{-1}, whereas for N < 0.01 s^{-1} it declines sharply with N. Similarly, the minimum D is approximately 1.17 for all N > 0.02 s^{-1}, whereas it declines sharply with N for N < 0.02 s^{-1}. Data from both lakes follow all these patterns in essentially the same way. Amongst the scale-specific versions of D, the relationship with N is strongest for D_F, where a very clear relationship is observed, which is best fitted by an exponential regression line (r = 0.902; *n* = 290; *p* << 0.001 for Esthwaite Water; r = 0.964; *n* = 104; *p* << 0.001 for

Blelham Tarn). This association is weaker, but still evident for D_I, although the shape of this plot has more in common with that of D than that of D_F (i.e., no relationship between D_I and N above small values of each parameter, rather than an exponential relationship). The relationship between D_C and N appears essentially non-existent. This implies that the layering structure at fine scales (<20 mm) is strongly associated with the buoyancy frequency (which is calculated as an average value for each two-meter depth bin, so can be thought of as a background or ambient value), but that this association weakens as the scale of the layering increases, so that layering structure at scales of >200 mm has no significant association with the background buoyancy frequency.

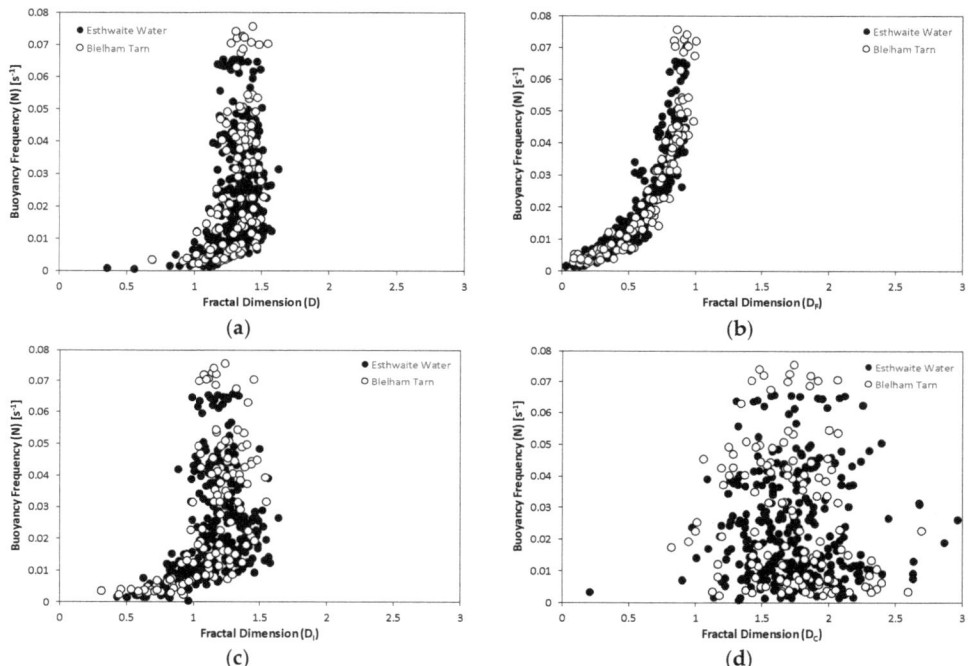

Figure 10. Plots of buoyancy frequency against (**a**) D; (**b**) D_F; (**c**) D_I; and (**d**) D_C. Data from all depths bins, all profiles and all dates combined.

The relationships between D (and its scale-specific variants) and the parameters which quantify turbulent stirring and mixing, L_T and (Thorpe scale-derived) K_Z show a consistent pattern (Figures 11 and 12). For both lakes, there is a strongly significant negative linear correlation between D and log(L_T) (r = −0.533; n = 290; p << 0.001 for Esthwaite Water; r = −0.442; n = 104; p << 0.001 for Blelham Tarn), and D and log(K_Z) (r = −0.304; n = 290; p << 0.001 for Esthwaite Water; r = −0.298; n = 104; p = 0.002 for Blelham Tarn), with larger D values (more layering structure) corresponding to smaller L_T values and vice versa, but there is also a great deal of scatter around this general trend. Similarly, there is a strongly significant negative linear correlation between D and log(K_Z) (r = −0.304; n = 290; p << 0.001 for Esthwaite Water; r = −0.298; n = 104; p = 0.002 for Blelham Tarn), implying that larger values of D are found where there is less turbulent diffusion. For both L_T and K_Z, the correlation is stronger with D_F (particularly for K_Z) and D_I (particularly for L_T) (i.e., at scales < 200 mm). As is the case with N, there is no significant association of L_T or K_Z with D_C. This implies that the active stirring and mixing processes represented in the turbulent overturns in the temperature profiles are most strongly associated with layering structure in the profile at fine and intermediate scales.

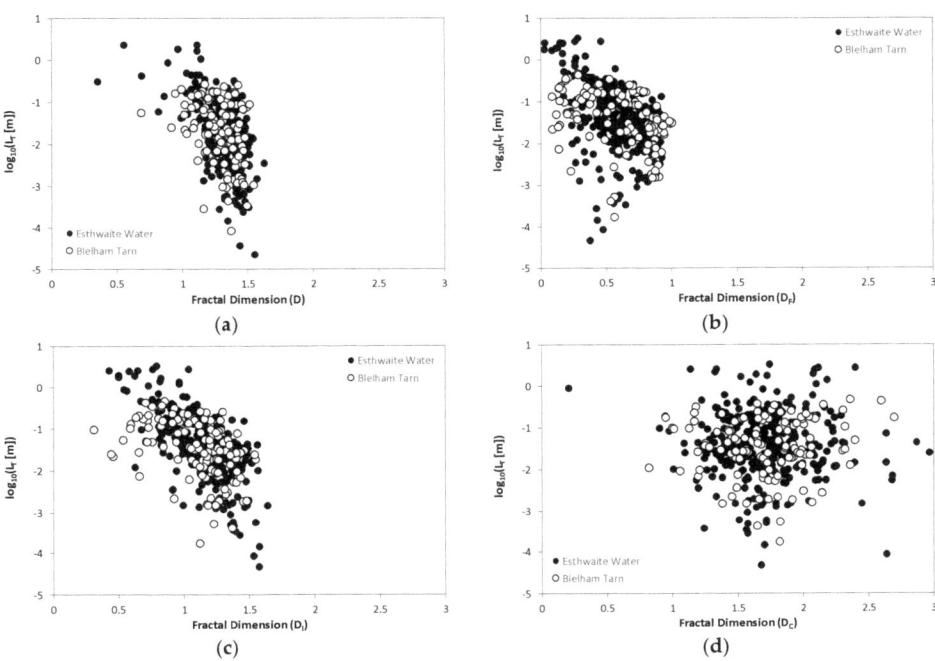

Figure 11. Plots of $\log_{10}(L_T)$ against (**a**) D; (**b**) D_F; (**c**) D_I; and (**d**) D_C. Data from all depths bins, all profiles and all dates combined.

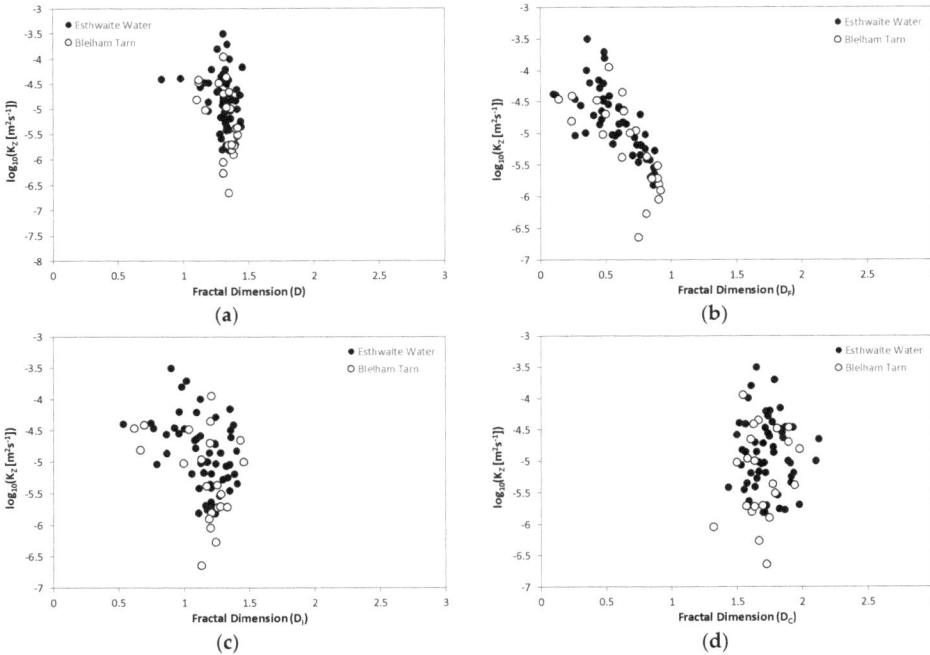

Figure 12. Plots of $\log_{10}(K_Z)$, calculated using the Batchelor method, against (**a**) D; (**b**) D_F; (**c**) D_I; and (**d**) D_C. Data from all depths bins, all profiles and all dates combined.

4. Discussion

The profiles of temperature and centered Thorpe displacement, together with the mixing layers indicated by the segmentation lines (i.e., the left-hand panels in Figures 5 and 6) show clearly the limited extent of active mixing during the stratified season, its tendency to be most common near the water surface and lake bed, and its increasing prevalence as the stratification breaks down as autumn sets in. This is all very much in agreement with normal expectations for turbulent mixing in lakes [6]. The mixing layer segmentation lines provide a clear visual idea of the proportion of the water column that is actively mixing and the extent to which the mixing layers are separated by non-mixing regions. The changes in the patterns of these aspects of the profiles are conflations of variations on timescales varying from sub-hourly to seasonal, so no attempts have been made to find any consistent trends in them. This is a good example of the way in which collection of microstructure profiles strongly under-samples the temporal variations in turbulence mixing activity.

With regard to the K_Z estimates, the Batchelor method is long-established, and its veracity has been demonstrated in a very wide range of field, laboratory and numerical studies. The Thorpe-scale method, on the other hand, has only been quite recently proposed, and deployed in the analysis of oceanographic, rather than limnological, data [19]. It seems appropriate, therefore, to assume that the Batchelor method values of K_Z are the "correct" ones, and to ask why the Thorpe-scale method values do not equate to them. From the plots in Figures 5 and 6, the two sets of values appear to agree better with each other in the top half of the water column (down to 6 m depth), i.e., above and within the upper part of the metalimnion, than they do below this. Regression of the two sets of values bears this out: for data from <6 m depth (from both lakes, all dates combined), r = 0.662 ($n = 144$; $p = 1.55 \times 10^{-19}$), whereas for data from below that depth, r = 0.250 ($n = 177$; $p = 8 \times 10^{-4}$). Moreover, the Thorpe scale method appears incapable of producing values at the larger end of the of values provided by the Batchelor method: for the Thorpe scale method, the range of $\log_{10}(K_Z)$ values across both lakes, and all depths and dates is -6.6 to -4.3, whereas for the Batchelor method, it is -6.6 to -1.5. In the upper 6 m of the water column, the values in the range -4.3 to -1.5 provided by the Batchelor method all occur in the two earliest profiles in the year (22 May and 16 June). These are unusually high values of K_Z for lakes (c.f. typical values quoted by, for example, [6]), but they are not outliers in the profiles from those dates, so there appears no reason to treat them as any less reliable than the other Batchelor method values. Moreover, they occur in regions where stratification is very weak, and are consistent with the intuitive concept that mixing will be rapid in these regions, because it is not very constrained by buoyancy forces.

The Thorpe scale method parameterization of K_Z assumes that conditions are in the energetic regime ($\varepsilon/\nu N^2 > 100$) defined by [20]. This may explain why the values it provides in the deeper part of the water column, below the thermocline, match the Batchelor method values less well than in the upper part of the water column. There is relatively little turbulent mixing in these deeper waters, and the buoyancy frequency is also generally higher than in the upper waters. It is concluded that, in the circumstances studied here, the Thorpe scale method provides a reasonably accurate and relatively straightforward method of estimating K_Z, which provides values that compare closely to those provided by the Batchelor method above the thermocline, except at times when the Batchelor method indicates high values of K_Z. From the data in this study, "high" in this context means $\log_{10}(K_Z) > -4.3$.

With regard to the pseudo-spectra of layering structure, and specifically the values of the parameter D presented as a convenient quantification of them, the data in Figures 8–10 show that there is a strong relationship between D and the buoyancy frequency N, which persists when the data are averaged across dates and lakes (Figure 8), and when seasonal and daily timescale variations are considered (Figure 9). The plot of D_F vs. N in Figure 10 shows that this relationship is particularly strong when considering the fine-scale layering structure. Conversely, the plot of D_C vs. N shows that buoyancy frequency has no consistent correlation with D at coarse scales. As noted above, the parameters that quantify turbulent stirring and mixing, L_T and K_Z, correlate best with D at intermediate scales, and indicate that there is more layering structure when there is less stirring or mixing.

The fine scale layering structure identified by the analysis presented here and quantified by D_F can be taken as equivalent to the fine-scale structure identified by [28] and to be associated with the dissipation of turbulent kinetic energy (ε), while the coarse scale structure quantified by D_C can be associated with the irreversible changes to the potential energy caused by turbulent mixing (b). The intermediate scale, quantified by D_I, is representative of variations that are actively stirring and mixing, and which could go either way—they could break down into finer scale structure and dissipate away, or merge and smooth out into coarser-scale structure and become irreversible changes in potential energy. If D_F, D_I and D_C are thought of in this way, Figures 10 and 12 can be interpreted to make a number of suggestions. Firstly, the coarse-scale layering structure (i.e., the vertical distribution of potential energy) recorded in a microstructure profile is essentially unrelated to the turbulent stirring (L_T) and mixing (K_Z) occurring at the time that the profile was taken (which implies that it is due to previous turbulent activity) and is independent of the average buoyancy frequency (i.e., the amount of coarse-scale layering structure in any given layer is independent of whether that layer as a whole is strongly or weakly stratified). Secondly, the fine-scale layering structure is very strongly correlated with buoyancy frequency (there is more fine-scale structure in more stratified layers) such that there is more fine-scale structure in more stratified layers. It is also correlated with the Thorpe scale and diffusivity coefficient, but less strongly and in a negative sense. This seems counter-intuitive initially—one would expect a fine-scale structure indicative of more turbulent dissipation to be less prevalent in regions of stronger stratification and less turbulent stirring and mixing. Our interpretation of this finding is that it is pre-dominantly a consequence of the fine-scale layering structure being smoothed out quickly in regions of lower stratification, because of the smaller density differences between layers involved, but to be more persistent in more strongly-stratified regions because of the greater density differences between layers. The relationships between the fractal dimension at intermediate scales D_I and N, L_T and K_Z suggest a situation intermediate to those of D_F and D_C, indicating that a mix of the drivers determining these relationships at coarse and fine scales are operating at this scale.

The layering structure analysis has the potential to be useful because it analyses an aspect of the data that is indicative of the history of turbulent mixing, not just the mixing that is occurring at the time of the profiling. Ways in which it might be useful are (1) to provide measures of longer term averages of K_Z than current methods of analysing microstructure profilers provide; or (2) to provide alternative estimates of the efficiency of turbulent mixing (i.e., the value of R_f, and thus of Γ) that can be used to triangulate values provided by other methods. To test the first of these, an additional method of measuring the long-term average K_Z is required—for example the temperature diffusion method of [21] or the very similar dye diffusion method used by, for example [38], against which values of D and its scale-specific versions can be assessed. Neither of these were available in this study, so this suggestion remains as a proposal for further study. However, for the sake of providing an indication of what K_Z values this method might provide, the relationships between D and K_Z in Figure 12 are noted. To test (2), and further investigate the extent to which the fractal dimension parameters introduced here can be used to indicate mixing efficiency, requires comparison with results from numerical modelling [13,23].

5. Conclusions

The thermal microstructure profiles analysed in this study show the behaviour expected in terms of stratification structure, turbulent mixing activity and vertical variation in the thermal diffusivity coefficient, K_Z. While the values of K_Z calculated using two different methods—Batchelor curve fitting to the temperature gradient spectrum; and calculation directly from measurements of N and L_T using the equation of [19] based on the parameterization of K_Z of [20]—show good agreement in many cases, they also differ strongly in other cases. Given that the Batchelor curve fitting method is very well established and its accuracy has been demonstrated in many previous studies of small lakes, it is recommended that the Thorpe-scale method, whilst attractive for its simplicity and directness, is only used to calculate K_Z values above the thermocline and during the strongly stratified period in mid to late summer in lakes such as those studied here.

The novel analysis of the layering structure and its pseudo-spectra presented here shows that they have some properties that are consistent across the datasets used here, and other properties that vary consistently with other parameters. Values of the parameter D—the slope of the pseudo-spectrum—vary most consistently with the buoyancy frequency, especially D_F, the fine-scale specific version of D.

The main limitation of the findings presented here is that, at present, the novel parameter derived, D, has no clear practical use. To address this, it is suggested that D, and its scale-specific variants D_F, D_I and D_C, as defined here, may be useful in two ways: firstly, to provide measures of longer term averages of K_Z than current methods of analysing microstructure profilers provide; and secondly, to provide alternative estimates of the efficiency of turbulent mixing that can be used to triangulate values provided by other methods. Testing of the ability of these parameters to be of use in these ways requires further work involving field measurements of time-integrated values of K_Z (using, for example dye or thermal diffusion methods) and numerical modelling of turbulent mixing using, for example, DNS methods [13,23].

Funding: This research was funded by the UK Natural Environment Research Council, grant numbers NE/F00995X/1 and NE/G010498/1.

Institutional Review Board Statement: Not applicable.

Informed Consent Statement: Not applicable.

Data Availability Statement: The data presented in this study are openly available in Pure at https://www.research.lancs.ac.uk/portal/en/datasets/search.html. (accessed on 26 September 2021).

Acknowledgments: I am grateful to Fanghua Li, Rebecca Messham and Eleanor Mackay for their roles in the data collection fieldwork, and to Ian Jones and Joshua Arnott for useful discussions about the data and its analysis.

Conflicts of Interest: The author declares no conflict of interest.

References

1. Caulfield, C.P. Layering, Instabilities, and Mixing in Turbulent Stratified Flows. *Ann. Rev. Fluid Mech.* **2021**, *53*, 113–145. [CrossRef]
2. Roget, E.; Lozovatsky, I.D.; Sanchez, X.; Figueroa, M. Microstructure measurements in natural waters: Methodology and applications. *Prog. Oceanogr.* **2006**, *70*, 126–148. [CrossRef]
3. MacIntyre, S. Vertical mixing in a shallow, eutrophic lake: Possible consequences for the light climate of phytoplankton. *Limnol. Oceanogr.* **1993**, *38*, 798–817. [CrossRef]
4. Piera, J.; Quesada, R.; Catalan, J. Estimation of nonlocal turbulent mixing parameters derived from microstructure profiles. *J. Mar. Res.* **2006**, *64*, 123–145. [CrossRef]
5. Goto, Y.; Yasuda, I.; Nagasawa, M. Comparison of Turbulence Intensity from CTD-Attached and Free-Fall Microstructure Profilers. *J. Atmos. Ocean. Tech.* **2018**, *35*, 147–162. [CrossRef]
6. Wuest, A. ALorke Small-scale hydrodynamics in lakes. *Ann. Rev. Fluid Mech.* **2003**, *35*, 373–412. [CrossRef]
7. Fine, E.C.; Alford, M.H.; MacKinnon, J.A.; Mickett, J.B. Microstructure Mixing Observations and Finescale Parameterizations in the Beaufort Sea. *J. Phys. Oceanogr.* **2021**, *51*, 19–35. [CrossRef]
8. Guo, S.-X.; Cen, X.-R.; Qu, L.; Lu, Y.-Z.; Huang, P.-Q.; Zhou, S.-Q. Quantifying Flow Speeds by Using Microstructure Shear and Temperature Spectral Analysis. *J. Atmos. Ocean. Tech.* **2021**, *38*, 645–656. [CrossRef]
9. Le Boyer, A.; Alford, M.H.; Couto, N.; Goldin, M.; Lastuka, S.; Goheen, S.; Nguyen, S.; Lucas, A.J.; Hennon, T.D. Modular, Flexible, Low-Cost Microstructure Measurements: The Epsilometer. *J. Atmos. Ocean. Tech.* **2021**, *38*, 657–668. [CrossRef]
10. Thorpe, S.A. Turbulence and mixing in a Scottish loch. *Phil. Trans. Roy. Soc. A* **1997**, *286*, 125–181. [CrossRef]
11. Brainerd, K.E.; Gregg, M.C. Surface mixed and mixing layer depths. *Deep-Sea Res. Part I* **1995**, *42*, 1521–1543. [CrossRef]
12. Moum, J.N. Energy-containing scales of turbulence in the ocean thermocline. *J. Geophys. Res.* **1996**, *101*, 14095–14109. [CrossRef]
13. Smyth, W.D.; Moum, J.N.; Caldwell, D.R. The Efficiency of Mixing in Turbulent Patches: Inferences from Direct Simulations and Microstructure Observations. *J. Phys. Oceanogr.* **2001**, *31*, 1969–1992. [CrossRef]

14. Thomson, R.E.; Fine, I.I. Estimating mixed layer depth from Oceanic Profile Data. *J. Atmos. Ocean. Tech.* **2003**, *20*, 319–329. [CrossRef]
15. Stevens, C.; Ward, B.; Law, C.; Walkington, M. Surface layer mixing during the SAGE ocean fertilization experiment. *Deep-Sea Res. II* **2011**, *58*, 776–785. [CrossRef]
16. Lozovatsky, I.D.; Roget, E.; Fernando, H.J.S.; Figueroa, M.; Shapovalov, S. Sheared turbulence in a weakly stratified upper ocean. *Deep Sea Res. Part I* **2006**, *53*, 387–407. [CrossRef]
17. Batchelor, G.K. Small-scale variation of convected quantities like temperature in turbulent fluid Part 1. General discussion and the case of small conductivity. *J. Fluid Mech.* **1959**, *5*, 113–133. [CrossRef]
18. Luketina, D.A.; Imberger, J. Determining turbulent kinetic energy dissipation from Batchelor curve fitting. *J. Atmos. Ocean. Tech.* **2001**, *18*, 100–113. [CrossRef]
19. Park, Y.H.; Lee, J.H.; Durand, I.; Hong, C.S. Validation of Thorpe-scale-derived vertical diffusivities against microstructure measurements in the Kerguelen region. *Biogeosciences* **2014**, *11*, 6927–6937. [CrossRef]
20. Shih, L.H.; Koseff, J.R.; Ivey, G.N.; Ferziger, J.H. Parameterization of turbulent fluxes and scales using homogeneous sheared stably stratified turbulence simulations. *J. Fluid Mech.* **2005**, *525*, 193–214. [CrossRef]
21. Jassby, A.; Powell, T. Vertical patterns of eddy diffusion during stratification in Castle Lake, California. *Limnol. Oceanogr.* **1975**, *20*, 530–543. [CrossRef]
22. Weck, J.; Lorke, A. Mixing efficiency in the thermocline of lakes observed from eddy correlation flux measurements. *J. Geophys. Res. Ocean.* **2017**, *122*, 291–305. [CrossRef]
23. Ivey, G.N.; Winters, K.B.; Koseff, J.R. Density Stratification, Turbulence, but How Much Mixing? *Ann. Rev. Fluid Mech.* **2008**, *40*, 169–184. [CrossRef]
24. Stretch, D.D.; Venayagamoorthy, S.K. Diapycnal diffusivities in homogeneous stratified turbulence. *Geophys. Res. Lett.* **2010**, *37*, L02602. [CrossRef]
25. Bluteau, C.E.; Jones, N.L.; Ivey, G.N. Turbulent mixing efficiency at an energetic ocean site. *J. Geophys. Res.* **2013**, *118*, 4662–4672. [CrossRef]
26. Maffioli, A.; Brethouwer, G.; Lindborg, E. Mixing efficiency in stratified turbulence. *J. Fluid Mech.* **2016**, *794*, R3. [CrossRef]
27. Salehipour, H.; Peltier, W.R.; Whalen, C.B.; MacKinnon, J.A. A new characterization of the turbulent diapycnal diffusivities of mass and momentum in the ocean. *Geophys. Res. Lett.* **2016**, *43*, 3370–3379. [CrossRef]
28. Venaille, A.; Gostiaux, L.; Sommeria, J. A statistical mechanics approach to mixing in stratified fluids. *J. Fluid Mech.* **2017**, *810*, 554–583. [CrossRef]
29. Vladoiu, A.; Bouruet-Aubertot, P.; Cuypers, Y.; Ferron, B.; Schroeder, K.; Borghini, M.; Leizour, S.; Ismail, B.S. Mixing efficiency from microstructure measurements in the Sicily Channel. *Ocean. Dyn.* **2019**, *69*, 787–807. [CrossRef]
30. Zhou, Q.; Taylor, J.R.; Caulfield, C.P. Self-similar mixing in stratified plane Couette flow for varying Prandtl number. *J. Fluid Mech.* **2017**, *820*, 86–120. [CrossRef]
31. Mackay, E.B.; Jones, I.D.; Folkard, A.M.; Barker, P. Contribution of sediment focussing to heterogeneity of organic carbon and phosphorus burial in small lakes. *Freshw. Biol.* **2012**, *57*, 290–304. [CrossRef]
32. Ramsbottom, A.E. *Depth Charts of the Cumbrian Lakes*, 1st ed.; Freshwater Biological Association: Windermere, UK, USA, 1976; 39p.
33. Soga, C.L.M.; Rehmann, C.R. Dissipation of turbulent kinetic energy near a bubble plume. *ASCE J. Hydraul. Eng.* **2004**, *130*, 441–449. [CrossRef]
34. Gargett, A.; Garner, T. Determining Thorpe scales from shiplowered CTD density profiles. *J. Atmos. Ocean. Tech.* **2008**, *25*, 1657–1670. [CrossRef]
35. Gill, A.E. *Atmosphere-Ocean Dynamics*, 1st ed.; International Geophysics Series; Academic Press: London, UK, 1982; Volume 30, 681p.
36. Imberger, J.; Boashash, B. Application of the Wigner–Ville Distribution to Temperature Gradient Microstructure: A New Technique to Study Small-Scale Variations. *J. Phys. Oceanogr.* **1986**, *16*, 1997–2012. [CrossRef]
37. Lozovatsky, I.D.; Fernando, H.J.S. Mixing efficiency in natural flows. *Phil. Trans. Roy. Soc. A* **2013**, *371*, 20120213. [CrossRef] [PubMed]
38. Heinz, G.; Ilmberger, J.; Schimmele, M. Vertical mixing in Überlinger See, western part of Lake Constance. *Aquat. Sci.* **1990**, *52*, 256–268. [CrossRef]

Article

Stratification in a Reservoir Mixed by Bubble Plumes under Future Climate Scenarios

David Birt [1,*], Danielle Wain [2,3,*], Emily Slavin [4], Jun Zang [1], Robert Luckwell [5] and Lee D. Bryant [1]

1. Department of Architecture & Civil Engineering, University of Bath, Claverton Down, Bath BA2 7AY, UK; jz235@bath.ac.uk (J.Z.); lb712@bath.ac.uk (L.D.B.)
2. 7 Lakes Alliance, 137 Main Street, Belgrade Lakes, ME 04918, USA
3. Colby College, University of Maine, Mayflower Hill Drive, Waterville, ME 04901, USA
4. Drinking Water Inspectorate, Area 1A, Nobel House, 17 Smith Square, London SW1P 3JR, UK; emily.slavin@defra.gov.uk
5. Bristol Water, Bridgwater Road, Bristol BS13 7AT, UK; Robert.Luckwell@bristolwater.co.uk
* Correspondence: djb95@bath.ac.uk (D.B.); danielle.wain@7lakesalliance.org (D.W.); Tel.: +44-791-093-2933 (D.B.); +1-207-205-6341 (D.W.)

Abstract: During summer, reservoir stratification can negatively impact source water quality. Mixing via bubble plumes (i.e., destratification) aims to minimise this. Within Blagdon Lake, a UK drinking water reservoir, a bubble plume system was found to be insufficient for maintaining homogeneity during a 2017 heatwave based on two in situ temperature chains. Air temperature will increase under future climate change which will affect stratification; this raises questions over the future applicability of these plumes. To evaluate bubble-plume performance now and in the future, AEM3D was used to simulate reservoir mixing. Calibration and validation were done on in situ measurements. The model performed well with a root mean squared error of 0.53 °C. Twelve future meteorological scenarios from the UK Climate Projection 2018 were taken and down-scaled to sub-daily values to simulate lake response to future summer periods. The down-scaling methods, based on diurnal patterns, showed mixed results. Future model runs covered five-year intervals from 2030 to 2080. Mixing events, mean water temperatures, and Schmidt stability were evaluated. Eight scenarios showed a significant increase in water temperature, with two of these scenarios showing significant decrease in mixing events. None showed a significant increase in energy requirements. Results suggest that future climate scenarios may not alter the stratification regime; however, the warmer water may favour growth conditions for certain species of cyanobacteria and accelerate sedimentary oxygen consumption. There is some evidence of the lake changing from polymictic to a more monomictic nature. The results demonstrate bubble plumes are unlikely to maintain water column homogeneity under future climates. Modelling artificial mixing systems under future climates is a powerful tool to inform system design and reservoir management including requirements to prevent future source water quality degradation.

Keywords: 3D modelling; stratification; bubble plumes

Citation: Birt, D.; Wain, D.; Slavin, E.; Zang, J.; Luckwell, R.; Bryant, L.D. Stratification in a Reservoir Mixed by Bubble Plumes under Future Climate Scenarios. *Water* **2021**, *13*, 2467. https://doi.org/10.3390/w13182467

Academic Editor: Lars Bengtsson

Received: 15 July 2021
Accepted: 25 August 2021
Published: 8 September 2021

Publisher's Note: MDPI stays neutral with regard to jurisdictional claims in published maps and institutional affiliations.

Copyright: © 2021 by the authors. Licensee MDPI, Basel, Switzerland. This article is an open access article distributed under the terms and conditions of the Creative Commons Attribution (CC BY) license (https://creativecommons.org/licenses/by/4.0/).

1. Introduction

During the summer months, increased atmospheric heating leads to many reservoirs stratifying as increased surface heating creates temperature differences in the water column [1,2]. Reservoir stratification is defined as when there is a temperature gradient within the water column [3]. Thermally stratified water bodies are stable and mixing is suppressed [4]; these physical effects can have considerable influence on the biological, chemical, and general ecosystem processes of the water body [5]. Correspondingly, summer stratification directly affects the water quality within reservoirs [6] via processes including benthic sediment oxygen demand and decomposition of organic matter which consume oxygen from the hypolimnion. When stratification persists for long enough, the inhibition of oxygen supply to the hypolimnion results in oxygen depletion and may result in hypoxic conditions (i.e., the near absence of oxygen at <2 mgL^{-1} dissolved oxygen). Whether a water body is hypoxic or not drives many critical biogeochemical processes including trace metal transport, phytoplankton dynamics, and the carbon cycle [3,7–9].

Artificial destratification (i.e., mixing) is often used in lakes and reservoirs to overcome negative effects of summer stratification [10]. A popular method used is bubble plumes [11]. The reported success of bubble plumes in reservoir management has been varied, suggesting that there is a lack of guidance regarding best operational practice [12]. Bubble plumes are usually installed at the deepest point of the water column and force compressed air into the bottom water, which rises and forms a plume [13]. Bubble plumes work to destratify the water column as the rising bubbles entrain the surrounding denser hypolimnion water. The denser water is then raised into the lighter epilimnion, promoting mixing [14]. This allows aeration via both atmospheric gas exchange and directly from the produced bubbles [15]. Many water utilities use bubble plume destratification systems to ultimately improve source water quality prior to draw-off [16]. A minimum airflow of 9.2 m^3 min^{-1} km^{-2} has been given as a threshold to ensure total destratification of a reservoir via these plumes [17,18]. The aim of destratification systems such as bubble plumes is to reduce cyanobacteria biomass and minimise concentrations of trace metals, such as soluble manganese, entering the water treatment works to improve the sustainability of treatment and reduce costs to the consumers [16,18]. Oxygenation of the hypolimnion has been shown to decrease concentrations of soluble reduced forms [19] of iron and manganese [20] which are released from sediments under low-oxygen conditions [21]. Additional mixing can affect phytoplankton by increasing the mixed depth to below the photic zone, thereby reducing irradiance which has been shown to reduce cyanobacteria blooms [22].

Currently, increased greenhouse gases (GHG), such as methane (CH_4) and carbon dioxide (CO_2) [23], are leading to a rapid rise in global temperatures. It has been shown that inland waters (e.g., reservoirs) may account for ~18% of CH_4 emissions globally [24,25]. With existing trends, future GHG release will further contribute to global temperature rise and its consequences. Climate change influences inland water bodies via alterations to air temperature and precipitation [26]. This introduces new elements that threaten to exacerbate water-quality issues related to reservoir stratification [27]. For many reservoirs, there will be alterations to the timing of stratification, potentially forming earlier and destratifying later, leading to enhanced periods of hypolimnetic anoxia and subsequent release of deleterious chemical species from the sediment [27–29]. This directly affects carbon fluxes and long-term dissolved organic matter trends by extending anoxic periods in the hypolimnion [27]. Rising global temperatures and increased anthropogenic eutrophication of freshwater systems are likely to promote favourable growth conditions for cyanobacteria. Increased water column stability can favour bloom-forming or positively buoyant species of cyanobacteria, some of which may produce taste and odour compounds or cyanotoxins and result in a deterioration in source water quality [30].

Prolonged stratification seasons predicted with climate change will require destratification systems to be efficient in both current and future climates. In addition, more stable water columns (i.e., stronger stratification) will be increasingly likely, which will require greater energy input from destratification systems to successfully mix lakes and reservoirs [28]. However, the relationship between increased temperature and stratification stability is not a linear effect and stratification can be strongly influenced by water-body morphometry and volume [5]. To date, research has largely focused on the effects of artificial mixing on source water quality in current climates. In this novel study, modelled future climate scenarios were used to estimate how effective current destratification systems are likely to be in the future.

During June 2017, much of western Europe was struck by a heatwave. Record-breaking temperatures were recorded across many countries, including the UK. On 21 June, temperatures in the UK reached 34.5 °C; at the time, this was the hottest June since

1976 [31]. Such heatwaves have been shown to cause strong stratification events in shallow polymictic lakes; several of which are similar in depth to the study site, Blagdon Lake, located in the southwestern UK [32].

This study presents in situ observations of temperature in the shallow, aerated Blagdon Lake during the 2017 heatwave. These observations are used as the basis for the development of a 3D hydrodynamic model, via the widely used AEM3D [33–37] which is available publicly with a yearly licence, to capture effects of extreme events on stratification such as the 2017 summer heatwave. With this calibrated model and down-scaled hourly future forcing data, this study examines how effective bubble plumes will be under future climate scenarios. Results show that existing issues with reservoir mixing interventions will likely continue into the future and managers will need to consider future proofing options.

2. Materials and Methods

2.1. Study Site

Blagdon Lake is a shallow drinking water reservoir, located in Somerset, England, and operated by Bristol Water Plc. The reservoir has a surface area of 1.78 km^2 with a mean depth of 4.75 m and a maximum depth of approximately 12 m (Figure 1). The lake was created when the River Yeo was dammed in 1905. Several small streams feed into Blagdon and these inflows have a combined catchment area of 21.8 km^2.

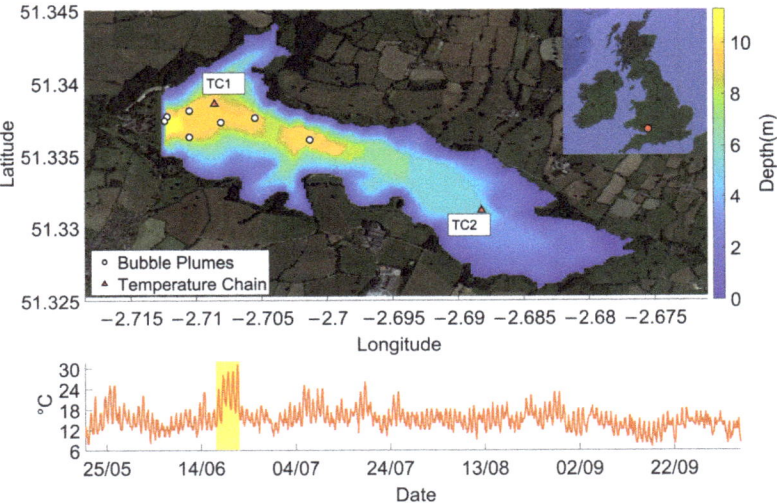

Figure 1. Global position and bathymetry of Blagdon Lake. The locations of the seven bubble plumes used to mix the reservoir and the two temperature chains deployed for this study are marked (white circles and red triangles, respectively; **top**). A time series of 2017 air temperature data is shown, including the June 2017 heatwave which is highlighted in yellow (**bottom**).

Artificial destratification was first implemented in Blagdon in the 1970s and since 2007 there have been seven bubble plumes installed (Figure 1). The bubble plumes are typically operated from April to September each year with the aim of destratifying the reservoir during the summer. Specifically, the bubble plumes were installed at Blagdon to address problems with soluble manganese concentrations and phytoplankton cell counts at the draw-off. Initially, five of the bubble plumes were positioned at 200 m, 250 m, 400 m, 600 m, and 850 m away from the draw-off tower at the dam. An additional two bubble plumes were placed nearer the draw-off tower to reduce soluble manganese concentrations entering the treatment plant. The bubble plumes contain no moving parts and have a 2-meter pipe containing a helical structure where compressed air bubbles can mix with the bottom water and generate a vertical plume that rises to the surface. These bubble plumes have a reported airflow of 0.011 m^{3}s^{-1} (Bristol Water pers. Comms).

Per data collected by Bristol Water, the bubble plumes appear to have reduced concentrations of soluble manganese at the Blagdon draw-off tower, with a 91.6% reduction from 2007 to 2008 of maximum observed soluble manganese. However, the effects of bubble plumes on phytoplankton cell counts at Blagdon, in particular cyanobacteria, appear less successful. Bloom-forming cyanobacteria are positively buoyant and have specific adaptations, such as gas vacuoles, that provide a competitive advantage under stratified conditions. Generally, the frequency of high counts of bloom-forming cyanobacteria at the Blagdon draw-off have increased since 2006. In 2014, for example, cyanobacteria cell counts at the draw-off exceeded 20,000 cells mL^{-1} on twelve occasions and peaked at 125,038 cells mL^{-1}, which related to a Microcystis bloom. On 26 June 2017, counts of Microcystis were elevated at the Blagdon draw-off, following the heatwave. Data provided by Bristol Water indicate that bubble plumes are not fully effective at reducing the buoyancy advantage of bloom-forming cyanobacteria during warm periods [38].

2.2. Observation Methodology

From 20th May to 5th October 2017, two temperature chains were deployed in Blagdon Lake to better understand the thermal regime in the reservoir over this time period. Each of the two chains were set at one-metre resolution and recorded data every ten minutes. One chain was placed in the deeper part of the reservoir (at depth of approximately 9 m), located within the zone immediately influenced by the bubble plumes located at latitude 51.33858 and longitude −2.70858; this chain is hereby referred to as temperature chain 1 (TC1). Per instructions from the water utility, this chain was placed as close as allowed to this intake zone. In order to evaluate the spatial extent of bubble plume mixing, the other temperature chain was placed further away in the shallower (at depth of approximately 5 m), non-aerated section of the reservoir located at latitude 51.33116 and at longitude −2.688. This shallower chain is referred to as temperature chain 2 (TC2). Over the 2017 observation period, a series of Secchi depths were also taken within Blagdon Lake at various points along the bubble plume transect; see Figure 1.

2.3. Data

2.3.1. Boundary Condition Data

To cover the forcing requirements of AEM3D, eight meteorological inputs were used. These include solar radiation (Wm^{-2}), cloud cover (decimal), air temperature (°C), atmospheric pressure (Pascals), precipitation (mday^{-1}), wind speed (ms^{-1}), wind direction (°) and relative humidity (decimal). Information about wind, air temperature, cloud cover and atmospheric pressure was taken from the weather station at Bristol Airport, which is proximal (around 5 km) to Blagdon Lake. Precipitation, solar radiation and relative humidity were sourced from the Filton weather station; this station is further from the lake (20 km) but offers a more comprehensive suite of measured variables than available for Bristol Airport. Weather station positions relative to Blagdon Lake are shown in Figure 2. The Met Office MIDAS Open UK Land Surface Stations Data was used to gather relative humidity, solar radiation and precipitation [39]. Remaining weather variables were sourced from the sub-daily Met Office Hadley Centre's Integrated Surface Database [40–43]. Whilst the most proximal sources were considered, the lack of direct meteorological data is a limitation that was taken into account in the evaluation of results but not considered a critical detriment. Mass balance information for the calibration and validation periods were sourced from outflow and reservoir capacity provided by Bristol Water. This calculated inflow was then separated between the six stream inflows into the lake, namely the Yeo, Butcombe, Rickford, Ubley, Copse and Holt Farm; the weighting applied to each inflow was based on the percentage of the lake's catchment each tributary contributed. Forcing data related to the seven bubble plumes installed in the reservoir was based on their 0.011 m^{3}s^{-1} flow rate. Modelling was focused on characterising the summer season to cover time periods when stratification was most likely to occur; as such, the bubble plumes were assumed to be on throughout the simulations.

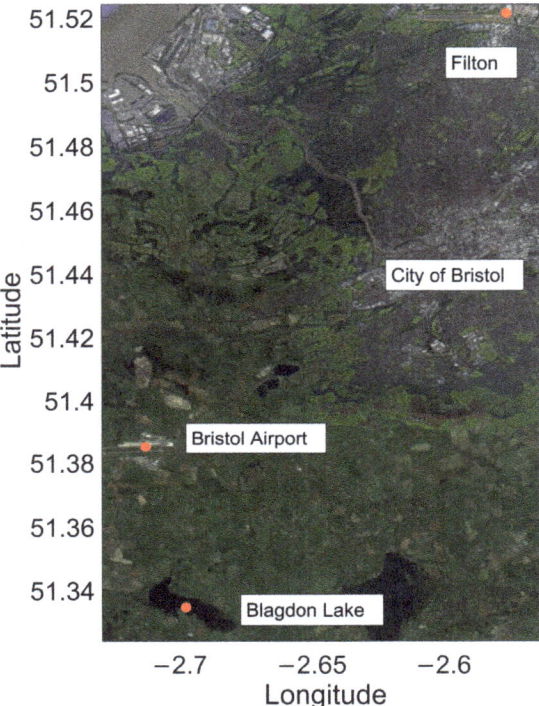

Figure 2. Global position of weather stations (Bristol Airport and Filton) used for the observational weather forcing file. The study site relative to the city and the weather stations is also shown.

2.3.2. Future Meteorological Data

Future forcing data were sourced from the UK Climate Projection 2018 (UKCP18) project which was designed to help inform adaptions as a result of climactic change [44]. The UKCP18 made use of models from the Coupled Model Intercomparison Project Phase 5 (CMIP5). These data sets were taken from regional climate model projections for the future climate of the UK extending from a 100-year period from 1981 to 2080 [45]. The climate projections were considered for the highest GHG emission scenario used by the IPCC, the representative pathway 8.5 (RCP 8.5). This representative pathway, named for the projected radiative forcing of 8.5 Wm^{-2} by 2100, predicts a future where high energy demand and high GHG emissions occur with little to no climate change policies to counteract this, thereby worsening climate change; due to this RCP 8.5 is considered the "worst case" scenario [46,47]. These projections were done using the UK Met Office's Hadley Centre Global Environmental model (HadGEM3), a coupled atmosphere-ocean climate model [48].

Twelve UKCP18 climate scenarios were down-scaled from twelve different HadGEM3-GC3.05 simulations from a grid size of 60 km to a higher resolution of 12 km via the HadREM3-GA705 model covering much of the British Isles on the Ordnance Survey's British National Grid [45]. The dataset has several separate projections of future climates that were used to force the future model runs; the scenarios used were 1, 4–13 and 15 since these had the required variables. These twelve scenarios offer distinct projections of climate variability due to climate change over the British Isles until 2080. From these projections, the following data were sourced at daily values with the following units:

downwards shortwave radiation (Wm^{-2}), northerly wind speed (ms^{-1}), easterly wind speed (ms^{-1}), relative humidity (%), cloud cover (%), atmospheric pressure (Pascals), precipitation (mmday^{-1}), maximum temperature (°C), and minimum temperature (°C) [44]. The future years consist of 360-day years, with twelve 30-day months. As this date format was incompatible with the model set-up, some days were repeated for an additional day after their occurrence or needed to be treated differently to get the climate scenarios into a standard date format. Depending on whether there was a leap year or not, one or two days were moved from February to later months. The final days of July, August, October and December were repeated; additionally, during a leap year May also had a repeated final day. This was done to space out repeated days throughout the year. Future model runs of Blagdon lake did not consider inflows or outflows. The HadGEM3 projects used were notably warmer than other CMIP5 model runs, though all were within the IPCC's stated range for future warming. This may contribute to higher water temperature being predicted by model runs [45].

2.4. Future Weather Down-Scaling

Predicted future data were obtained as daily values; in order to get sub-daily values, these data sets needed to be down-scaled temporally. The down-scaled methodology was evaluated using observed weather data, where daily averages of the observed data were produced before applying the various methods used in the down-scaling; this was also used to calibrate several of the required parameters. Down-scaling results were compared with the original observations. Numerous approaches were used for down-scaling, which will be detailed in the following sections. The data sets are openly available [49].

2.4.1. Temperature

The temporal down-scaling of air temperature data was done by using maximum and minimum temperature for any single given day in the future. These are all measured in °C. As air temperature follows a diurnal pattern related to the day-night cycle, air temperature can be described numerically with a sine function and an exponential function based on the time of day. The daytime air temperature function is described by the equation:

$$T_{a(t)} = T_{n(min)} + ((T_{n(max)} - T_{n(min)}) * S_{(t)}) \quad (1)$$

where $T_{a(t)}$ describes the air temperature at time t. $T_{n(max)}$ and $T_{n(min)}$ denote the daily maximum and minimum temperatures for any given day. The $S_{(t)}$ refers to the following sine pattern at time t:

$$S_{(t)} = sin(\pi \frac{t - SM + \frac{DL}{2}}{DL + 2P}) \quad (2)$$

DL is the day length at the field site in hours; this required the times of sunrise and sunset to be known. SM defines the time of solar maximum and P is the delay between the time of SM and the $T_{n(max)}$. These are all measured in hours.

For periods after sunset, the temperature is instead measured as an exponential decline curve based on the sunset temperature from the current day down to lowest temperature of the next day. This curve is based on lowest temperatures occurring just before dawn. The night temperature function can be written as:

$$T_{a(t)} = \frac{T_{n+1(min)} - (T_{ss} * e^{(-\frac{24-DL}{\tau})}) + ((T_{ss} - T_{n+1(min)}) * e^{(-\frac{t-ss}{\tau})})}{1 - e^{(-\frac{24-DL}{\tau})}} \quad (3)$$

T_t and t maintain the same definitions as previous; these are measured as °C and hours. T_{ss} and ss refer to the temperature and time of the day's sunset, respectively; these are measured as °C and hours. $T_{n+1(min)}$ is the minimum temperature the following day. τ is a time coefficient, in hours, that is calibrated for the field site using observational weather data and selecting the τ value that produced the smallest error and bias towards underestimation or overestimation. Prior to using this method with modelled future data, it was calibrated for the field site with observed data [50].

2.4.2. Relative Humidity

Relative Humidity (RH) was calculated as a decimal, from the ratios of actual vapour pressure in air (VPA) and the saturated vapour pressure (e_s). These are measured in kPa [50]. Equations used treat RH as a percentage; these can be calculated with the following equations:

$$e_s = 6.107 * e^{\frac{(17.4*T_a)}{(239+T_a)}} \tag{4}$$

$$VPA = 6.107 * e^{\frac{(17.4*T_d)}{(239+T_d)}} \tag{5}$$

$$RH = 100 * \frac{VPA}{e_s} \tag{6}$$

T_a and T_d are the air temperature and dew-point temperature measured in °C. As these are a function of temperature, RH also follows a diurnal pattern. In order to down-scale this RH, the initial daily value of T_a and the initial daily value of RH were used to produce a daily T_d. This was done numerically with the following rearranged equation based on Equations (4)–(6). This produced two separate estimations for VPA, one made by subbing in various values for T_d within a sensible temperature range for the UK (-10 °C to 30 °C), then subtracting the two estimations. The chosen T_d value was based on the result closest to zero.

$$0 = (6.107 * e^{\frac{(17.4*T_d)}{(239+T_d)}}) - \frac{e_s * RH}{100} \tag{7}$$

Once this daily T_d was estimated, a final VPA was calculated with the dew-point value. e_s was calculated with the sub-daily modeled T_a and from these, an estimated sub-daily RH on the same time step was calculated using Equation (6). At times, this equation produced results of greater than 100%; these values were ignored and set to 100% as beyond this limit the model will not accept the forcing data [50].

2.4.3. Downwards Shortwave Radiation

The diurnal solar radiation pattern was based on the total global radiation, denoted as R_{sum}; this value was calculated by multiplying the daily modelled incoming radiation (in units of watts) by the number of seconds in a day to joules. From this total insolation, sub-daily values were estimated with a sine function [50]. Hourly incoming solar radiation (R_t) was estimated as a function of solar declination (°), solar elevation (°), latitude (°), and day length (hours), as defined by the following set of Equations (9)–(12) [50]. Solar declination, denoted by δ, was calculated via the following equation [51]:

$$\delta = 23.45 * sin((284+n) * \frac{360}{365}) \tag{8}$$

where n is the day number. Seasonal offset and amplitude of the sine wave, SD and CD, respectively, were calculated with the following equations:

$$SD = sin(L) * sin(\delta) \tag{9}$$

$$CD = cos(L) * cos(\delta) \tag{10}$$

L is the latitude of the site in question. These can then be used to obtain the sine of the solar elevation, $sin\ \beta$, as calculated by:

$$sin(\beta) = SD + (CD * cos(\pi \frac{t-SM}{12})) \tag{11}$$

The solar maximum, *SM*, and the time of day, *t*, are in hours. For this method of estimation of solar radiation, a linear increase of the atmospheric transmissivity with the sine of solar height was assumed. R_t can then be calculated with the following equation:

$$R_t = R_{sum} * sin(\beta) * \frac{(1 + (C * sin(\beta)))}{DSBE * 3600} \tag{12}$$

C is a meteorological variable characterising dependence of transmissivity on solar height equal to about 0.4 [52] and *DBSE* is the integral of solar radiation from sunrise to sunset. This then produces an R_t in a sine pattern; periods at night when the insolation was below zero were adjusted to zero values [50].

2.4.4. Longwave Radiation

Downwards long-wave radiation (L_d) was chosen to be down-scaled, as opposed to down-scaling cloud cover, as L_d is more diurnal due to being a function of T_a. The Stefan-Boltzmann equation presented below is traditionally used to determine long-wave radiation [53]:

$$L_d = E_{eff} \sigma T_{eff}^4 \tag{13}$$

E_{eff} is effective emissivity (dimensionless), T_{eff} is the temperature from the atmosphere (measured in K within the equation) above and σ is the Stefan-Boltzmann constant (5.670367×10^{-8} kg s^{-3} k^{-4}). The parameterisation of L_d normally uses surface T_a and humidity measurements. Initially, the effective emissivity of a clear sky was needed. The Angstrom equation from 1918 was used to calculate clear sky emissivity as it has been shown to estimate L_d on a clear day [53]:

$$e_{clr} = 0.83 - (0.18 * 10^{-0.067 e_a}) \tag{14}$$

where e_{clr} is the clear sky emissivity (dimensionless) and e_a is vapour pressure. After the clear sky emissivity was calculated, an additional equation was used to calculate emissivity based on cloud cover. This was based on the Unsworth and Monteith equation from 1975 [54] as this is established as performing well for estimating cloud cover influence. These values of cloud cover, taken as a decimal, were obtained from daily predicated values from the twelve future climate scenarios considered [53].

$$E_{eff} = (1 - 0.84 Cf) e_{clr} + 0.84 Cf \tag{15}$$

Cf is the value of cloud cover. This was kept constant throughout the day using the daily value from future modelled weather data predictions. This was then placed into Equation (13) along with the daily down-scaled T_a to produce a time series of sub-daily L_d.

2.4.5. Wind Speed and Direction

Wind speed varies both cyclically and randomly in time. This often forms as random variance around a more regular diurnal cycle relating to atmospheric pressure and geostrophic wind. The deterministic approach is to use a wave function that varies from a minimum to a maximum wind speed over the course of the day [55].

Wind direction (°) and wind speed (ms^{-1}) were considered together, with wind speed down-scaled first and the wind direction calculated from that. With one wind value available for each day, an applicable method was used [56]. Firstly, wind speed was considered in its eastward and northward elements. These were then both down-scaled assuming a cosine function for wind speed, where maximum wind speed occurs later in the day and lower wind speed occurs earlier on in the day. These are based on the following equation [56]:

$$W_t = W_a + (\frac{1}{2} * W_a * cos(\frac{\pi * (t - H_{max})}{12})) \tag{16}$$

W_t references to wind speed at time t. W_a is the average wind speed. H_{max} is a time of maximum wind, as estimated from the observed wind speed. This was found to occur at midday. These sub-daily northward and eastward wind speeds were then used to calculate geostrophic wind speed and wind direction [56].

2.4.6. Air Pressure

When comparing observed values with their daily averaged values, it was found that daily variations of air pressure were minor when compared to timescales longer than a day. Due to this, a daily average of observed values was considered sufficient for down-scaling. Air pressure is measured in Pascals.

2.4.7. Precipitation

Rain in the study region is prone to periodic spikes in rainfall at shifting times. Due to this, a static rate across the entire day was used to capture general periods of high and low rainfall. As such, a daily average of observed values was considered sufficient for down-scaling. Precipitation rate is measured in $mday^{-1}$.

2.5. Down-Scaling Performance

The performance of these methods based on modern weather can be seen in Figure 3 and Table 1. The down-scaled temperature produced deviations more regularly from mode temperature towards the minimum and maximum. The root mean squared error (RMSE) produced an error of 1.6 °C. Errors were introduced in winter months when the sinusoidal temperature pattern was less dominating, when the down-scaled temperature always assumes a sinusoidal pattern. Error was also introduced during the night as warmer nights can have colder temperature allocated to them if a colder temperature occurred within the day. However, there is a high coefficient of correlation at 0.96, showing the method appropriately captured sub-daily patterns.

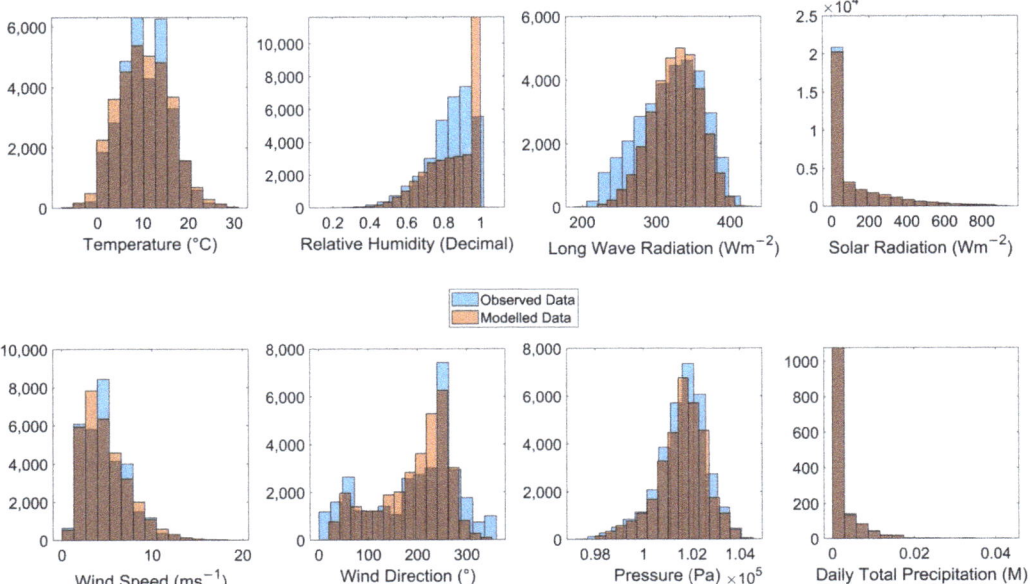

Figure 3. Frequency plots of the eight meteorological elements (temperature, relative humidity, long wave and solar radiation, wind speed and direction, atmospheric pressure and precipitation) used in the model comparing the observed meteorological factors with the corresponding down-scaled values.

Table 1. Root mean squared error (RMSE) and the Correlation Coefficients for comparison between the weather monitoring data and the down-scaled hourly data.

Meteorological Factor	RMSE	Correlation Coefficient
Air Temperature	1.6 (°C)	0.96
Relative Humidity	0.09 (decimal)	0.78
Long-Wave Radiation	26.87 (Wm^{-2})	0.8
Solar Radiation	53.54 (Wm^{-2})	0.95
Wind Speed	2.06 (ms^{-1})	0.66
Wind Direction	55.23 (°)	0.79
Atmospheric Pressure	253.52 (Pa)	0.97
Daily Total Precipitation	0 (m)	1

Down-scaled *RH* overestimations occurred at higher humidity. When the cut-off filter of over 1 was applied, this overestimation was downplayed but is still evident in Figure 3. At lower *RH*, the down-scaled method performed much better. The RMSE from a time series comparison is 0.09; this is improved by placing a cap on humidity values. The correlation coefficient of 0.78 shows sufficient capturing of the general temporal trends.

Downwards L_d was the only meteorological factor where proximal observed data was not available so was not included in the down-scaling method. Due to this, T_a, *Cf* and *RH* were used instead as a proxy. The calculated values based on observations will be referred to as the "observed L_d" for simplicity. Both observed and modelled L_d centred around the same modal value, but the modelled had a narrower range of values with a lower maximum value and a higher minimum value. Modelled L_d under-predicted the frequency of lower values. The RMSE of 26.87 Wm^{-2} and a coefficient of correlation of 0.8 show that they capture the overall pattern sufficiently.

Solar radiation is among the best performing down-scaled methods within the study. The modelled solar radiation under-predicted times of no short-wave radiation, likely due to the hourly time step obfuscating exact times of sunrise and sunset. The method has a comparatively low RMSE of 53.54 Wm^{-2} and a high coefficient of correlation of 0.95. Differences may primarily be effects of *Cf* on solar radiation which is not considered within the down-scaling.

Wind speed methodology produced a similar range of results, but the centre of the down-scaled wind speed frequency was lower. However, there is a general agreement of the frequency for the other speeds. The RMSE is large compared to other meteorological elements at 2 ms^{-1}. The correlation coefficient was also the weakest, at 0.66. These results suggest that, despite a similar range and frequency of values on a similar time scale, the method does not perform as well at capturing hourly wind speeds compared to other meteorological parameters.

Down-scaled wind direction matches the mode wind direction at 270° within the observations and captures another small frequency peak around 90°. The down-scaled method does not represent northerly winds well (around 0°). There is a large RMSE of 55.2° and the coefficient of correlation is 0.79. The down-scaled values captured the frequency spread of wind direction well but did not perform as well on hourly comparisons.

Surface pressure has a comparatively low RMSE of 253.52 Pa; this down-scaled parameter had the highest correlation coefficient of 0.97. Due to using an average of the air pressure, as it does not vary massively over a daily time period, agreement was very close.

When considering the performance of down-scaled precipitation, daily total precipitation was used. The down-scaling worked remarkably well with a RMSE within a rounding error of 0 and coefficient of correlation of 1.

Across all the down-scaled meteorological factors, there was found to be suitable agreement between observed and down-scaled weather parameters. All shared a similar range of values capturing the variability experienced, and many had low RMSE and high correlation. Thus, the down-scaled parameters were considered to sufficiently represent the required modelling needs.

2.6. AEM3D

2.6.1. Model Description

AEM3D, a coupled 3D model of hydrodynamics and ecology, was used for this study. This model is established for considering various hydro-environments and capturing many related physical and biogeochemical processes [33,35,37,57], though this study only focused on physical process modelling. It has often been employed in reservoir studies [33], including the evaluation of dammed rivers [36] and reservoirs of various sizes [35]. The model allows for the prediction of mixing requirements [36], temperature arrangements [36] and management methods [35]. The volume of research done shows that the model has a wide range of applicability [33–37]. This model works by coupling ELCOM and CAEDYM routines [33] which enables AEM3D to be a hydrodynamic model and/or fully coupled with a biogeochemistry module [33,35]. The model uses a z-grid system [57]. The solver for the hydrodynamics, ELCOM, solves in 3D with hydrostatic, Boussinesq, Reynolds-averaged, unsteady, viscous Naiver-Stokes equations [33]. For the vertical turbulence closure of Reynolds stresses, and corresponding turbulent fluxes, a 1D mixed layer model is utilized [34]. The biogeochemical element, CAEDYM, includes an array of algorithms to incorporate various production and cycling processes [33]; specifically, this module contains descriptions for primary production, secondary production, oxygen dynamics and nutrient cycling [34]. When new wet cells, i.e., cells within the model where water is present, are filled, as a default the surrounding cells are averaged to inform these new cells. Options exist such that non-temperature factors can be set to zero instead as the new wet cell fills. When water level drop is sufficient, wet cells will empty of water and convert to dry cells [57]. AEM3D also includes a module that simulates bubble plumes, allowing for characterisation of aeration-induced mixing in Blagdon Lake for the current study [57].

2.6.2. Model Set-Up

For running model calibration and validation, the full length of observations from 20th May to 6th October 2017 was used. For the future runs, focus was placed on the summer stratification period and 20th May through the end of August was considered, performed at five-year intervals from 2030 to 2080. This was done to capture a sufficient period during which the lake exhibited ephemeral stratification, whilst also optimising computational time. Measured 2017 temperature chain data on 20th May were used as initial conditions for each model run, though future results were processed using data from June onwards. The domain chosen for the model used grid cells with 25-m sides and with 1-m height in the vertical for the more stable epilimnion and hypolimnion; 0.5 m was used in the vertical to capture the more dynamic metalimnion zone. These regions within the model domain were estimated based on observed temperatures.

Secchi depth (Z_{SD} in m) casts obtained during the 2017 observation period were used to establish a range of possible light attenuation (K) estimates based on the Poole and Arkins (1929) formulation [58].

$$K = \frac{1.7}{Z_{SD}} \qquad (17)$$

To best match the dynamics of the measured data, inflows were included in the model for calibration. Some editing of the inflow boundary conditions was required. Stream inflows ultimately needed to be routed from the edge of the reservoir into its deeper region since, with varying water levels, inflows set directly to the reservoir shoreline become invalid; as the water level dropped, these wet cells become dry. Due to the shape of Blagdon, some of the inflow streams cannot be directly routed to the deeper sections of the reservoir so were ultimately moved to new locations, proximal to the actual position but with a more direct path to the lake's interior in order to facilitate long-term simulations that capture the varying water levels. Though used for model calibration, once the model was validated inflows were not included in future-prediction model runs for the sake of minimising error related to uncertainty in estimating changes in demand for water and stream inflows.

Bubble plumes were assigned to grid cells within the model domain closest to their actual locations in the reservoir (Figure 1). Due to the long-term operation of bubble plumes in Blagdon Lake, reaching back over four decades, there is no available data for the natural mixing regime during the summer stratification period. This means that quantifying the actual influence of the bubble plumes on the mixing regimes compared to the natural regime is difficult. To quantify the bubble plumes' effect on the reservoir, a series of model runs were done with the bubble plume module within the model turned on and off.

Figure 4 shows model run results and highlights that the bubble plumes had minimal influence on the physical mixing regime within the model. A localized mixing effect and increased heat transference are shown around the bubble plume at 350 m from the dam, suggesting the model underestimated the bubble plumes' influence and/or they have a limited effect. The airflow of the mixers in Blagdon lake is estimated to at 2.57 m^3 min^{-1} km^{-2}, below the given threshold of 9.2 m^3 min^{-1} km^{-2}, suggesting they might not be as effective as recommended by design standards.

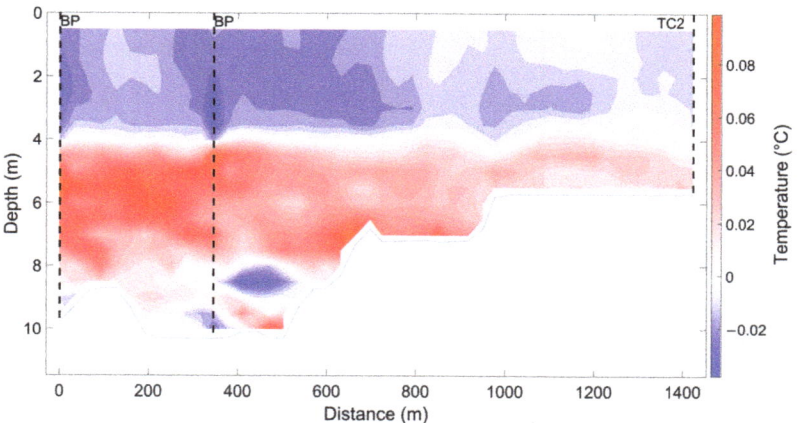

Figure 4. Differences in mean water temperature over the period of in situ observations between a run where the AEM3D bubble plume module is turned on and then off. Black dotted lines show the positions of bubble plumes and temperature chains. These were run for June in 2017.

2.7. Calibration and Validation

2.7.1. Performance Criteria

For calibration and validation of the model developed and used for this study, the RMSE between the observed values and the modelled outputs were calculated. For this operation, temperature was chosen as the evaluated parameter as it is the most representative of the overall physical environment of the reservoir.

$$RMSE = \sqrt{\sum_{i=1}^{n} \frac{(\hat{x} - x)^2}{n}} \tag{18}$$

n represents the number of considered points, $\hat{x}$ is the modelled data and x is the observed data.

Mean bias error (MBE) will be considered when analysing the performance of stability calculations to understand how well the model captures the mixing. This is shown in the equation below which uses the same definitions as Equation (18).

$$MBE = \frac{1}{n}\sum_{i=1}^{n}(\hat{x} - x) \qquad (19)$$

2.7.2. Approach

For calibration and validation of the model, the 2017 observed temperatures were used; TC1 and TC2 modelled locations were based on their real-world locations, shown in Figure 1. The calibration period used was from the start of temperature measurements, May 20th, to the end of June. The first recorded temperatures were used as initial conditions. The remaining observation period was used for validation. Based on sensitivity analysis runs, wind stress coefficient, heat transfer coefficient and albedo (defined by the fraction of incident solar radiation reflected back [59]) were chosen as model calibration parameters. Light attenuation was also used as a calibration parameter. A series of automated model runs varying these values was undertaken where MATLAB was used to alter the values and re-run the model. Albedo and the heat transfer coefficient have been used for model calibration in similar work [35]. The base values for heat transfer and wind stress coefficients within the model are both 0.0013; these coefficients were varied along a similar range from 0.001 to 0.0016. Albedo was varied from 8%, the model's base value, to 6%, based on northern altitude lakes [60]. Light attenuation was calculated from the Poole equation based on a range of Secchi depths taken from the site, 2 m to 5 m, and the established ratios between the various light attenuation factors contained within the base setup of the model were adjusted accordingly. The runs with the lowest RMSE were chosen to use for validation and used for model runs considering future climates.

2.8. Schmidt Stability

Schmidt stability is a widely used indicator of a water body's resistance to mixing, commonly used with both observed and simulated temperatures [61–66]. Schmidt stability describes the amount of mechanical work needed per unit surface area to mix a stratified water column into a homogeneous state [67]. According to Schmidt stability theory, larger values denote strong stratification and near-zero values indicate an isothermal, fully mixed structure [62]. Schmidt stability (S), measured in Jm^{-2}, is described in Equation (20) [68].

$$S = \frac{g}{A_0}\int_{z_0}^{z_m}(z - z^*)(\rho_z - \rho^*)A_z dx \qquad (20)$$

Area (m^2) is represented by A_0 for surface area and A_z for surface area at depth z (m). ρ_z stands for the density (kgm^{-3}) at any given depth, where ρ^* is the volume-weighted mean density within the water column. The depth at which this mean density occurred is denoted by z^* (m). Gravity-induced acceleration is indicated by g (ms^{-2}).

Acknowledged downsides to S analysis include that, primarily, it is limited by a 1D assumption which simplifies both the stratification and the final homogeneous state of the full water column. This 1D calculation, which commonly uses measurements taken from the deepest point within a reservoir [66,69], can be problematic in larger water bodies or reservoirs where there are large mass-balance changes for which horizontal heterogeneity cannot be assumed.

S was used to diagnose when stratification was present and was not present. A threshold was based on 10% of the maximum daily S of the observed summer, 28.26 Jm^{-2}. This placed the threshold at 2.83 Jm^{-2}; this will be used when considering the results.

2.9. Mann-Kendall Trend Test

The Mann–Kendall trend test [70,71] allows for distinguishing statistically significant trends, both increasing and decreasing, over a temporal data set. This statistical test has often been used to examine trends within hydrology and meteorological time series and has been used within studies looking at climate change effects [72,73]. In this study, a significance level of $\alpha = 0.05$ was considered, with the analysis being done in MATLAB.

3. Results
3.1. Lake Observational Data

The observational datasets from the in situ temperatures chains are shown in Figure 5. Due to operational issues with the thermistor at 8 m on TC1, it was omitted when considering results.

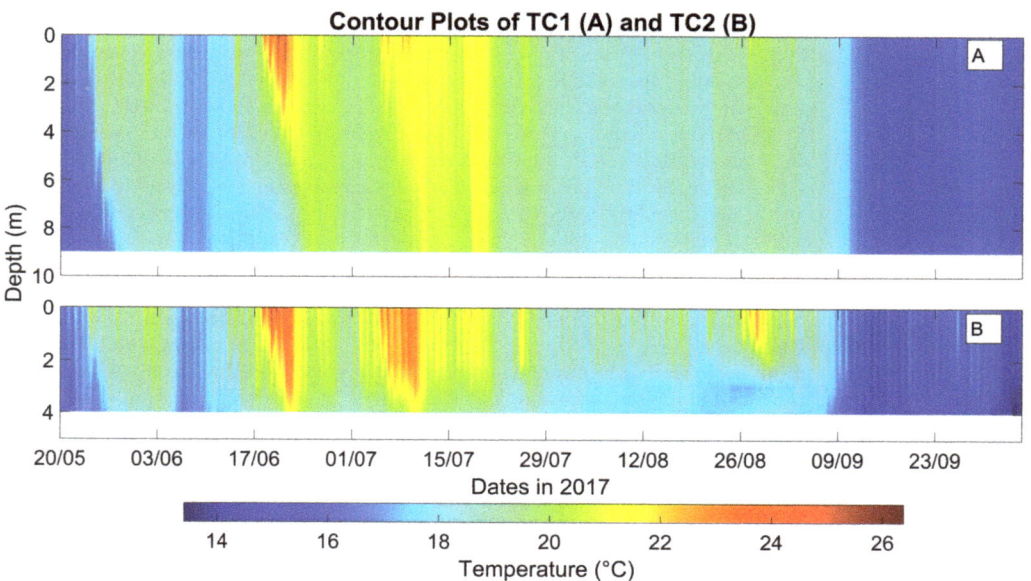

Figure 5. Contour plots of the observed data from temperature chain 1 (TC1) (panel **A**) and temperature chain 2 (TC2) (panel **B**); locations of these temperature chains shown in Figure 1. Temperature data were collected from the 20th of May to the 5th of October 2017.

Temperature chain observations show that surface water temperatures increased to between 24 °C and 26 °C from 18–21 June 2017. Consequently, stratification of the water column, characterised by S, was observed with a temperature difference of 8 °C. The stratification during the June 2017 heatwave was thus found to increase the amount of energy required to mix the reservoir, reaching near 40 Jm^{-2}, as shown in Figure 6.

The water column stratified despite the operation of the bubble plumes in the reservoir and data indicate that stratification occurred at other times during the 2017 summer, not just during the heatwave. Blagdon appears to be polymictic, with seven mixing events (water column S above the threshold seen in Figure 6) observed in the 2017 summer; for these calculations, daily averaged water temperatures were considered. At TC1, there were three notable stratification events: 24–29 May, the heatwave (14–24 June), and 6–11 July, as seen in Figure 5. The stratification events in May and July 2017 both had a maximum temperature difference of around 4 °C at TC1. At TC2, the water column was much shallower and subsequently stratified more intensely than TC1 from 22 August–3 September 2017, where the water temperature difference peaked at 6 °C. Temperature difference at TC2 also peaked at 4.7 °C, 7 °C and 5.5 °C for the previously described stratification in May, June, and July 2017, respectively. There were additional smaller events in early June and two later in July.

3.2. Model Performance

The validation of the model produced a RMSE of 0.53 °C over all the comparable depths of observation. Heating and wind transfer coefficient were estimated at 0.0011 with albedo at 6%. Light attenuation factors within the model were based on Secchi depth of 2 m. The model did very well at simulating a shallower lake and did much to capture the thermal structure within the lake, as highlighted in Figure 6. The model also performed well at simulating the evolution of lake temperature over summer, capturing episodes of ephemeral stratification throughout this period.

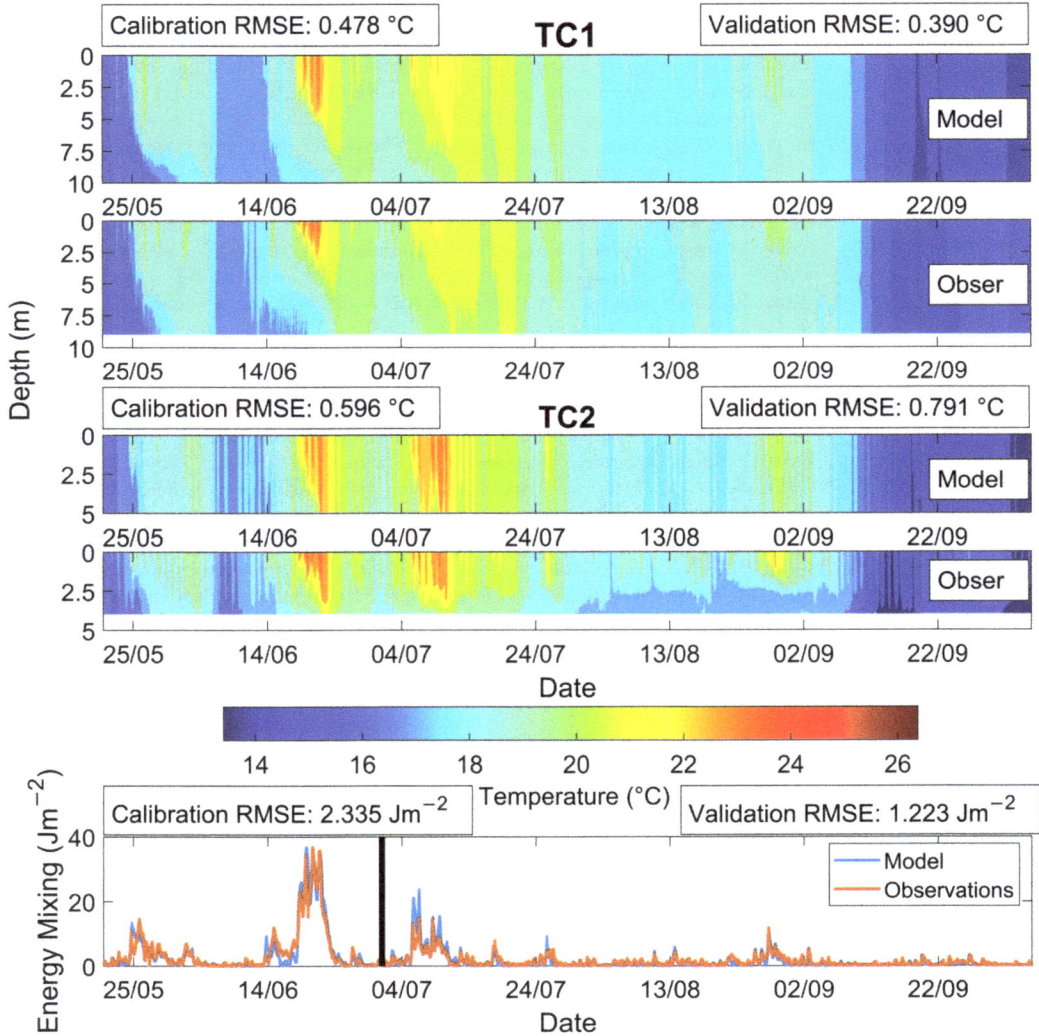

Figure 6. Assessment of model performance. Contour plots show comparisons between modelled results and observations for temperature (**top** panels). Also shown is a plot of Schmidt stability (S) over time between the observed and modelled runs (**bottom** panel). RMSE between these observations and model data is also shown. Calibration and validation periods are separated by a vertical black line.

The estimated S is based on the bathymetry input to the model and TC1 observations, the deeper of the temperature chains. Within the model, the corresponding cell to the location of TC1 was used to make results more directly comparable with the observations. The comparable S values are shown in Figure 6, with a corresponding RMSE of 2.335 Jm^{-2} during the calibration period and RMSE 1.223 Jm^{-2} for the validation period. MBE estimations support the visual indication that the model was successful at predicting mixing, as both periods have small near-zero values. MBE values were 0.118 Jm^{-2} and 0.005 Jm^{-2}, respectively, for the calibration and validation periods. Additionally, correlation coefficients were calculated from the two S time series to consider changes in S over time. Both periods showed high correlation at 0.96 for the calibration period and 0.9 for the validation period, highlighting that the model is capturing the peaks in S and, therefore, reflecting when the reservoir is forming layers. Ultimately, these results indicate that the model is appropriately capturing the trends within the lake.

The model predicted a deeper thermocline than the observed values, which was more noticeable at TC2 (shallower site; Figure 6). The model also underestimated the magnitude of short-term heating events within the surface layer of the lake, which was shown in the validation and calibration period. During the June heatwave, the model under-predicts the surface water temperature by a maximum of 1.99 °C and a mean under-prediction of 0.69 °C. Comparing the simulated and observed TC2, the model under-predicted surface water temperatures and over-predicted bottom water temperatures. Simulations showed diurnal heating penetrated to the sediment, which was not present in the observations and was the largest source of error at a RMSE of 1.2 °C. Between mid-September and early October, when the reservoir became mixed again, simulations underestimated the temperature at every depth, often around 1 °C. This is present in both simulated TC1 and TC2 locations.

Figure 7 depicts how utilizing the down-scaled meteorological factors for forcing the model yielded comparatively accurate results with a RMSE of 1 °C. This method produced larger deviations from observations than the observed forced model shown in Figure 6. However, as supported by the S calculations, model results did capture many of the periods of increased S, though seemingly missing a small event in late August. During the heatwave in June, the down-scaled forcing model under-predicts S to a greater extent than the model using the observed forcing (as evidenced by the lower temperatures and increased mixing shown in Figure 7, compared with Figure 6). This deviation likely explains the larger RMSE in S of 3 Jm^{-2} and a bias of -1.06 Jm^{-2}. However, during the heatwave there is a noticeably large error with a mean difference of 9.72 Jm^{-2} between observations and modelled results. There is a high correlation coefficient between the model and the observations at 0.82, suggesting shared periods of raised and lowered energy requirements. In mid-August, there is another notable deviation where the model over-predicts temperature increases and corresponding mixing. This might be due to the use of estimated long-wave radiation, since no direct or near observations were available. Another consideration is that the down-scaled runs tend to have peak temperatures at approximately 19:00; while this behaviour can be observed in the in situ data, it is infrequent.

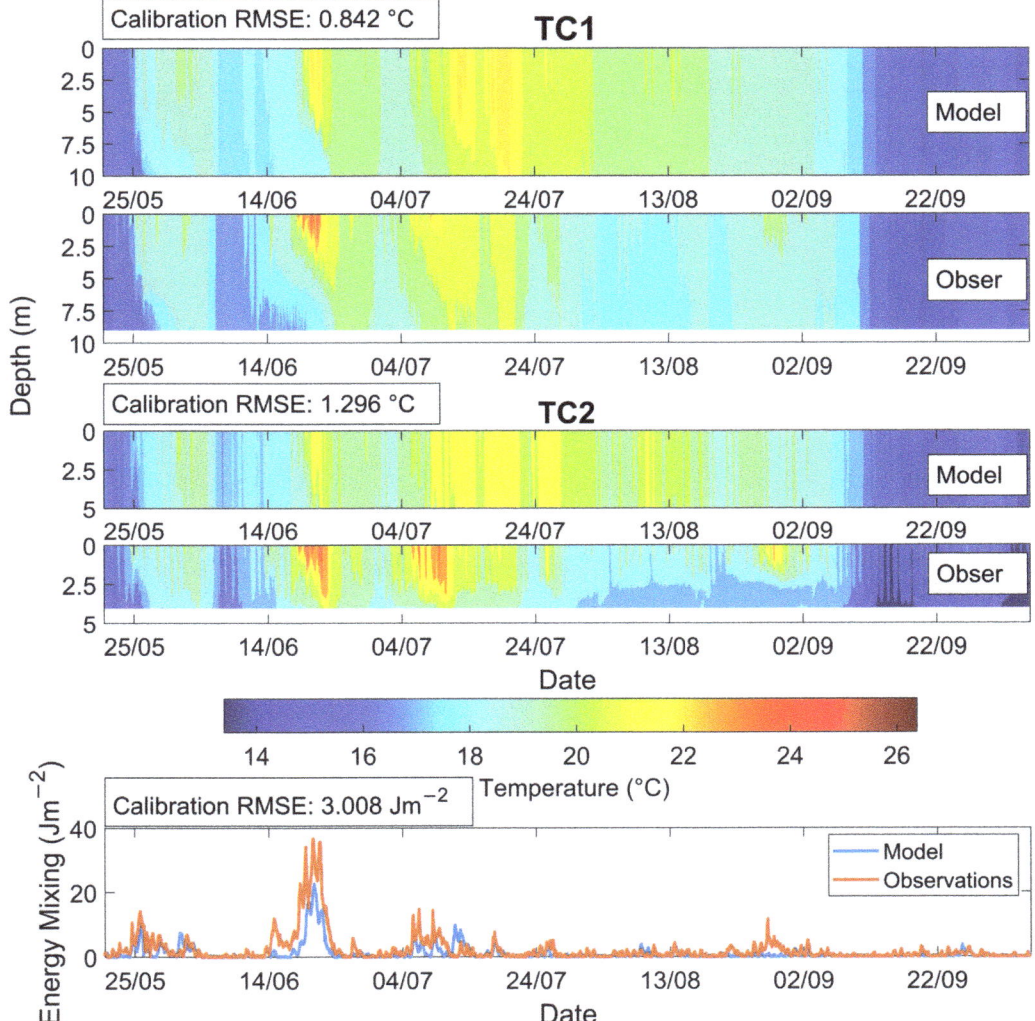

Figure 7. Assessment of model performance when forcing data was averaged daily then placed through the down-scaling procedure. Contour plots show comparisons between modelled with down-scaled forcing data and observations for temperature (**top** panels). Also shown is a plot of S over time between the observed and modelled runs (**bottom** panel). RMSE between these observations and down-scaled model data is also shown.

3.3. Future Lake Regimes under Climate Change

Figure 8 shows mean water temperature, S (based on the deepest part of the reservoir and the model bathymetry) and number of mixing events for every iteration of the model, covering twelve different future climate projection scenarios in five-year intervals from 2030 to 2080. Every climate scenario shows that, with the warming atmosphere from the increasing GHG, there is increased mean water temperature with time. When a linear trend is fitted, all scenarios display a positive trend with an increase in mean water temperature of around 0.05°C each year; trend lines are shown in Figure 8 with corresponding slopes in Table 2. Most of these scenario runs were found to exhibit statistically significant trends, excluding scenarios 1, 5, 9 and 15. Results suggest that these four scenarios' trends of increased temperature with time could be due to chance as opposed to linked relationship.

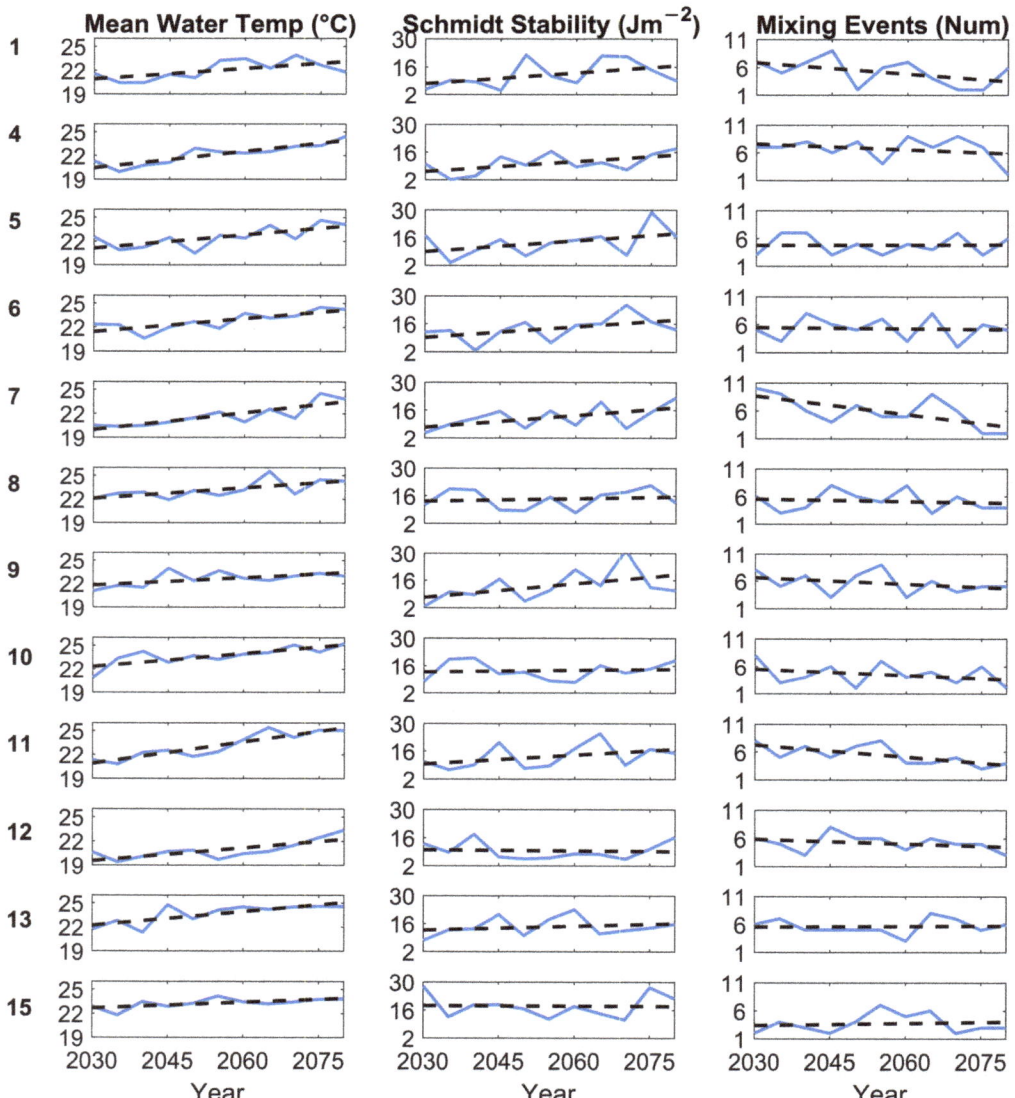

Figure 8. Compiled results from the twelve used UK Climate Projection 2018 (UKCP18) modelled scenarios across the range of years 2030 to 2080 with increasing greenhouse house gases with time, including the mean reservoir temperature, S and number of mixing events each year. Dashed black lines are trend lines.

Table 2. *p*-values (calculated with the Mann-Kendall test) and slope of fitted linear trend for mean water temperature, *S* and mixing events from the 12 different climate scenarios across the range of years the model was run for. Bold text shows statistically significant *p*-values of below 0.05, indicating a relationship between the modelled parameters and time not likely attributed to chance.

Scenario	Mean Water Temperature (*p*-Values)	Schmidt Stability (*p*-Values)	Mixing Events (*p*-Values)
1	0.062	0.350	0.186
4	**0.002**	0.119	0.876
5	0.087	0.276	1.000
6	**0.020**	0.062	0.876
7	**0.003**	0.119	**0.043**
8	**0.043**	0.876	0.755
9	0.087	0.119	0.392
10	**0.008**	0.755	0.436
11	**0.003**	0.350	**0.043**
12	**0.008**	0.755	0.276
13	**0.013**	0.276	1.000
15	0.062	0.755	0.697

Scenario	Mean Water Temperature (Slope)	Schmidt Stability (Slope)	Mixing Events (Slope)
1	0.044	0.182	−0.067
4	0.069	0.170	−0.035
5	0.056	0.183	0.002
6	0.055	0.177	−0.007
7	0.070	0.197	−0.109
8	0.045	0.043	−0.015
9	0.032	0.232	−0.040
10	0.055	0.028	−0.038
11	0.090	0.147	−0.071
12	0.054	−0.016	−0.027
13	0.057	0.066	0.004
15	0.024	−0.006	0.013

S values show a mix of positive and negative trends, with scenarios 12 and 15 displaying a trend of decreased *S* and, subsequently, decreased levels of mixing required to maintain destratification. The other scenarios show a positive trend, meaning that more energy input would be needed to mix the reservoir. None of the scenarios exhibits a statistically significant trend of increasing *S* and corresponding stratification. Scenario 9 has the largest trend line slope of 0.23 Jm^{-2} per year.

There were two significant trends in reducing mixing events for scenarios 7 and 11. Scenarios 1, 4 and 6–12 showed slight negative trends on the order of magnitude of 0.01 events per year; Scenario 7 was higher, on the order of 0.1 events per year. The other scenarios showed positive trends of similar values. The trend lines are centred on about six mixing events occurring per year, similar to the 2017 observations which showed seven such events. Across nearly all the modelled years and scenarios, ephemeral mixing remains a constant feature with multiple mixing events throughout the summer.

3.4. Future Stratification Extent

As seen in Figure 9, across nearly all model runs as time progresses, barring scenario 15, the linear trends show an increasing number of days where stratification is present during the simulated summers. Most scenarios show an increase of between 0.2 and 0.7 days per year, though scenario 8 has a notably flat trend. Scenario 15, the only one with a negative trend, is also similarly flat. The number of days of stratification varies between 20 days and 90 days, highlighting that there are approximately three weeks or more of stratification within every modelled year for each scenario. As shown by Table 3, the vast majority of scenarios showed no significant trend in increasing period of stratification. Scenarios 1, 6 and 11 showed a statistically significant trend where the number of stratified days increased with time over the modelled years.

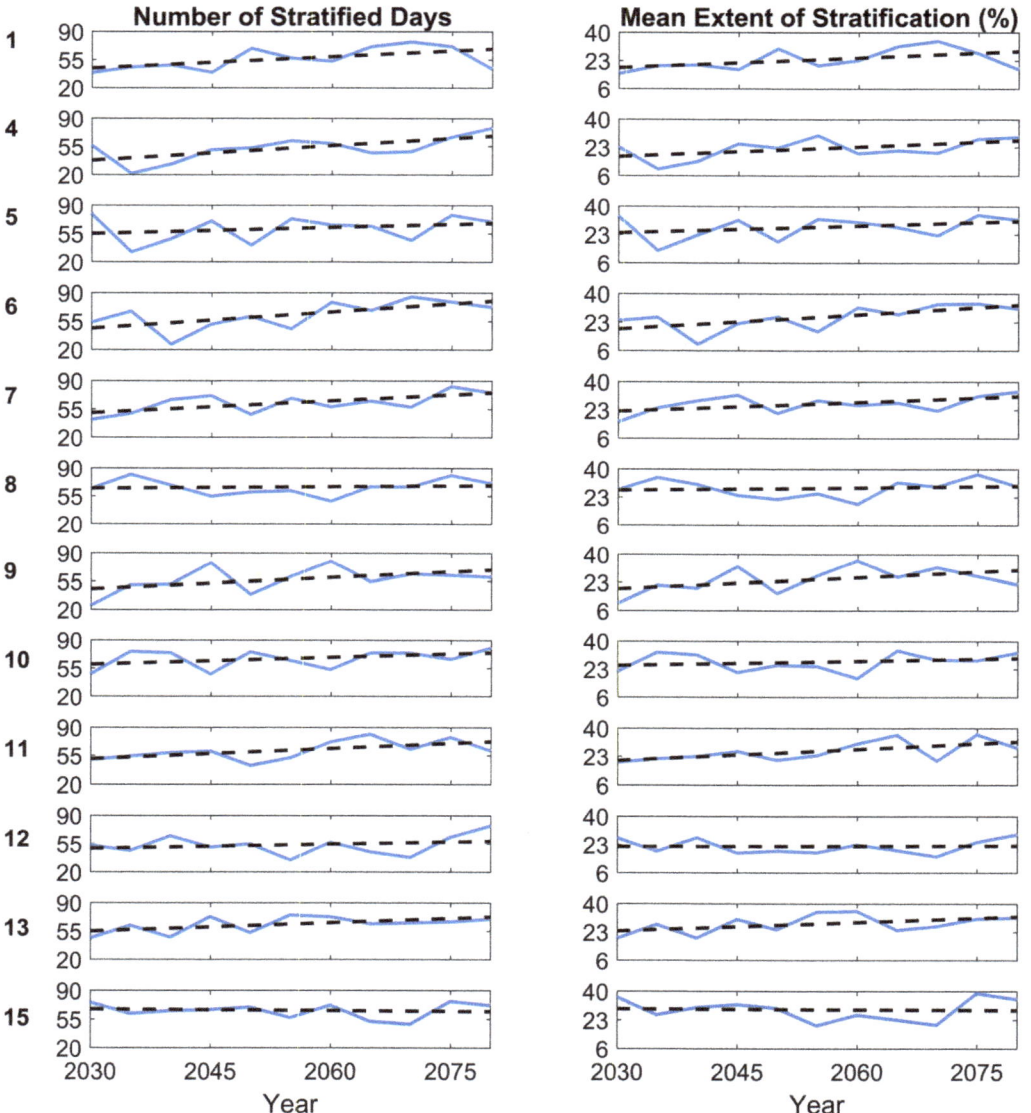

Figure 9. Compiled results from the twelve UKCP18 modelled scenarios used across the range of years, including the number of days of stratification (**left** column) and the mean surface area, as a percentage of the reservoir, that was stratified (**right** column).

Table 3. *p*-values (calculated with the Mann-Kendall test) and slope of fitted linear trends for mean water temperature, *S* and mixing events from the 12 different climate scenarios across the range of years for which the model was run. Bold text shows statistically significant *p*-values of below 0.05, indicating a relationship between the modelled parameters and time not likely attributed to chance.

Scenario	Period of Stratification (*p*-Values)	Extent of Stratification (*p*-Values)
1	**0.043**	0.087
4	0.062	0.213
5	0.755	0.533
6	**0.043**	**0.020**
7	0.087	0.119
8	0.640	0.755
9	0.119	0.276
10	0.436	0.640
11	**0.043**	**0.043**
12	0.640	0.755
13	0.119	0.087
15	1.000	0.640

Scenario	Period of Stratification (Slope)	Extent of Stratification (Slope)
1	0.480	0.199
4	0.602	0.194
5	0.253	0.139
6	0.672	0.286
7	0.495	0.179
8	0.071	0.053
9	0.470	0.228
10	0.288	0.087
11	0.429	0.228
12	0.184	0.012
13	0.352	0.173
15	−0.058	−0.017

The mean extent of stratification refers to the mean surface area, as a percentage of the reservoir, that is stratified. Similar to the duration of stratification, only scenario 15 exhibits a slightly negative trend for mean extent of stratification with time; these negative changes were on the order of 0.02% each year, as seen in Figure 9. For scenarios with positive trends, the largest increase shown is 0.3% per year for scenario 6. The mean extent of stratification throughout the simulated summer varies between 6% and 40%. As shown in Table 3, scenarios 6 and 11 displayed a statistically significant trend of increased mean stratification extent over the simulated years.

While there is not a wealth of statistically significant increases in the mean percentage of the reservoir stratification or in the total period the reservoir is stratified (Figure 9), it should be noted that no model run remained well mixed throughout the entirety of the simulation period; some degree of stratification always remained. Though there was large variation in maximum daily *S*, none of the scenarios assessed had a value below the threshold 2.83 Jm^{-2}. Some runs with more extreme results did also show evidence of monomictic lake behaviour. This is evident in Figure 10, where several model runs show a low number of mixing events and may remain strongly stratified (as indicated by the warmer colour hues).

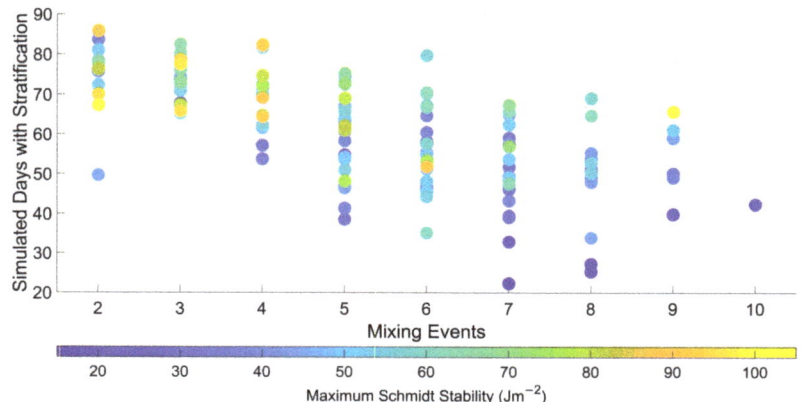

Figure 10. Comparison of mixing events and number of simulated days where stratification was present, where S is above the threshold. The scatter plot colour scheme depicts the maximum daily S from the model runs varying from minimum (cooler, blue tones; minimum of 15 Jm^{-2}) to maximum (warmer, yellow tones; maximum of 105 Jm^{-2}).

4. Discussion

In situ observations, displayed in Figure 5, show that Blagdon Lake will stratify during a heatwave such as that observed in June 2017. Furthermore, other stratification events were also observed in May, July and August 2017 when considering the entire reservoir. This shows that the bubble plumes were not able to prevent stratification within the reservoir over the duration of the summer. It should be noted, however, that the bubble plumes have been considered largely successful at reducing soluble manganese concentrations in the Blagdon in-take water to the treatment plant.

As shown from the results in Figure 6, this study provides a robust example of the effectiveness of the AEM3D model for simulating hydrodynamic environments. Even with indirect forcing, the model was able to represent ephemeral stratification within a shallow reservoir. The model does remarkably well at resolving the energy requirements to mix the reservoir. Results also showed that the developed model appropriately characterised the thermal structure and much of the mixing/stratification within the reservoir.

There were mixed results based on the methodology for down-scaling daily meteorological factors into hourly datasets. Some factors with stronger diurnal patterns such as T_a, RH and solar radiation produced much closer agreement (Figure 3). There were less accurate fits for wind where this diurnal pattern is much less prevalent. Nevertheless, all the down-scaled parameters had strong correlation coefficients shown in Table 1. Methods were selected that made use of the available future modelled weather; this is one area that could be improved upon, e.g., if a more diverse future data set was to be used. When the down-scaled forcing data were used for forcing, the model produced results in line with in situ lake temperature measurements (Figure 7). The shift in time of the warmest daily temperature might delay the estimate for lake temperature maximum. This may limit the model applicability to resolve issues on smaller time steps. However, the reasonable temperature error and correlation in energy requirements over the length of the simulation, seen in Figure 7, show it is applicable when discussing trends over the course of the year. Additionally, the model was found to underestimate the stability (S in Figure 7), capturing 47% of the mean energy requirement during the June heatwave. The down-scaled run underestimated the surface water temperature during this heatwave but performed better during other stratified periods. This should be considered when interpreting the results.

Despite the increased heat content within the reservoir (Figure 9), the lake maintained polymixis and mixed numerous times over the modelled summers, though scenarios 7 and 11 showed a statistically significant decreasing trend of mixing events. As these scenarios were modelled over summer periods, the model is possibly missing shorter stratification events before and/or after the summer and a true monomictic regime cannot be confirmed. Additionally, there are a few runs that stand out as outliers, where a more monomictic nature with higher S and a lower number of mixing events, is displayed during the simulation period (Figure 10). The model did highlight some evidence of increased extent and length of stratification (Table 3 and Figure 9) as scenarios 6 and 11 showed evidence of a statistically significant trends in these factors. Other studies have provided some evidence of polymictic lakes stratifying for longer periods [74–76]. There was no statistically significant trend for increasing energy requirements to mix the reservoir, though most of the scenarios showed a positive trend. It should be noted that the actual energy requirement is likely underestimated, as seen in Figure 7.

Across all runs, none remained well mixed throughout the simulation period, with some stratified periods always remaining present. This is highlighted in Figure 10 where all the model runs had S above the chosen threshold of 2.83 Jm^{-2} during the simulation, supporting the view that, in the future, the reservoir will continue to be at risk of stratification and the associated water quality impacts. It is thought that the current ephemeral stratification persists as the increased atmospheric heating is not sufficient to stabilise the water column enough to prevent mixing. Thus, when mixing occurs the increased heat content is mixed down into the hypolimnion; as such, when the next stratification event occurs, the relative temperature difference between the epilimnion and hypolimnion is not meaningful different to conditions experienced in the present and hence the stability does not meaningfully increase with time.

Model runs for this study were performed based on the RCP 8.5 climate projection, which projects an increase in GHG production from the present and is often considered the "worst case" scenario for climate change, with the temperature increase reaching 4.3 °C by 2100. With a continuation of the current stratification regime under this "worst case" scenario, this would imply any results based on modelling using less-extreme temperature increases will also show little to no modification to the current stratification regime.

The use of discontinuous runs means that the model results will not take the retention of heat from previous years into account. This was primarily done to optimise the available computational times. The amount of heat retained year to year cannot be quantified as observations are not sufficiently long enough. However, the reservoir has a dynamic outflow and the residence time is around 315 to 430 days. As the reservoir can completely flush in less than a year, this yearly heat retention may not be a significant issue.

The results of the study indicate that the bubble plume mixer arrangement, as currently operated within the Blagdon Lake, will continue to be unable to prevent stratification moving forwards, as shown by summer 2017 heatwave data in Figure 5 and future model results in Figure 10. Results suggest than an artificial mixing system that can effectively prevent stratification from developing during current conditions will likely also be effective at preventing stratification for several decades. Additionally, with scenarios 6 and 11 showing an increase in the areal extent of stratification, it is possible that some alterations to the areal extent of mixing via bubble plumes might need to be considered.

Ultimately, the results here demonstrate that modelling is an effective tool for predicting whether existing reservoir aeration systems are currently effective and if they are likely to remain so in the future. While stratification does not meaningfully change under the modeled future scenarios, the over-heating of the reservoir may pose other concerns. Conditions may start to favour algal growth at higher temperatures; this may require additional management focus and adapted strategies in order to counteract this potential future water quality threat. Furthermore, increased water temperature may cause shifts in microbial communities, corresponding oxygen consumption and biogeochemical cycling (e.g., trace metal release from sediments) [77].

5. Conclusions

A shallow drinking water reservoir, artificially mixed by bubble plumes, was observed for summer 2017 and, using the 3D hydrodynamic model AEM3D, modeled with downscaled hourly future climate data. The calibrated model did well at representing the shallow lake based on a summer of observations with a RMSE of 0.53 °C. In situ measurements from 2017 and supporting model results highlight that, despite the operation of bubble plumes, the reservoir still stratified during the summer. Supporting source water quality data suggest that the plumes have been successful at addressing issues related to soluble manganese but not cyanobacteria blooms.

Twelve climate scenarios from RCP 8.5, often thought to be the worst-case scenario prediction, were considered using 5-year intervals from 2030 to 2080. These were downscaled using various sub-daily methods which exhibited mixed results. Using the bubble plume arrangement at Blagdon, eight of these scenarios showed statistically significant trends of increasing mean water temperature increasing with the warming climate. On average, across all the scenarios this increase is around 2.7 °C between 2030 and 2080. Only two scenarios showed a statistically significant trend of decreasing mixing events and no scenarios showed a meaningful trend of changing energy requirements to mix the reservoir. The majority of the evidence suggests that, while the reservoir warmed, the present stratification regime will likely continue with no significant climate-induced changes to the current stratification regimes. Though some scenarios showed some evidence of gradual shift to monomitic conditions, scenarios 6 and 11 showed a significant increase in the extent (between 0.2 and 0.3 % of the reservoir per year) and length of stratification (between 0.4 and 0.7 days per year).

The results show that under the current bubble plume arrangement and operation, issues with stratification will remain; every model run showed the presence of stratification, with maximum S of 15 Jm^{-2} or greater. To prevent further stratification-related issues with source water quality, reservoir managers need to consider what mitigation options might be required to be more effective in the future. Here, in situ observations demonstrated that bubble plumes are currently not fully effective and, under modelled scenarios, their ineffectiveness will likely continue. The bubble plumes are currently being operated below the given threshold of airflow of 9.2 m^3 min^{-1} km^{-2} for total mixing; targeting this threshold could be a better approach for future proofing. Modelling can be an effective tool for reservoir managers to proactively assess the effectiveness of current reservoir management schemes as well as necessary mitigation in the future. This approach will support evidence-based decisions on whether further mitigation and/or enhancement of current management strategies are required.

Future work could examine similar aeration systems in future climates to see how these will perform in comparison to results obtained for Blagdon Lake. Consideration should be placed on how to improve best practice for such artificial destratification systems to safeguard against future water quality risk. A wider range of climate scenarios could also be considered to establish a broader scope of possible responses to future climatic change.

Author Contributions: Conceptualization, D.B., D.W., E.S., J.Z. and L.D.B.; methodology, D.B., D.W. and E.S; software, D.B.; validation, D.B.; formal analysis, D.B. and E.S.; investigation, D.B., E.S. and R.L.; resources, D.W., E.S., J.Z., R.L. and L.D.B.; data curation, D.B. and E.S.; writing—Original draft preparation, D.B.; writing—Review and editing, D.W., E.S., L.D.B. and J.Z.; visualization, D.B.; supervision, J.Z., L.D.B. and D.W.; project administration, D.B., D.W., E.S., J.Z. and L.D.B.; funding acquisition, D.W. All authors have read and agreed to the published version of the manuscript.

Funding: This study and D.B. was supported by the Engineering and Physical Sciences Research Council in the UK via grant EP/L016214/1 awarded for the Water Informatics: Science and Engineering (WISE) Centre for Doctoral Training, which is gratefully acknowledged. E.S. was supported by a NERC GW4+ Doctoral Training Partnership studentship from the Natural Environment Research Council via grant NE/L002434/1.

Institutional Review Board Statement: Not applicable.

Informed Consent Statement: Not applicable.

Data Availability Statement: Publicly available data sets were analyzed in this study: Air Temperature, Pressure, Cloud Cover, Wind Speed and Direction can be found here: [https://catalogue.ceda.ac.uk/uuid/61392a37ab614f349a4c20df4d08871c] (accessed on 6 December 2019). Solar Radiation, Precipitation and Relative Humidity (within a general weather observations) can be found here: [https://catalogue.ceda.ac.uk/uuid/dbd451271eb04662beade68da43546e1] (accessed on 5 July 2019). Sunrise and sunset for Blagdon were found here: [timeanddate.com/sun/@2655407] (accessed on 2 March 2020). Future Climate Projections can be found here: [https://catalogue.ceda.ac.uk/uuid/589211abeb844070a95d061c8cc7f604] (accessed on 18 June 2020). The MATLAB scripts, modelled and processed observed data presented in this study are openly available in University of Bath Research Data Archive at [https://doi.org/10.15125/BATH-01036] (accessed on 21 February 2021).

Acknowledgments: The authors would like to thank NERC, WISE and GW4+ for supporting the work. Thanks also goes to Bristol Water for supporting field monitoring and research on their reservoirs. We would also like to thank the reviewers for their feedback helping to improve the manuscript.

Conflicts of Interest: The authors declare no conflict of interest.

References

1. Stephens, R.; Imberger, J. Reservoir destratification via mechanical mixers. *J. Hydraul. Eng.* **1993**, *119*, 438–457. [CrossRef]
2. Koue, J.; Shimadera, H.; Matsuo, T.; Kondo, A. Evaluation of thermal stratification and flow field reproduced by a three-dimensional hydrodynamic model in Lake Biwa, Japan. *Water* **2018**, *10*, 47. [CrossRef]
3. Woolway, R.I.; Maberly, S.C.; Jones, I.D.; Feuchtmayr, H. A novel method for estimating the onset of thermal stratification in lakes from surface water measurements. *Water Resour. Res.* **2014**, *50*, 5131–5140. [CrossRef]
4. McGinnis, D.; Wuest, A. *Lake Hydrodynamics*; Yearbook of Science & Technology; McGraw Hill: London, UK, 2005.
5. Kraemer, B.M.; Anneville, O.; Chandra, S.; Dix, M.; Kuusisto, E.; Livingstone, D.M.; Rimmer, A.; Schladow, S.G.; Silow, E.; Sitoki, L.M. Morphometry and average temperature affect lake stratification responses to climate change. *Geophys. Res. Lett.* **2015**, *42*, 4981–4988. [CrossRef]
6. Elçi, S. Effects of thermal stratification and mixing on reservoir water quality. *Limnology* **2008**, *9*, 135–142. [CrossRef]
7. Nürnberg, G.K. Quantifying anoxia in lakes. *Limnol. Oceanogr.* **1995**, *40*, 1100–1111. [CrossRef]
8. Mishra, S.; Kumar, A. Estimation of physicochemical characteristics and associated metal contamination risk in the Narmada River, India. *Environ. Eng. Res.* **2021**, *26*, 190521. [CrossRef]
9. Kumar, S.S.; Kumar, A.; Singh, S.; Malyan, S.K.; Baram, S.; Sharma, J.; Singh, R.; Pugazhendhi, A. Industrial wastes: Fly ash, steel slag and phosphogypsum- potential candidates to mitigate greenhouse gas emissions from paddy fields. *Chemosphere* **2020**, *241*, 124824. [CrossRef]
10. Gantzer, P.A.; Bryant, L.D.; Little, J.C. Controlling soluble iron and manganese in a water-supply reservoir using hypolimnetic oxygenation. *Water Res.* **2009**, *43*, 1285–1294. [CrossRef]
11. Hill, D.F.; Vergara, A.M.; Parra, E.J. Destratification by mechanical mixers: Mixing efficiency and flow scaling. *J. Hydraul. Eng.* **2008**, *134*, 1772–1777. [CrossRef]
12. Chen, S.; Carey, C.C.; Little, J.C.; Lofton, M.E.; McClure, R.P.; Lei, C. Effectiveness of a bubble-plume mixing system for managing phytoplankton in lakes and reservoirs. *Ecol. Eng.* **2018**, *113*, 43–51. [CrossRef]
13. Schladow, S.G. Lake destratification by bubble-plume systems: Design methodology. *J. Hydraul. Eng.* **1993**, *119*, 350–368. [CrossRef]
14. Moshfeghi, H.; Etemad-Shahidi, A.; Imberger, J. Modelling of bubble plume destratification using DYRESM. *J. Water Supply Res. Technol.—AQUA* **2005**, *54*, 37–46. [CrossRef]
15. Sahoo, G.B.; Luketina, D. Response of a tropical reservoir to bubbler destratification. *J. Environ. Eng.* **2006**, *132*, 736–746. [CrossRef]
16. Austin, D.; Scharf, R.; Chen, C.F.; Bode, J. Hypolimnetic oxygenation and aeration in two Midwestern USA reservoirs. *Lake Reserv. Manag.* **2019**, *35*, 266–276. [CrossRef]
17. Lorenzen, M.; Fast, A.; Wermes, I.M. *A Guide to Aeration/Circulation Techniques for Lake Management*; Environmental Protection Agency, Office of Research and Development: Washington, DC, USA, 1977.
18. Hanson, D.; Austin, D. Multiyear destratification study of an urban, temperate climate, eutrophic lake. *Lake Reserv. Manag.* **2012**, *28*, 107–119. [CrossRef]
19. Davison, W. Iron and manganese in lakes. *Earth-Sci. Rev.* **1993**, *34*, 119–163. [CrossRef]
20. Ma, W.X.; Huang, T.L.; Li, X. Study of the application of the water-lifting aerators to improve the water quality of a stratified, eutrophicated reservoir. *Ecol. Eng.* **2015**, *83*, 281–290. [CrossRef]
21. Zaw, M.; Chiswell, B. Iron and manganese dynamics in lake water. *Water Res.* **1999**, *33*, 1900–1910. [CrossRef]

22. Heo, W.M.; Kim, B. The effect of artificial destratification on phytoplankton in a reservoir. *Hydrobiologia* **2004**, *524*, 229–239. [CrossRef]
23. Kiehl, J.T.; Briegleb, B.P. The relative roles of sulfate aerosols and greenhouse gases in climate forcing. *Science* **1993**, *260*, 311–314. [CrossRef]
24. Maeck, A.; DelSontro, T.; McGinnis, D.F.; Fischer, H.; Flury, S.; Schmidt, M.; Fietzek, P.; Lorke, A. Sediment trapping by dams creates methane emission hot spots. *Environ. Sci. Technol.* **2013**, *47*, 8130–8137. [CrossRef]
25. Kumar, A.; Yang, T.; Sharma, M.P. Greenhouse gas measurement from Chinese freshwater bodies: A review. *J. Clean. Prod.* **2019**, *233*, 368–378. [CrossRef]
26. Stocker, T.; Qin, D.; Plattner, G.K.; Tignor, M.; Allen, S.; Boschung, J.; Nauels, A.; Xia, Y.; Bex, V.; Midgley, P. (Eds.) Summary for Policymakers. In *Climate Change 2013: The Physical Science Basis*; Contribution of Working Group I to the Fifth Assessment Report of the Intergovernmental Panel on Climate Change; Cambridge University Press: Cambridge, UK; New York, NY, USA, 2013; Volume 1542, pp. 33–36. [CrossRef]
27. Williamson, C.E.; Saros, J.E.; Vincent, W.F.; Smol, J.P. Lakes and reservoirs as sentinels, integrators, and regulators of climate change. *Limnol. Oceanogr.* **2009**, *54*, 2273–2282. [CrossRef]
28. Paerl, H.W.; Huisman, J. Climate change: A catalyst for global expansion of harmful cyanobacterial blooms. *Environ. Microbiol. Rep.* **2009**, *1*, 27–37. [CrossRef]
29. Bryant, L.D.; Hsu-Kim, H.; Gantzer, P.A.; Little, J.C. Solving the problem at the source: controlling Mn release at the sediment-water interface via hypolimnetic oxygenation. *Water Res.* **2011**, *45*, 6381–6392. [CrossRef]
30. Paerl, H.W.; Paul, V.J. Climate change: Links to global expansion of harmful cyanobacteria. *Water Res.* **2012**, *46*, 1349–1363. [CrossRef]
31. Sánchez-Benítez, A.; García-Herrera, R.; Barriopedro, D.; Sousa, P.M.; Trigo, R.M. June 2017: The Earliest European Summer Mega-heatwave of Reanalysis Period. *Geophys. Res. Lett.* **2018**, *45*, 1955–1962. [CrossRef]
32. Wilhelm, S.; Adrian, R. Impact of summer warming on the thermal characteristics of a polymictic lake and consequences for oxygen, nutrients and phytoplankton. *Freshw. Biol.* **2008**, *53*, 226–237. [CrossRef]
33. Tranmer, A.W.; Marti, C.L.; Tonina, D.; Benjankar, R.; Weigel, D.; Vilhena, L.; McGrath, C.; Goodwin, P.; Tiedemann, M.; Mckean, J. A hierarchical modelling framework for assessing physical and biochemical characteristics of a regulated river. *Ecol. Model.* **2018**, *368*, 78–93. [CrossRef]
34. Romero, J.R.; Antenucci, J.P.; Imberger, J. One-and three-dimensional biogeochemical simulations of two differing reservoirs. *Ecol. Model.* **2004**, *174*, 143–160. [CrossRef]
35. Ulańczyk, R.; Kliś, C.; Absalon, D.; Ruman, M. Mathematical Modelling as a Tool for the Assessment of Impact of Thermodynamics on the Algal Growth in Dam Reservoirs–Case Study of the Goczalkowice Reservoir. *Ochrona Srodowiska i Zasobów Naturalnych* **2018**, *29*, 21–29. [CrossRef]
36. Zamani, B.; Koch, M. Comparison between two hydrodynamic models in simulating physical processes of a reservoir with complex morphology: Maroon Reservoir. *Water* **2020**, *12*, 814. [CrossRef]
37. Ryu, I.; Yu, S.; Chung, S. Characterizing Density Flow Regimes of Three Rivers with Different Physicochemical Properties in a Run-Of-The-River Reservoir. *Water* **2020**, *12*, 717. [CrossRef]
38. Environment Agency. *Nitrate Vulnerable Zone (NVZ) Designation, 2017 Eutrophication (Lakes) Evidence of Eutrophication 2017*; Technical Report Blagdon Lake, Environment Agency: Rotherham, UK, 2016.
39. Met Office. *Met Office MIDAS Open: UK Land Surface Stations Data (1853-Current)*; Centre for Environmental Data Analysis: Didcot, UK, 2019.
40. Dunn, R.J.H.; Willett, K.M.; Parker, D.E.; Mitchell, L. Expanding HadISD: Quality-controlled, sub-daily station data from 1931. *Geosci. Instrum. Methods Data Syst.* **2016**, *5*, 473–491. [CrossRef]
41. Dunn, R.J.H.; Willett, K.M.; Thorne, P.W.; Woolley, E.V.; Durre, I.; Dai, A.; Parker, D.E.; Vose, R.S. HadISD: A quality-controlled global synoptic report database for selected variables at long-term stations from 1973–2011. *Clim. Past* **2012**, *8*, 1649–1679. [CrossRef]
42. Smith, A.; Lott, N.; Vose, R. The integrated surface database: Recent developments and partnerships. *Bull. Am. Meteorol. Soc.* **2011**, *92*, 704–708. [CrossRef]
43. Met Office Hadley Centre; National Centers for Environmental Information—NOAA. *HadISD: Global Sub-Daily, Surface Meteorological Station Data, 1931-2018, v3.0.0.2018f*; Centre for Environmental Data Analysis: Didcot, UK, 2019.
44. Met Office Hadley Centre. *UKCP18 Regional Projections on a 12km Grid over the UK for 1980–2080*; The Centre for Environmental Data Analysis, STFC Rutherford Appleton Laboratory, Harwell Campus: Didcot, UK, 2018.
45. Lowe, J.A.; Bernie, D.; Bett, P.; Bricheno, L.; Brown, S.; Calvert, D.; Clark, R.; Eagle, K.; Edwards, T.; Fosser, G. *UKCP18 Science Overview Report*; Met Office Hadley Centre: Exeter, UK, 2018.
46. Meinshausen, M.; Smith, S.J.; Calvin, K.; Daniel, J.S.; Kainuma, M.L.T.; Lamarque, J.F.; Matsumoto, K.; Montzka, S.A.; Raper, S.C.B.; Riahi, K. The RCP greenhouse gas concentrations and their extensions from 1765 to 2300. *Clim. Chang.* **2011**, *109*, 213–241. [CrossRef]
47. Riahi, K.; Rao, S.; Krey, V.; Cho, C.; Chirkov, V.; Fischer, G.; Kindermann, G.; Nakicenovic, N.; Rafaj, P. RCP 8.5—A scenario of comparatively high greenhouse gas emissions. *Clim. Chang.* **2011**, *109*, 33–57. [CrossRef]

48. Sexton, D.M.H.; McSweeney, C.F.; Rostron, J.W.; Yamazaki, K.; Booth, B.B.B.; Murphy, J.M.; Regayre, L.; Johnson, J.S.; Karmalkar, A.V. A perturbed parameter ensemble of HadGEM3-GC3. 05 coupled model projections: Part 1: Selecting the parameter combinations. *Clim. Dyn.* **2021**, *56*, 3395–3436. [CrossRef]
49. Birt, D. *Results and Supporting Datasets for "Stratification in a Reservoir Mixed by Bubble Plumes under Future Climate Scenarios"*; University of Bath: Bath, UK, 2021.
50. Ephrath, J.E.; Goudriaan, J.; Marani, A. Modelling diurnal patterns of air temperature, radiation wind speed and relative humidity by equations from daily characteristics. *Agric. Syst.* **1996**, *51*, 377–393. [CrossRef]
51. George, A.; Anto, R. Analytical and experimental analysis of optimal tilt angle of solar photovoltaic systems. In Proceedings of the 2012 International Conference on Green Technologies (ICGT), Trivandrum, India, 18–20 December 2012; pp. 234–239.
52. Spitters, C.J.T.; Toussaint, H.; Goudriaan, J. Separating the diffuse and direct component of global radiation and its implications for modeling canopy photosynthesis Part I. Components of incoming radiation. *Agric. For. Meteorol.* **1986**, *38*, 217–229. [CrossRef]
53. Flerchinger, G.N.; Xaio, W.; Marks, D.; Sauer, T.J.; Yu, Q. Comparison of algorithms for incoming atmospheric long-wave radiation. *Water Resour. Res.* **2009**, *45*, 1–13. [CrossRef]
54. Unsworth, M.H.; Monteith, J.L. Long-wave radiation at the ground I. Angular distribution of incoming radiation. *Q. J. R. Meteorol. Soc.* **1975**, *101*, 13–24. [CrossRef]
55. Donatelli, M.; Bellocchi, G.; Habyarimana, E.; Confalonieri, R.; Micale, F. An extensible model library for generating wind speed data. *Comput. Electron. Agric.* **2009**, *69*, 165–170. [CrossRef]
56. Guo, Z.; Chang, C.; Wang, R. *A Novel Method to Downscale Daily Wind Statistics to Hourly Wind Data for Wind Erosion Modelling BT-Geo-Informatics in Resource Management and Sustainable Ecosystem*; Springer: Berlin/Heidelberg, Germany, 2016; pp. 611–619.
57. Hodges, B.; Dallimore, C. *Aquatic Ecosystem Model: AEM3D: v1.2 User Manual*, v1.2 ed.; HydroNumerics: Docklands, VIC, Australia, 2021; pp. 1–194.
58. Idso, S.B.; Gilbert, R.G. On the universality of the Poole and Atkins Secchi disk-light extinction equation. *J. Appl. Ecol.* **1974**, *11*, 399–401. [CrossRef]
59. Twomey, S. Pollution and the planetary albedo. *Atmos. Environ.* **1974**, *8*, 1251–1256. [CrossRef]
60. Rouse, W.R.; Oswald, C.J.; Binyamin, J.; Spence, C.; Schertzer, W.M.; Blanken, P.D.; Bussières, N.; Duguay, C.R. The role of northern lakes in a regional energy balance. *J. Hydrometeorol.* **2005**, *6*, 291–305. [CrossRef]
61. Lawson, R.; Anderson, M.A. Stratification and mixing in Lake Elsinore, California: An assessment of axial flow pumps for improving water quality in a shallow eutrophic lake. *Water Res.* **2007**, *41*, 4457–4467. [CrossRef]
62. Winder, M.; Schindler, D.E. Climatic effects on the phenology of lake processes. *Glob. Chang. Biol.* **2004**, *10*, 1844–1856. [CrossRef]
63. Wu, B.; Wang, G.; Jiang, H.; Wang, J.; Liu, C. Impact of revised thermal stability on pollutant transport time in a deep reservoir. *J. Hydrol.* **2016**, *535*, 671–687. [CrossRef]
64. Vinçon-Leite, B.; Lemaire, B.J.; Khac, V.T.; Tassin, B. Long-term temperature evolution in a deep sub-alpine lake, Lake Bourget, France: how a one-dimensional model improves its trend assessment. *Hydrobiologia* **2014**, *731*, 49–64. [CrossRef]
65. Minns, C.K.; Moore, J.E.; Doka, S.E.; St. John, M.A. Temporal trends and spatial patterns in the temperature and oxygen regimes in the Bay of Quinte, Lake Ontario, 1972–2008. *Aquat. Ecosyst. Health Manag.* **2011**, *14*, 9–20. [CrossRef]
66. Noges, P.; Noges, T.; Ghiani, M.; Paracchini, B.; Grande, J.P.; Sena, F. Morphometry and trophic state modify the thermal response of lakes to meteorological forcing. *Hydrobiologia* **2011**, *667*, 241–254. [CrossRef]
67. Schmidt, W. Über Die Temperatur-Und Stabili-Tätsverhältnisse Von Seen. *Geogr. Ann.* **1928**, *10*, 145–177. [CrossRef]
68. Idso, S.B. On the concept of lake stability 1. *Limnol. Oceanogr.* **1973**, *18*, 681–683. [CrossRef]
69. Rolland, D.C.; Bourget, S.; Warren, A.; Laurion, I.; Vincent, W.F. Extreme variability of cyanobacterial blooms in an urban drinking water supply. *J. Plankton Res.* **2013**, *35*, 744–758. [CrossRef]
70. Mann, H.B. Nonparametric tests against trend. *Econom. J. Econom. Soc.* **1945**, *13*, 245–259. [CrossRef]
71. Kendall, M.G. *Rank Correlation Methods*; Charles Griffin & Co.: London, UK, 1975.
72. Gocic, M.; Trajkovic, S. Analysis of changes in meteorological variables using Mann-Kendall and Sen's slope estimator statistical tests in Serbia. *Glob. Planet. Chang.* **2013**, *100*, 172–182. [CrossRef]
73. Kumar, A.; Taxak, A.K.; Mishra, S.; Pandey, R. Long term trend analysis and suitability of water quality of River Ganga at Himalayan hills of Uttarakhand, India. *Environ. Technol. Innov.* **2021**, *22*, 101405. [CrossRef]
74. Boehrer, B.; Schultze, M. Stratification of lakes. *Rev. Geophys.* **2008**, *46*, 1–27. [CrossRef]
75. Hetherington, A.L.; Schneider, R.L.; Rudstam, L.G.; Gal, G.; DeGaetano, A.T.; Walter, M.T. Modeling climate change impacts on the thermal dynamics of polymictic Oneida Lake, New York, United States. *Ecol. Model.* **2015**, *300*, 1–11. [CrossRef]
76. Woolway, R.I.; Merchant, C.J. Worldwide alteration of lake mixing regimes in response to climate change. *Nat. Geosci.* **2019**, *12*, 271–276. [CrossRef]
77. Bryant, L.D.; Little, J.C.; Bürgmann, H. Response of sediment microbial community structure in a freshwater reservoir to manipulations in oxygen availability. *Fems Microbiol. Ecol.* **2012**, *80*, 248–263. [CrossRef] [PubMed]

Article

Dissolved Oxygen in a Shallow Ice-Covered Lake in Winter: Effect of Changes in Light, Thermal and Ice Regimes

Galina Zdorovennova [1,*], Nikolay Palshin [1], Sergey Golosov [1], Tatiana Efremova [1], Boris Belashev [2], Sergey Bogdanov [1], Irina Fedorova [3], Ilia Zverev [1], Roman Zdorovennov [1] and Arkady Terzhevik [1]

[1] Northern Water Problems Institute, Karelian Research Centre RAS, 185000 Petrozavodsk, Russia; npalshin@mail.ru (N.P.); sergey_golosov@mail.ru (S.G.); efremova@nwpi.krc.karelia.ru (T.E.); sergey.r.bogdanov@mail.ru (S.B.); iliazverev@mail.ru (I.Z.); romga74@gmail.com (R.Z.); ark1948@list.ru (A.T.)
[2] Institute of Geology, Karelian Research Centre RAS, 185000 Petrozavodsk, Russia; belashev@krc.karelia.ru
[3] Institute of Earth Sciences, Saint-Petersburg State University, 199034 Saint Petersburg, Russia; i.fedorova@spbu.ru
* Correspondence: zdorovennova@gmail.com

Abstract: Oxygen conditions in ice-covered lakes depend on many factors, which, in turn, are influenced by a changing climate, so detection of the oxygen trend becomes difficult. Our research was based on data of long-term measurements of dissolved oxygen (2007–2020), water temperature, under-ice solar radiation, and snow-ice thickness (1995–2020) in Lake Vendyurskoe (Northwestern Russia). Changes of air temperature and precipitation in the study region during 1994–2020 and ice phenology of Lake Vendyurskoe for the same period based on field data and FLake model calculations were analyzed. The interannual variability of ice-on and ice-off dates covered wide time intervals (5 and 3 weeks, respectively), but no significant trends were revealed. In years with early ice-on, oxygen content decreased by more than 50% by the end of winter. In years with late ice-on and intermediate ice-off, the oxygen decrease was less than 40%. A significant negative trend was revealed for snow-ice cover thickness in spring. A climatic decrease of snow-ice cover thickness contributes to the rise of under-ice irradiance and earlier onset of under-ice convection. In years with early and long convection, an increase in oxygen content by 10–15% was observed at the end of the ice-covered period, presumably due to photosynthesis of phytoplankton.

Keywords: shallow lake; ice-covered period; ice phenology; snow-ice cover thickness; dissolved oxygen; water temperature; under-ice irradiance; radiatively driven convection; climate change

1. Introduction

Dissolved oxygen (DO) involved in chemical and biological processes is one of the most important parameters of aquatic ecosystems [1]. A decrease in the oxygen content worsens the quality of water and the habitat of aquatic organisms. The development of hypoxia and anoxia [2] can lead to fatal consequences, such as mass death of fish [3]. Moreover, under anaerobic conditions, the release and accumulation of greenhouse gases can occur [4]. These factors determine the ecological significance of studying the oxygen regime of lakes.

During the open water period, oxygen conditions in lakes are largely formed through the gas exchange with the atmosphere and gas redistribution over the water column due to wind-wave mixing and convective movements. In the presence of photosynthesis, oxygen is released and organic matter is formed, which then settles to the bottom and creates the prerequisites for oxygen deficiency in the following winter season [5,6].

Continuous ice cover almost completely suppresses the exchange of heat, gas, and momentum between the water mass and the atmosphere. Snow on ice sharply reduces the penetration of photosynthetic radiation into the water column. Inhibited by low illumination, photosynthesis cannot be a significant source of oxygen as long as the snow layer on the ice exceeds 10–20 cm [7–10]. After the snow has melted and under-ice illumination has increased, a significant increase of oxygen due to photosynthesis of phytoplankton can be expected [11,12]. River runoff, underground springs, and melt water can be additional sources of oxygen in ice-covered lakes, but they can play a significant role in the oxygen budget only in small shallow lakes [13,14]. Thus, in the absence of significant sources of oxygen, its content decreases during the winter, mainly due to bacterial decomposition of organic matter, respiration of organisms, and chemical reactions near the surface of bottom sediments [5,6,13–19].

Analysis of the data of long-term oxygen measurements shows that the degree of winter oxygen depletion differs in the same lake in different years (see, for example, [11,13,14,16,20–22]). What factors influence the rate of oxygen consumption and production in an ice-covered lake in different years? And how does modern climate variability affect these factors?

First of all, the duration of the ice-covered period determines the duration of winter hypoxia and anoxia; hence, the dates of ice-on and ice-off should be taken into account. These dates on a particular lake being largely determined by meteorological conditions, so, air temperature, precipitation, wind speed, and solar radiation can be considered among the main factors influencing ice phenology [23–25].

The rate of bacterial decomposition of organic matter, and, consequently, the rate of oxygen consumption, depends on the water temperature [15], which can therefore be considered as a predictor of DO conditions. The temperature of water in lakes in winter, in turn, depends on the date of ice-on and heat transfer from bottom sediments. A prolonged pre-ice period contributes to the loss of heat accumulated by the water mass and bottom sediments in summer, and the winter water temperature in this case is lower. On the contrary, early formation of ice ensures the preservation of the heat accumulated in summer, and the water temperature during the winter season is higher [26].

The thickness and structure of the snow-ice cover can be considered as a factor that can have a limiting effect on the photosynthesis of phytoplankton and, hence, on the release of oxygen. Snow not only effectively absorbs solar radiation, but can also turn into white ice or slush over black ice surface [7,8,27]. White ice and slush also effectively absorb solar radiation [7,8]. Therefore, climatic changes thickness of snow, white ice and slush can cause long-term changes in phytoplankton environmental conditions and oxygen production.

Light, temperature, and hydrodynamic conditions favorable for the growth of phytoplankton and release of oxygen are usually observed in boreal lakes in late winter, during radiatively driven convection [28]. Convective currents keep non-mobile phytoplankton within the photic zone and transport nutrients from the bottom layers to the surface [9,10,21,29,30]. Therefore, the duration of radiatively driven convection can be considered as a factor influencing oxygen increase at the end of the ice-covered period.

Thus, oxygen conditions in ice-covered lakes are determined by many factors, which, in turn, may be influenced by climate change: weather conditions, dates of ice-on and ice-off, duration of the ice season and under-ice convection, under-ice solar radiation, water temperature, thickness and structure of the snow-ice cover.

In this study, we analyzed the variability of DO content in a small boreal Lake Vendyurskoe (northwest Russia, southern Karelia) in winters 2007–2020. We investigated the effect of long-term changes of climatic parameters (air temperature and precipitation) in 1994–2020 on the lake's thermal, light, and ice regimes. The purpose of this work is to investigate the influence of various factors on the rate of oxygen consumption and production in an ice-covered lake under modern climatic conditions.

2. Materials and Methods

2.1. Study Site

Lake Vendyurskoe is located in the northwest of Russia, southern Karelia (62°10′ N, 33°10′ E) (Figure 1a). The limnological description of Lake Vendyurskoe and characteristics of the climate in the study region are given in [31,32]. This is a typical lake of glacial origin; such lakes are widespread in Fennoscandia. The lake surface area is 10.4 km^2, length—7.0 km, width—1.5–2.0 km, water volume—5.5 × 10^7 m^3, mean and maximum depths are 5.3 m and 13.4 m. Despite its relatively large average and maximum depths, Lake Vendyurskoe is a typical polymictic lake. During the open water period, it can be fully mixed up to 5–6 times [33]. There are several small inlet streams and one outlet, but their discharge is low. Precipitation and evaporation have a decisive influence on changes in the volume of the lake water mass during a year. Thermal stratification in summer and in winter leads to oxygen deficiency in the bottom layers of the central basin of the lake [5,6,33]. Ice-on occurs from early November to mid-December, and ice-off occurs from late April to the third week of May [34].

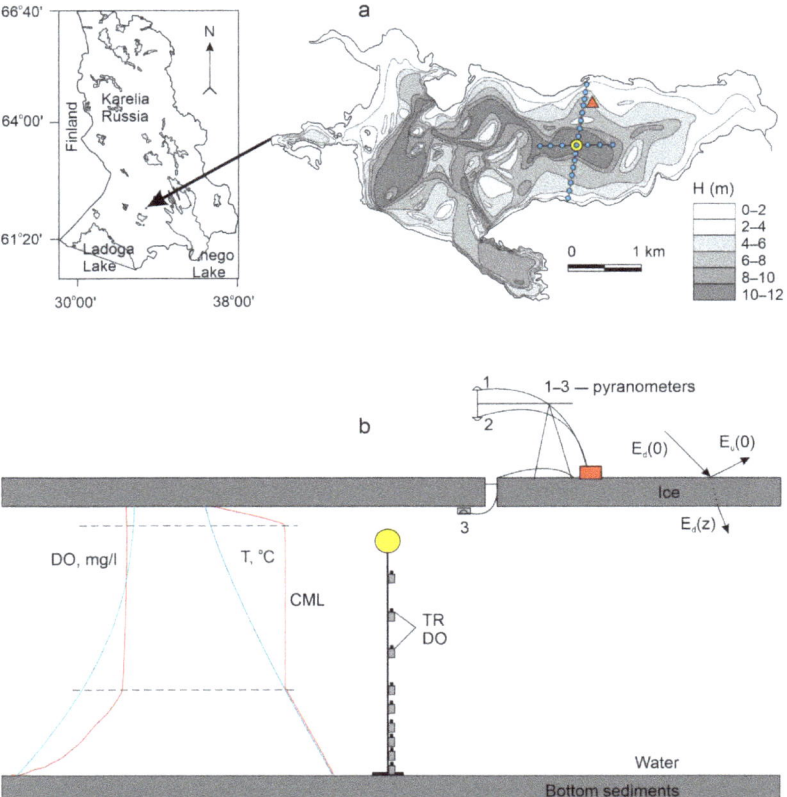

Figure 1. (**a**) Geographic location (black arrow) and bathymetry map of Lake Vendyurskoe, and locations of measurement Scheme 1995. (**b**) Setup of the observation site, sketch of the thermal and DO structures during spring under-ice convection and of radiative transfer through the ice. Blue and red lines indicate vertical DO and temperature profiles in winter and during radiatively driven convection, respectively. CML—convectively mixed layer. $E_d(0)$ and $E_d(z)$—downwelling planar irradiances at the ice surface and under the ice, $E_u(0)$—upwelling irradiances at the ice surface.

2.2. Climate Data

To analyze the climate variability in the study region in 1994–2020, we used data from the Petrozavodsk weather station (WMO station ID: 22820, coordinates 61°49′ N and 34°16′ E), which is located 75 km to the south-east from Lake Vendyurskoe. Data on air temperature and precipitations were obtained from the "Reliable Forecast" website [35] and the website of the All-Russian Scientific Research Institute of Hydrometeorological

Information—World Data Center (VNIIGMI-WDC) [36]. Average monthly air temperature at the Petrozavodsk station for the baseline period of 1961–1990 was obtained from the website of the North Eurasian Climate Center [37].

To characterize the variability of the regional climate of southern Karelia in 1994–2020, we used two parameters—air temperature and precipitation. Since the ice-covered period on Lake Vendyurskoe lasted from November to May, we analyzed the variability of the selected climatic parameters for the months from November of the previous year to May of the next year. We calculated the deviation of the air temperature of November, December, January, February, March, April, and May relative to the corresponding months of the baseline period 1961–1990 (Table S1) and also analyzed the air temperature trends for each of these months in order to identify the climatic reasons for changes in the dates of ice-on and ice-off (Table S2).

To describe how severe or mild each winter season was, we calculated the average temperature for the period from November of the previous year to April of the next year, and then compared it to the baseline values (the rightmost column in Table S1). In this case, we did not take into account the temperature in May, since freezing temperatures in this month are extremely rare and cannot have a significant effect on the severity or mildness of the winter.

To track how the conditions of ice formation and ice melting changed in the winter seasons of 1994–2020 and to identify the climatic triggers for changes in the thickness and structure of the snow-ice cover, we calculated the number of days with an average positive daily air temperature (Days_T_{AIR} > 0) and the number of days with precipitation (Days_P) in November–April for each winter season. If precipitation happened on a day with a positive average daily air temperature, then we assumed the precipitation was liquid or mixed; then the number of such days (Days_P_{LM}) was calculated.

2.3. Under Ice Radiation

Measurements of solar radiation were carried out on different dates in spring months in 1995–2000, 2002–2018, and 2020; the duration of radiation measurements differed markedly from year to year—from one to 14 days, with measurement interval from one to five minutes. Dates of the first and last days of measurements in different years are shown in Table S3. The downwelling $E_d(0)$ and upwelling $E_u(0)$ planar irradiances at the ice surface and downwelling irradiances $E_d(z)$ under the ice were measured with Star-shaped pyranometers (Theodor Friderich & Co, Meteorologische Gerate und Systeme, Germany, resolution 0.2 W/m^2); two pyranometers were mounted on a tripod at one meter above ice surface, and the third one was fixed on a foam plate under the ice surface (Figure 1b). In this study, we used only data from the third pyranometer. Data from the first and second pyranometers were analyzed in [32]. For each day of radiation measurements, the maximum and average values of $E_d(z)$ were calculated for the period from 8 a.m. to 8 p.m. Table S3 shows the values of $E_d(z)$ for the first and last days of radiation measurements for each year.

2.4. Ice Regime of Lake Vendyurskoe. Measurements and Modelling

Measurements of snow-ice cover thickness were carried out on the first and last days of radiation measurements in spring months of 1995–2000, 2002–2018 and 2020 at 22 stations (Figure 1a), then the average and standard deviation of snow, total ice, white and black ice, and slush were calculated over all stations (Table S3). We then calculated the snow-ice cover thickness for mid-April (H_{APRIL_MEAS}) for each year of measurements using linear interpolation.

Dates of ice-on and ice-off were determined using TR-chain data (Table 1; see procedure in Section 2.5).

Table 1. Estimates of the statistical significance of trends (1994–2020) by Mann–Kendall test (minus means the trend is not significant, plus means the trend is significant at the level less than 5%).

Parameters/Characteristics	Z (h)	b1	b0	Confidence Interval b1 95%
Days_P	1.0409(−)			
Days_T_{AIR} > 0	2.0504(+)	1.1053	−2160.2	[0.091 1.996]
Days_P_{LM}	2.0115(+)	0.6667	−1307.8	[0.0 1.3586]
Ice on (field)	1.243(−)			
Ice on (model)	0.7249(−)			
Ice off (field)	−1.5152(−)			
Ice off (model)	−1.1696(−)			
$H_{IS\_APRIL\_MEAS}$	−3.4599(+)	−0.9161	1894.2	[−1.333–0.4562]
H_{ICE_MODEL}	−1.7516(−)			

Since field data have gaps for the dates of ice-on and ice-off in some years (Table 1), we also investigate the ice regime of Lake Vendyurskoe using the one-dimensional thermohydrodynamic model FLake [38,39]. This model was developed jointly by the German Meteorological Service, the Institute of Limnology of the Russian Academy of Sciences, the Leibnitz-Institute of Water Ecology and Inland Fisheries (Germany), and the Northern Water Problems Institute of the Russian Academy of Sciences (Russia). The FLake model is used to solve a wide range of limnological problems, serves as a basic tool for developing models of the functioning of aquatic ecosystems and the formation of water quality. The standard meteorological parameters, namely short-wave solar radiation, air temperature, wind speed, absolute humidity, cloudiness, atmospheric pressure and precipitation, are used to reproduce the thermal regime and mixing conditions in a lake. In our case, meteorological data from the ERA 5 reanalysis [40] were used to study the ice regime of Lake Vendyurskoe for the period from 1994 to 2019.

According to FLake model calculations, the dates of ice-on and ice-off on Lake Vendyurskoe in 1994–2019 were determined, and the ice thickness was calculated for each day of each ice season. Then, we calculated the average ice thickness (H_{ICE_MODEL}) for each winter season of 1995–2019.

2.5. Water Temperature and Dissolved Oxygen Measurements

Measurements of water temperature (1995–2020) and DO content (2007–2020) were carried out in the central deep-water basin of Lake Vendyurskoe, where the temperature chain was installed (Figure 1b). The TR-1 temperature sensors (Aanderaa Instruments, Norway, accuracy 0.15 °C) were used in 1995–2006, and RBR temperature and DO loggers (RBR, Canada, temperature sensor accuracy 0.002 °C, oxygen range 0–150%, accuracy 1%) were used in 2007–2020. Sensors were fixed on a cable every 0.02–1.5 m, with a smaller spacing towards the bottom (Figure 1b). The measurement interval was 1–3 h in 1995–2006 and one minute in 2007–2020.

During most of the research years, the chain was installed on the lake bottom before ice-on (October–early November) and was removed from the lake after ice-off (May–June). However, in some years the chain was installed after the ice formation and was removed from the lake before the ice degradation. For these years, we could not determine the dates of ice-on (1997, 2000–2002, 2006) and ice-off (1998, 2001–2003, 2007).

According to TR-chain data, the dates of ice-on and ice-off and the start of the spring radiatively driven convection were determined (Table 1). The dates of ice-on were determined by the moment when the water temperature near the bottom starts to increase and DO concentration starts to decrease. The dates of ice-off were determined by the moment of a sharp drop in the bottom temperatures and sharp increase of DO concentrations (Figure S1). The start of the radiatively driven convection was determined by the appearance of daily temperature variability in the upper chain temperature sensor and formation of the convectively mixed layer (CML).

We used the column-average water temperature on the day of ice-on (T_{W_ICE-ON}) and the near bottom water temperature at the end of the winter season (T_{WB_MAX}) (Table 1) as factors that reflect the interannual variability of the lake's thermal regime and can affect the oxygen consumption rate in winter.

The oxygen content over the water column was calculated for the first day of the ice period C_0 and then for each day of the ice period C_t; then the ratio C_t/C_0 was calculated for each day of the ice period (Table 1). This indicator declined during the winter, reflecting the total oxygen consumption in the lake. For each winter season the extent of oxygen depletion during winter (C_t/C_{0_MIN}) was calculated. In some years, a pronounced increase in oxygen was observed by the end of the ice-covered period, $C_t/C_{0\_END} - C_t/C_{0\_MIN} = C_t/C_{0\_PLUS}$, which reflected an increase in the oxygen content in the lake. We suppose that the most likely reason for this increase was the release of oxygen as a result of photosynthesis of phytoplankton, since this increase was observed before the ice-off and could not be associated with gas exchange with the atmosphere.

2.6. Statistical Data Analysis

Trend analysis was applied to: (1) air temperature in November, December, January, February, March, April, and May in 1994–2020, (2) number of days with positive air temperature (Days_$T_{AIR} > 0$), with precipitation (Days_P), and with liquid or mixed precipitation (Days_P_{LM}) in November–April in 1994–2020, (3) dates of ice-on and ice-off on Lake Vendyurskoe in 1995–2020, both measured and simulated, and (4) measured snow-ice thickness in mid-April ($H_{IS\_APRIL\_MEAS}$) and simulated ice thickness averaged over the winter season (H_{ICE_MODEL}) in 1995–2019.

The level of statistical significance of the trends was assessed by Student's t-statistics and the nonparametric Mann–Kendall test [41]. Sen's method was used to determine the slope of the trend and its confidence intervals [41].

We investigated the regression between the days with positive air temperature (Days_$T_{AIR} > 0$) and liquid or mixed precipitation (Days_P_{LM}) in November–April and maximal thickness of the snow-ice cover of Lake Vendyurskoe in mid-April.

We analyzed the influence of the thickness of the snow-ice cover, as well as the thickness of snow, white ice, and slush on under-ice radiation in spring.

Then, we analyzed the relationship between the extent of oxygen depletion during winter (C_t/C_{0_MIN}) and such factors as ice-on date, average water column temperature on the day of ice-on (T_{W_ICE-ON}), water temperature in the bottom layer at the end of the winter season (T_{WB_MAX}), and the number of days with positive air temperature during November–April as an indicator of the severity of the winter.

To understand which factors determine the oxygen increase in late winter, we looked for links between the increase in DO content (C_t/C_{0_PLUS}) and factors such as the date of onset of radiatively driven convection and its duration, the number of days with positive air temperatures (Days_$T_{AIR} > 0$) and days with liquid or mixed precipitation (Days_P_{LM}) in March and April. The last two parameters could reflect the effect of weather conditions on snow-ice thickness and under-ice irradiance, assuming that the more days with thaws or days with liquid or mixed precipitation, the thinner the ice and the greater the under-ice irradiance. Statistical analysis of field and model data was carried out using the software packages STATISTICA and MatLab.

3. Results

3.1. Climatic Variability of Air Temperature and Precipitation at the Petrozavodsk Weather Station in 1994–2020

Average air temperatures of January, February, March, April, May, November, and December in the 1961–1990 baseline period were −11.4 °C, −9.8 °C, −4.4 °C, +1.5 °C, +8.4 °C, −2.5 °C, and −7.6 °C, respectively. Table S1 shows the deviations of average monthly air temperatures of these months in 1994–2020 from the corresponding values of the baseline period. The months most often warmer than the baseline was January

(22 years), April (20 years), December (19 years), May (18 years), and February (16 years). Positive deviation of 5 °C or more from the baseline (bold in Table S1) most often occurred in December (8 years), January (7 years), and February (8 years). March was warmer than the baseline by 5 °C or more only in two years. In November, April, and May, the deviation from the baseline did not exceed 5 °C, with maximum values of 4.9 °C, 4.2 °C, and 4.3 °C, respectively. The greatest positive deviation from the baseline was recorded in January 2020 (9.4 °C). Positive trends were found for all months, but only November and May air temperature trends were statistically significant (Table S2).

The average temperature for the period including January, February, March, April, November, and December in 1961–1990 was −5.7 °C. Seven winter seasons in 1994–2020 were colder than this value (numbers with asterisks in the rightmost column in Table S1); other winter seasons in the considered period were warmer than the baseline. The period from November to April was the warmest in 2007–2008, 2013–2014, 2014–2015, and 2019–2020, with average temperatures of −2.2 °C, −1.6 °C, −2.1 °C, and −1.0 °C.

The number of days with precipitation in November–April in 1994–2020 did not demonstrate any significant trend (Table 1). The number of days with a positive average daily air temperature (Table 1, Figure 2a) and the number of days with liquid or mixed precipitation (Table 1, Figure 2b) in November–April in 1994–2020 increased significantly during the considered period.

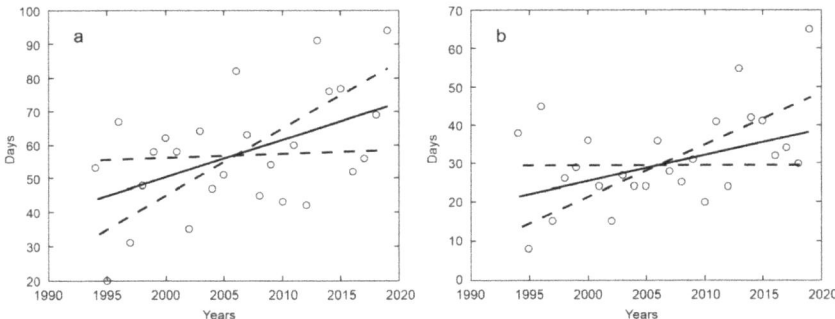

Figure 2. Number of days with positive air temperature (**a**), and with liquid or mixed precipitation (**b**) in November–April in 1994–2020. Solid lines—linear trends, dashed lines—95% confidence interval.

3.2. Ice Phenology and Thickness of the Snow and Ice Cover of Lake Vendyurskoe: Field Data and Model Calculations

The dates of ice-on and ice-off on Lake Vendyurskoe in 1994–2020 varied widely. Unstable ice formed on the lake, starting from the end of October, and stable ice formed from 7 November to 12 December. Ice-off took place between 29 April and 19 May (Table 2). Thus, the dates of stable ice formation varied in different years within 5 weeks, and the dates of ice-off—within 3 weeks. Simulations yielded ice-on dates 5–20 days earlier than the observed dates in some years; simulated ice-off dates in some years were also 5–20 days ahead of the field data, but in some years simulated ice-off lagged 3–12 days behind. Both field data and model calculations show a positive trend for ice-on dates and a negative trend for ice-off dates, but these trends are not significant (Table 1, Figure S2).

Table 2. Dates of ice-on, ice-off and start of spring under-ice convection, water temperature on ice-on date (T_{W_ICE-ON}), maximal bottom water temperature at the end of winter (T_{WB_MAX}), duration of convection, minimal C_t/C_0 during winter (C_t/C_{0_MIN}), values of C_t/C_0 on the last day of the ice season (C_t/C_{0_END}), and increase in C_t/C_0 during convection (C_t/C_{0_PLUS}) according to field measurements.

Winter Season	Ice-On	T_{W_ICE-ON}	T_{WB_MAX}	Ice-Off	Convection Start	Convection, Days	C_t/C_{0_MIN}	C_t/C_{0_END}	C_t/C_{0_PLUS}
1994–1995	7 November	1.2	5.16	19 May	15 April	34	n/d	n/d	n/d
1995–1996	7 November	0.5	5.3	14 May	15 April	29	n/d	n/d	n/d
1996–1997	12 December	0.6	4.3	14 May	16 April	28	n/d	n/d	n/d
1997–1998	n/d	n/d	4.8	11 May	21 April	20	n/d	n/d	n/d
1998–1999	10 November	2.1	5.0	30 April	1 April	30	n/d	n/d	n/d
1999–2000	15 November	0.6	4.8	29 April	1 April	29	n/d	n/d	n/d
2000–2001	n/d	n/d	n/d	n/d	n/d	n/d	n/d	n/d	n/d
2001–2002	n/d	n/d	5.5	3 May	28 March	37	n/d	n/d	n/d
2002–2003	n/d	n/d	4.3	13 May	11 April	33	n/d	n/d	n/d
2003–2004	18 November	1.5	n/d	n/d	n/d	n/d	n/d	n/d	n/d
2004–2005	17 November	1.8	n/d	n/d	7 April	32	n/d	n/d	n/d
2005–2006	20 November 4 December	2.1 0.2	n/d	23 November n/d	8 April	n/d	n/d	n/d	n/d
2006–2007	n/d	n/d	n/d	n/d	n/d	n/d	n/d	n/d	n/d
2007–2008	15 November	0.63	5.18	10 May	30 March	41	0.58	0.74	0.16
2008–2009	10 December	0.47	4.26	8 May	16 April	22	0.59	0.61	0.02
2009–2010	12 November 5 December	0.52 0.17	4.20	30 November 1 May	31 March	31	0.61	0.79	0.18
2010–2011	21 November	0.5	4.65	2 May	7 April	25	0.61	0.72	0.11
2011–2012	17 November 12 December	1.7 0.15	4.74	1 Dec 5 May	16 April	19	0.59	0.70	0.11
2012–2013	1 December	0.13	4.50	3 May	14 April	19	0.55	0.60	0.05
2013–2014	1 December	n/d	n/d	n/d	n/d	n/d	n/d	n/d	n/d
2014–2015	23 October 13 November	2.3 1.4	4.80	3 November 18 November	11–22 March 6–30 April	37	0.51	0.70	0.19
2015–2016	21 November 16 November 12 December	0.05 1.85 0.63	4.65	30 April 4 December 4 May	27 March	38	0.61	0.81	0.20
2016–2017	6 November	1.6	5.85	18 May	6 April	42	0.47	0.63	0.16
2017–2018	23 November	0.7	5.16	7 May	10 April	26	0.49	0.63	0.14
2018–2019	26 November	0.29	4.87	2 May	31 March	25	0.50	0.60	0.10
2019–2020	7 November 12 November 16 November 23 November	1.0 0.7 1.1 1.0	4.76	11 November 14 November 18 November 6 May	10–31 March 4–14 April 22 April–5 May	46	0.54	0.61	0.07

Model calculations of ice thickness in 1994–2020 for the entire ice period and field measurements of ice thickness at different points in the winter seasons are shown in Figure S3. For early winter, the model reproduces the ice thickness well, but for the spring period of some years the discrepancies can reach 10–15 cm (e.g., spring 1995 and 2008).

Statistical analysis of time series of the ice thickness of Lake Vendyurskoe in mid-April in 1995–2020 showed the presence of a significant negative linear trend ($H_{IS\_APRIL\_MEAS}$, Table 1, Figure 3); the average rate of decrease of ice thickness for this period was 1.1 cm per year. A model calculation of the average ice thickness of Lake Vendyurskoe for the winter period (H_{ICE_MODEL}) also shows a negative trend, but the trend is not significant.

For all years of measurements, snow thickness on the ice of Lake Vendyurskoe in April rarely exceeded 10 cm (Table S3), and in some years there was no snow during the measurements. In spring, the thickness of snow on ice sometimes changed quickly, decreasing as a result of intense melting, or increasing as a result of a fresh snowfall. In April, ice on Lake Vendyurskoe consisted of two layers, i.e., white ice and black ice. Black ice was usually thicker than white ice. In some years; however, white ice thickness was greater (2005, 2012 and 2017). In some years in April (2000, 2004, 2009–2012 and 2015), a 3–11 cm slush layer formed on the surface of white ice.

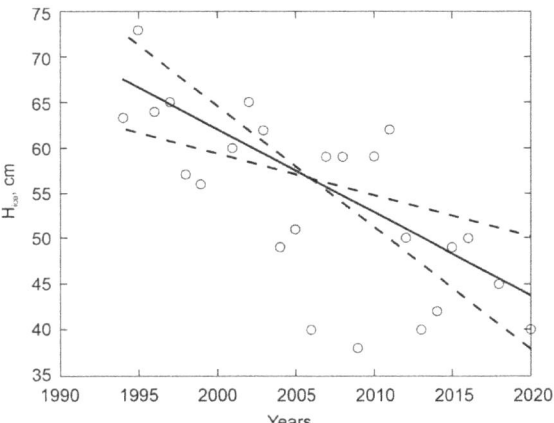

Figure 3. Maximal snow-ice cover thickness on Lake Vendyurskoe in spring in 1994–2000, 2002–2018, and 2020 (field data). Solid line—linear trend, dashed lines—95% confidence interval.

A significant negative relationship was found between the snow-ice cover thickness in April and the number of days with positive air temperatures in November–April ($p = 0.0015$), as well as with the number of days with liquid or mixed precipitation in the same period ($p = 0.0103$): the more such days there were, the thinner the snow-ice cover was in April (Figure 4).

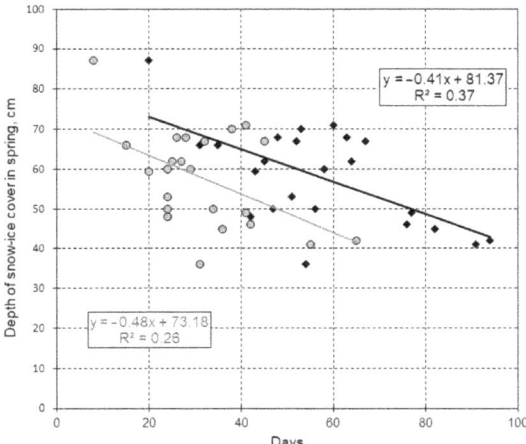

Figure 4. The thickness of the snow-ice cover of Lake Vendyurskoe in spring in 1995–2000, 2002–2018, and 2020 as a function of the number of days with positive air temperature in November–April (Days_T_{AIR} > 0, black diamonds), and the number of days with liquid or mixed precipitation (Days_P_{LM}, grey circles) in the same period. Bold lines are the regression lines.

3.3. Under-Ice Solar Radiation in Spring in 1994–2020

During the period of spring measurements, the daytime maxima of the incident radiation on the snow-ice cover surface $E_d(0)$ reached 600–800 W/m² on sunny days [32]. Depending on the thickness of snow and ice, as well as on meteorological conditions, the fluxes of under-ice radiation $E_d(z)$ varied over a wide range from values close to zero to 350 W/m² and more (Table S3). After the fall of fresh snow, the daytime maximum of under-ice radiation fluxes sharply declined to almost zero. Then, as the snow melted, it increased again. Since the weather conditions in spring are very unstable, a pronounced variability in the optical characteristics of the snow-ice cover of the lake and the fluxes of under-ice radiation was noted. A generalized analysis of the data from the spring period for all years of measurements showed a significant negative relationship between under-ice radiation and the snow-ice cover thickness during melting stage (Figure 5).

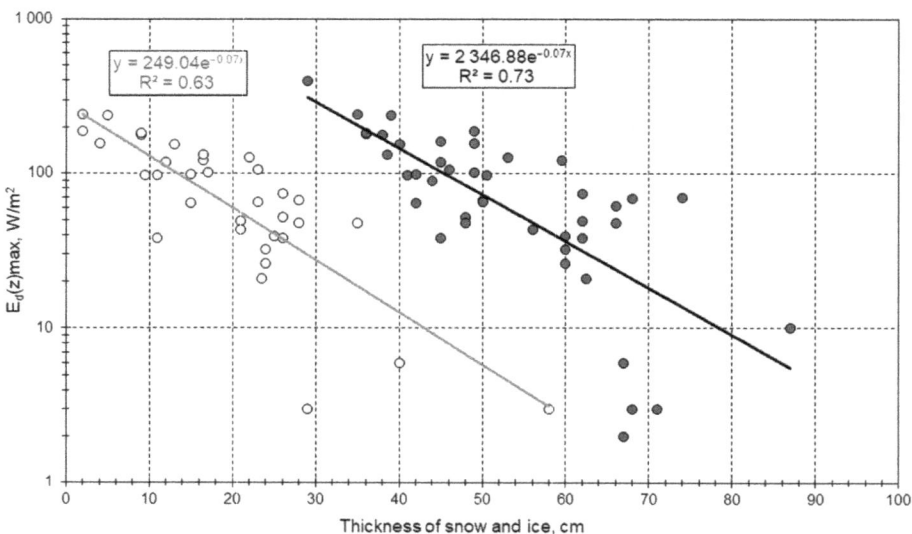

Figure 5. Solar radiation at the ice–water interface (daily maximum, $E_d(z)max$) as a function of snow-ice cover thickness (dark grey circles), and thickness of snow, slush, and white ice (light grey circles) in spring in 1995–2000, 2002–2018, and 2020. Bold lines are the regression lines.

3.4. Water Temperature in Winter in 1994–2020

According to the thermistor chain data, the mean water column temperature in the central part of the lake at ice-on varied from 0.13 to 2.3 °C (Table 2). In some years, warm windy weather caused ice to fall apart after ice-on, for example, in 2009, 2011, 2014, 2015, and 2019 (see Figure S1 for 2009). In most cases, ice formation after a transient ice-off occurred at a lower water temperature. For example, in autumn 2009 the first ice-on was on November 12, when the average temperature over the water column was 0.52 °C; the second ice formation was on December 5, at water temperature of 0.17 °C (Table 2, Figure S1a). By the end of the winter season, water temperature in the bottom layers of the central basin of the lake reached 4.20–5.85 °C. Spring under-ice convection began in the interval from March 10 to April 21 and lasted for 19 to 46 days. In some years, convection started very early, and then halted after cooling and snowfall, as happened, e.g., in April 2015 and 2020, and then, as the snow melted, convection continued to develop. For example, in the spring of 2020, under-ice convection began on March 10. The lower boundary of CML on March 20 was well identified at a depth of 3 m, and the temperature in this layer was 1.55 °C. During the 3rd–4th weeks of March, there was practically no snow on the ice surface and sunny weather activated convection. The temperature of the CML and the depth of its lower boundary increased rapidly and reached 2.26 °C and 6 m by March 31. On April 1 and 2, after snow fell, the solar radiation flux at the lower ice boundary dropped sharply to 5–10 W/m^2, and the convection slowed down. During the first week of April, the snow melted, but the weather being cloudy, the flux of under-ice radiation did not exceed 50–60 W/m^2. The depth of the CML lower boundary decreased to 3–4 m on April 6. During the second half of April, convection continued, sometimes increasing, sometimes weakening, depending on weather conditions and the presence of snow on ice, and by May 3 the temperature of CML rose to 4 °C, and the depth of its lower boundary was about 7.5 m.

The depth of the convective layer in different years reached 5.5–10.5 m, and its temperature was 2.5–4.0 °C. In some years (1994, 1996, 1997, 1999, 2016, 2020), an increase in water temperature above 4 °C was observed in the surface layers of the lake during the last days of the ice-covered period (before ice-off).

3.5. Dissolved Oxygen Content Change in the Winter Seasons of 2007–2020

In the pre-ice period (October, November), the water mass was quasi-homogeneous due to intensive wind mixing and was quite well saturated with oxygen (>90%). The DO content in the water column gradually grew as its solubility increased with decreasing water temperature. Maximal DO concentrations of >12.0 mg/L (saturation over 90%) were observed in the surface layer of the lake every autumn.

During the first 2–3 days after ice-on, DO concentration usually decreased by 0.5–1.0 mg/L (see, for example, Figure S1b for early winter in 2009). During the winter, the oxygen concentration in the surface layers of the water column gradually decreased, and by the end of the winter it was 9.0–9.5 mg/L at a saturation of 75–80%. In the bottom layers of the central deep-water basin, a rapid decrease of DO was observed during the first weeks of the ice season, and by the end of the second month of the ice period DO concentration in the 0.5 m bottom layer decreased to 2 mg/L. The thickness of the bottom anaerobic zone (DO concentration less than 2 mg/L) can reach two meters at the end of the ice-covered period in some years. The anaerobic zone in the deep-water part of the lake usually persists until ice-off because the bottom layer of the central basin is rarely involved in convective mixing. After ice-off in spring, complete mixing of the water column of Lake Vendyurskoe usually occurs, and the bottom anaerobic zone is destroyed.

In the first few days after ice-on, the rate of oxygen decline is maximal in the bottom layers of central deep-water basin and can reach 0.5–0.7 (mg/L)/day (Figure S1b).

Then, in the winter stagnation stage, the rate of DO decline in the surface layer did not exceed 0.01 (mg/L)/day. With depth, the rate of DO consumption gradually increased and reached 0.03–0.05 (mg/L)/day at 2–4 m above the bottom. The highest rates of DO decline are observed in the 0.2–0.3 m bottom layer, where they reached 0.1–0.4 (mg/L)/day during the first winter weeks.

In spring, as convection develops, the oxygen concentration gets equalized over CML. When convection develops, the underlying oxygen-depleted water is drawn into CML. As a consequence, the oxygen concentration in CML decreases.

Daily DO fluctuations with a distinct daytime maximum and a nighttime minimum were observed in spring in the surface layer of the lake during all winter seasons. For example, in the spring of 2020, such fluctuations were observed since the last ten days of March and continued until ice-off. The amplitude of these fluctuations can in some years reach 0.9 mg/L. After snowfall, under-ice radiation decreases, and the amplitude of daily oxygen fluctuations decreases too; as under-ice radiation increases, the amplitude of daily oxygen fluctuations also grows noticeably.

The decrease in the oxygen content of the water column during winter ranged from 53 to 39% in different years; the corresponding C_t/C_{0_MIN} value varied from 0.47 to 0.61 (Table 2, Figure 6). With the onset of under-ice convection, the decrease in the oxygen content ended, remained at approximately the same level for several days, and then began to increase (by 10–15% in some years).

To understand which factors influence the rate of oxygen decrease in Lake Vendyurskoe in winter, we analyzed the relationship between the minimum values of C_t/C_{0_MIN} during winter and the following factors: ice-on date, water temperature on the ice-on day, maximal bottom water temperature at the end of the winter, and the number of days with a positive average daily temperature in November–April.

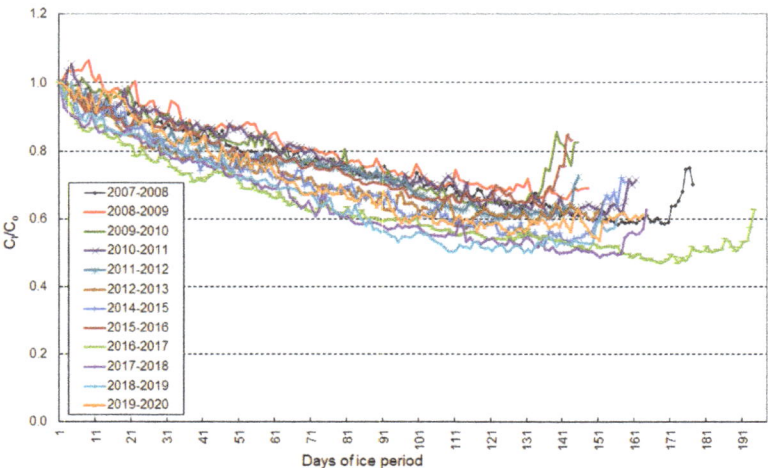

Figure 6. Change of C_t/C_0 in Lake Vendyurskoe during the ice-covered period in different years.

The values of C_t/C_{0_MIN} correlated the most significantly with the date of ice-on (p = 0.0256, Figure 7a), and with the temperature of the bottom water at the end of the winter (p = 0.0113, Figure 7b): the earlier the ice formed and the higher the bottom water temperature at the end of the winter, the lower the C_t/C_{0_MIN} ratio. The highest values of C_t/C_{0_MIN} were observed in years with intermittent ice-off at the beginning of the winter (Table 2). A less pronounced relationship (R^2 = 0.17) was found between C_t/C_{0_MIN} and the water temperature on the day of ice-on: the higher this temperature was, the lower the value of C_t/C_{0_MIN} was. No correlation was found between the degree of oxygen depletion in the lake in winter and the number of days with positive air temperatures in November–April.

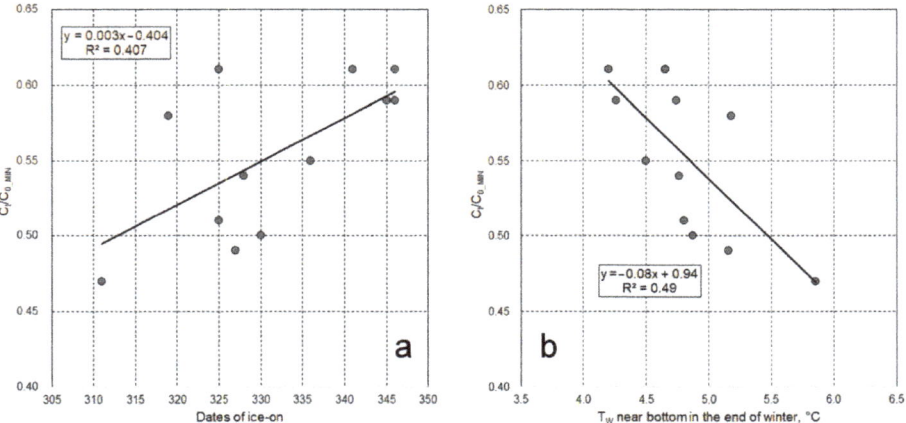

Figure 7. The value of C_t/C_{0_MIN} at the end of the winter as a function of ice-on dates (**a**), and bottom water temperature at the end of the winter (**b**).

To find out which factors contribute to the increase in the DO content at the end of the winter season, we analyzed the relationship between C_t/C_{0_PLUS} and the onset date of convection (Figure 8a) and its duration (Figure 8b), the number of days with a positive air temperature in March–April (Figure 8c), and the number of days with liquid or mixed precipitation in the same period (Figure 8d).

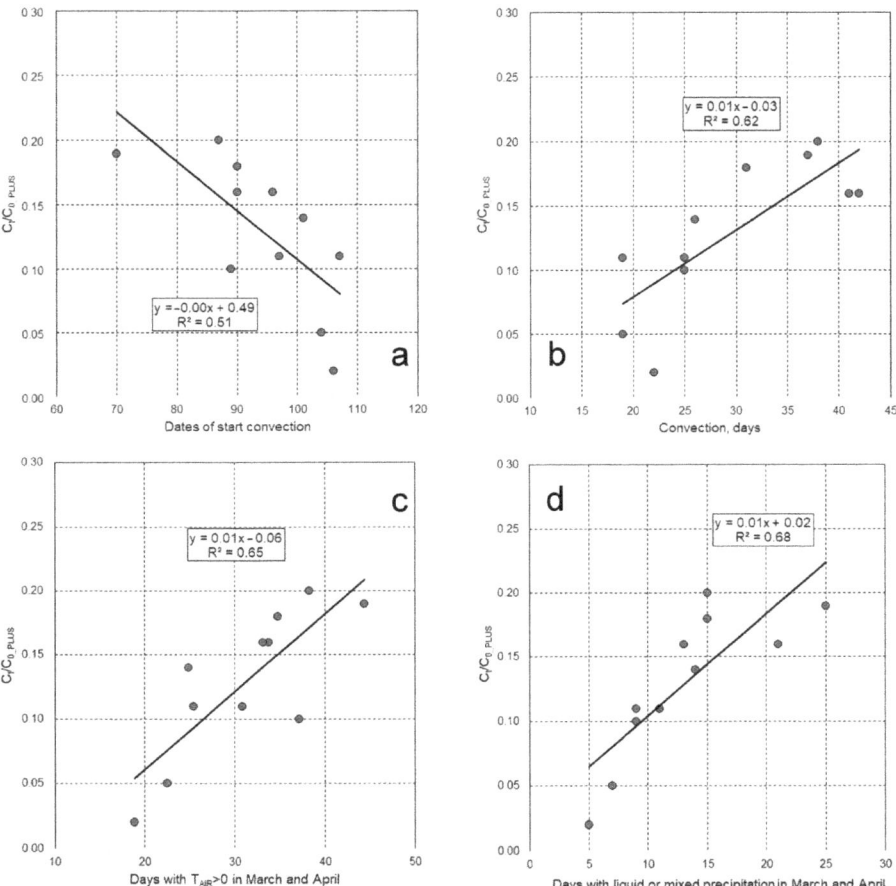

Figure 8. C_t/C_{0_PLUS} at the end of the winter as a function of the date of under-ice convection onset (**a**), the duration of convection (**b**), the number of days with a positive air temperature in March–April (**c**), and the number of days with liquid or mixed precipitation in March–April (**d**).

There are good correlations between C_t/C_{0_PLUS} and dates of start convection ($p = 0.0139$) and duration of convection ($p = 0.0042$). The greatest increase in the oxygen content in the lake occurred in years with an early onset and long duration of under-ice convection. The lowest values of C_t/C_{0_PLUS} were observed in years with a short convection period.

There is also a good correlation between C_t/C_{0_PLUS} and the number of days with a positive average daily temperature in March and April (0.0029) and several days with liquid or mixed precipitation ($p = 0.0019$).

4. Discussion

We analyzed the climate variability in southern Karelia (northwestern Russia) in 1994–2020 and revealed that the air temperature and precipitation in the colder half of the year (November–May) varied markedly during this period. The largest and most frequent positive deviation from the 1961–1990 baseline was recorded for the winter months (December, January, and February) (Table S1). Positive trends were detected for all of these months, but the trend was significant only in November and May (Table S2). During 19 of the 26 winter seasons in the series, the average air temperature in November–April period was higher than the baseline. The number of days with a positive average daily air temperature in November–April as well as the number of days with liquid or mixed precipitation increased significantly during 1994–2020 (Table 1).

The revealed trends in air temperature and precipitation in the winter seasons of 1994–2020 in southern Karelia are consistent with the climate tendency in other regions of the temperate zone and high latitudes of the Northern Hemisphere over the past decades. Positive trends in precipitation in winter (November–March) during 1921–2015 have been shown for Northern Eurasia (statistically significant) and North America (statistically insignificant) [42]. A rapid increase in air temperature (over 2 °C in 1847–2013 for annual air temperature, which amounts to 0.14 °C/decade) with the most pronounced increase in November, December, and January has been shown for Finnish territory [43]. A statistically significant increase in air temperature and precipitation in winter in 1948–2016, as well as a decrease in the proportion of precipitation falling as snow (i.e., the ratio of snowfall to total precipitation) associated with warming has been recorded for the territory of Canada [44]; the authors of said study emphasize that annual and winter precipitation may continue to increase everywhere in Canada over the course of the 21st century.

Climatic changes in air temperature and precipitation affect the ice and thermal regimes of lakes. Shortening of the ice period is observed and modelled on many temperate lakes [23–25,45–51]. An important consequence of shorter ice periods is a decrease in the duration of winter hypoxia and anoxia and, hence, improvement of oxygen conditions in the colder half of the year [39,52,53].

The changes revealed in air temperature and precipitation in southern Karelia in 1994–2020 were reflected in a pronounced variability of the dates of ice-on and ice-off on Lake Vendyurskoe. Both field data and FLake model calculations show a positive trend for ice-on dates and a negative trend for ice-off dates, although these trends are not statistically significant. Large gaps in the field series of ice-on and ice-off dates do not allow us to analyze the trend of the of ice-covered period duration for Lake Vendyurskoe. We can only state that the duration of the ice-covered period on this lake in 1994–2020 varied widely from 145 days (winter season 2011–2012) to 193 days (winter seasons 1994–1995 and 2016–2017).

One important feature of the change of the ice regime of Lake Vendyurskoe in early winter was revealed that could enhance its oxygen conditions in the subsequent winter. The increase in air temperature in November and December causes ice to become less stable in the initial period of winter. To wit, since 2005, intermediate ice-offs have become more frequent in the early winter (Table 2). In some years, there were two or three intermediate ice-off events. After an intermediate ice-off, the water temperature most often continued to decline until the ice set in again that lead to a lower water temperature and smaller bacterial oxygen consumption in the subsequent winter. In addition, aeration and oxygenation of the bottom layers of the lake occurred after an intermediate ice-off (Figure S1b). Presumably, the organic matter accumulated in the bottom layers of the lake were partially oxidized during the temporary existence of ice, and the rate of oxygen consumption in the subsequent ice-covered period decreased.

The climate variability has also affected the thickness of the lake's snow-ice cover, and, consequently, the under-ice illumination, which is an important factor for phytoplankton growth and oxygen release. A climatic decrease in snow-ice cover thickness is observed on different lakes in the temperate zone and high latitudes [46,51].

A statistically significant decrease in the thickness of the snow-ice cover of Lake Vendyurskoe during the spring periods of 1994–2020 (by 1.1 cm per year) was revealed (Table 1, Figure 3). We assume that this decrease is associated both with an increase in air temperature, namely, an increase in the number of days with positive air temperatures and liquid or mixed precipitation in winter and spring months (Figure 4). It should be kept in mind that ice and snow thickness measurements on Lake Vendyurskoe in spring were carried out in different stages of snow and ice melting in different years. This complicates the joint analysis of spring ice and snow thickness with climatic parameters. Measurements in spring during the period of intensive thawing do not allow estimating the average ice thickness over the winter season and investigating its climatic variability. FLake model calculations indicate that the average ice thickness over the winter season has most likely also decreased in 1994–2019 (Figure S3), but the trend of this parameter is not significant (Table 1).

The climatic decrease in the thickness of the snow-ice cover should lead to a long-term increase in the under-ice illumination. A decrease in the thickness of snow on the ice surface is often considered to be a major factor in increase in under-ice irradiation [7,9]. The critical role of the snow cover for the limitation of under-ice irradiation, primary production and oxygen release was shown for a lake of the Mongolian Plateau [12]. A significant inverse relationship between under-ice irradiation and snow depth in winter has been revealed for four lakes in Canada [9].

The ongoing climate changes of Northwestern Russia in the winter months, namely, an increase in the number of thaw days and an increase in days with liquid and mixed precipitation in November–April (Table 1), should contribute to a decrease in snow thickness. Indeed, there was often no snow at all when measurements were taken on Lake Vendyurskoe in spring (Table 2).

Another important component of the snow-ice cover of lakes in the temperate zone is white ice, which forms in winter as a result of snowfall on the surface of black ice [8]. Frequent thaws, alternating freeze–thaw cycles, and liquid or mixed precipitation also contribute to the formation of white ice and slush on the surface of black ice. White ice and slush are noticeably less transparent than black ice [7,8]. Therefore, formation of white ice and slush on the surface of black ice must have a limiting effect on under-ice irradiation. Some studies indicate that increased precipitation in winter, namely increased snowfall, inhibits the growth of black ice and contributes to the growth of white ice [8,27,51].

For each year of measurements in 1994–2020, the ice cover of Lake Vendyurskoe included both white and black ice. In some spring periods white ice was thicker than black ice; from time to time, slush was observed on the surface of white ice. Thus, despite the noticeable climatic decrease in the thickness of the snow-ice cover of Lake Vendyurskoe in spring in 1994–2020, the increase of under-ice radiation may be smoothed out by changes in the ice structure (namely, an increase in the thickness of white ice and slush), which can occur due to an increase in air temperature and change of precipitation.

During the period of intense spring thawing, a rapid decrease in the thickness of snow and ice occurs. Measurements of under-ice radiation and the thickness of snow and ice, carried out over several days during intense melting, make it possible to trace how the under-ice radiation increases depending on the decrease in the thickness of snow and ice (Figure 5). A joint analysis of the observational data on the thickness of snow, white and black ice, and under-ice radiation during the intensive spring thawing period showed that not only the complete disappearance of snow, but also a decrease in the thickness of white ice plays an important role in increasing under-ice radiation (see, for example, 1999, 2000, and 2004 in Table S3).

Long-term measurements of oxygen at autonomous stations create unique opportunities for studying high-frequency DO variability, providing new knowledge of the processes occurring in ice-covered lakes, which can affect the rate of DO consumption or production [9,11–13,16,20,22,23,39,54–58].

We used data from long-term high-frequency DO measurements in 2007–2020, and measurements of under-ice irradiation and water temperature in 1994–2020, which allowed us to study the change in oxygen content over the winter and reveal the interannual variability of oxygen conditions in the lake in relation to changes in light, water temperature and ice.

We found that the strongest oxygen depletion during the winter season in Lake Vendyurskoe occurred in years with early ice-on and high bottom water temperature at the end of the winter season. This is due to earlier ice-on, longer ice period, and, hence, longer period of oxygen absorption. In addition, high water temperature promotes oxygen consumption near bottom sediments [5,6,15]. Therefore, bottom-layer water temperature in a lake in winter can be an important predictor of the rate of oxygen consumption.

We saw an increase in oxygen content at the end of the winter season in all years of measurements (Table 2). Since this increase occurred before the destruction of ice, it could not be associated with the resumption of gas exchange with the atmosphere. We assume that the most likely reason for this increase may be the release of oxygen as a result of photosynthesis of phytoplankton. Aquatic ecosystems can have high productivity in winter [21,30,59,60]; consequently, some increase in DO content due to photosynthesis of phytoplankton can be expected [11,12,14].

Daily oscillations of oxygen in the surface layer of Lake Vendyurskoe in spring can be considered an indicator of the circadian rhythms of phytoplankton. Some studies argue that the circadian rhythms of oxygen presumably reflect the metabolism of the lake—a combination of production and destruction processes [12,22,57]. On the other hand, oxygen fluctuations over a 24-h period, which were observed in the surface layers of the ice-covered Lake Valkea-Kotinen (Finland) during the development of spring under-ice convection, were attributed by the authors to diurnal changes in water temperature due to radiative heating [56].

The greatest increase in the oxygen content in the lake occurred in years with an early onset and long duration of under-ice convection. The lowest values of C_t/C_{0_PLUS} were observed in years with a short convection period. The positive relationships we found for the duration of convection and the date of its onset with the increase in oxygen at the end of winter have a logical basis, since during the spring under-ice convection favorable light, temperature and hydrodynamic conditions are formed for the development of phytoplankton [9–11,29] and, hence, for the release of oxygen.

The correlation revealed between the increase of oxygen and the number of days with thawing and with liquid or mixed precipitation reflects the relationship between snow-ice thickness and under-ice solar radiation, since the more days with a thaw, the smaller the snow-ice thickness. With thinner ice, more solar radiation penetrates under the ice, creating favorable conditions for the development of phytoplankton and the production of oxygen as a result of photosynthesis.

5. Conclusions

A warming of the regional climate of southern Karelia (Northwest Russia) in 1994–2020 expressed itself in an increase in air temperature and the number of days with a thaw, and days with liquid or mixed precipitation in winter months. These changes have had an impact on the ice regime of the small shallow Lake Vendyurskoe, as well as on thermal, light, and oxygen conditions in this lake in the ice-covered period.

The main features of the ice regime of the lake in the modern climatic conditions are a substantial variability of ice-on (5 weeks) and ice-off dates (3 weeks), intermittent ice cover at the beginning of winter, as well as a noticeable thinning of the snow-ice cover in spring (a significant trend) in 1994–2020 and a change of its structure (low snow thickness, increase in the proportion of white ice and slush formation in some years). The positive trends detected in the dates of ice-on and negative trends in the dates of ice-off according to field data and FLake model calculations are statistically insignificant.

A significant negative trend of snow-ice cover thickness in the 1994–2020 springs is shown. This gives us grounds to assume that the long-term trend of snow-ice cover thinning potentially prolongs the period with favorable conditions for the development of radiatively driven convection, which, in turn, creates favorable light, temperature and mixing settings for the growth of phytoplankton and the associated oxygen release. It should be noted that the changes in the structure of the snow-ice cover observed in some years (an increase in the proportion of white ice and slush), which can reduce its transparency, may neutralize the effect of its thinning on the growth of under-ice radiation and DO production.

The seasonal DO decrease in the lake in winter is determined by two factors influencing by regional climate changes: ice-on date, and bottom-layer water temperature. Peak DO consumption can exceed 50% during winters with early ice-on and higher bottom-layer water temperature. In contrast, late ice-on, intermittent ice cover in early winter, and low water temperature result in a wintertime DO decrease by to less than 40%. It can be predicted that the future climatic shift to later ice-on dates will entail a reduction in oxygen consumption in lakes in winter.

The increase in oxygen content in the lake at the end of the ice-covered period, which in Lake Vendyurskoe is most likely associated with photosynthesis of phytoplankton, is the most significant in years with early onset and prolonged convection. The duration of convection can be expected to increase in the future, creating favorable conditions for the development of under-ice plankton and improving oxygen conditions in lakes in winter.

Supplementary Materials: The following are available online at https://www.mdpi.com/article/10.3390/w13172435/s1, Figure S1: Water temperature (a) and DO concentration (b) at different depths in the central deep-water part of Lake Vendyurskoe from 1 November to 19 December 2009. Numbers in the legend indicate the distance of sensor above the bottom (m). Depth of station is 11.3 m. Solid thick blue lines indicate the ice-covered periods, blue dots indicate the period of unstable ice; Figure S2: Dates of ice-on (a) and ice-off (b) on Lake Vendyurskoe in 1994–2020. Solid lines—linear trends (not significant). Black circles—observational data, grey diamonds—model FLake calculation; Figure S3: Ice thickness on Lake Vendyurskoe in 1994–2020. Grey line – model calculation, black circles—observational data, white diamonds—modelled average ice thickness for all winter seasons; Table S1: Deviations of average monthly air temperature from the baseline (ΔT, °C) in different winter seasons (Petrozavodsk meteorological station data). Bold means excess over the baseline by 5 °C or more. Rightmost column: average temperature from November to May. Asterisks mean the average air temperature from November to April was colder than in 1961–1990; Table S2: Estimates of the statistical significance for the trends (1994–2020) in air temperature by least squares method. $p < 0.05$ means the trend is significant (bold); Table S3: The dates of measurements, thickness of snow, slush, total ice, white ice, and black ice, averaged over 22 stations and r.m.s., and under ice irradiance Ed(z), maximal and average from 8 a.m. to 8 p.m. in spring in different years. Dash—no data.

Author Contributions: Conceptualization, G.Z. and N.P.; methodology, G.Z., N.P. and A.T.; software, B.B., T.E., S.G. and I.Z.; validation, S.G., I.Z., T.E. and G.Z.; formal analysis, B.B., T.E., N.P. and G.Z.; investigation, R.Z. and N.P.; resources, N.P., G.Z. and I.F.; data curation, R.Z. and N.P.; writing—original draft preparation, G.Z., A.T. and S.B.; writing—review and editing, G.Z., A.T., S.B., N.P., R.Z. and I.F.; visualization, B.B., T.E. and G.Z.; supervision, G.Z.; project administration, I.F. and G.Z.; funding acquisition, I.F. and G.Z. All authors have read and agreed to the published version of the manuscript.

Funding: The study was carried out as part of the state assignment of the Northern water problems Institute of Karelian research Center of RAS. Field studies were carried out with partial financial support from the RFBR, grant number 18-05-60291.

Data Availability Statement: All data created or used during this study are available by request to authors.

Acknowledgments: The authors are grateful to Andrey Mitrokhov (NWPI KarRC RAS) for many years of participation in field research and Olga Kislova for her help in translating the article into English.

Conflicts of Interest: The authors declare no conflict of interest.

References

1. Odum, E.; Barret, G. *Fundamentals of Ecology*, 5th ed.; Brooks Cole: Florence, KY, USA, 2004; p. 624.
2. Nürnberg, G.K. Quantifying anoxia in lakes. *Limnol. Oceanogr.* **1995**, *40*, 1100–1111. [CrossRef]

3. Greenbank, J. Limnological conditions in ice-covered lakes, especially as related to winter-kill of fish. *Ecol. Monogr.* **1945**, *15*, 343–392. [CrossRef]
4. Miettinen, H.; Pumpanen, J.; Heiskanen, J.J.; Aaltonen, H.; Mammarella, I.; Ojala, A.; Levula, J.; Rantakari, M. Towards a more comprehensive understanding of lacustrine greenhouse gas dynamics—Two-year measurements of concentrations and fluxes of CO_2, CH_4 and N_2O in a typical boreal lake surrounded by managed forests. *Boreal Environ. Res.* **2015**, *20*, 75–89.
5. Terzhevik, A.; Golosov, S.; Palshin, N.; Mitrokhov, A.; Zdorovennov, R.; Zdorovennova, G.; Kirillin, G.; Shipunova, E.; Zverev, I. Some features of the thermal and dissolved oxygen structure in boreal, shallow ice-covered Lake Vendyurskoe, Russia. *Aquat. Ecol.* **2009**, *43*, 617–627. [CrossRef]
6. Terzhevik, A.; Palshin, N.I.; Golosov, S.D.; Zdorovennov, R.E.; Zdorovennova, G.E.; Mitrokhov, A.V.; Potakhin, M.S.; Shipunova, E.A.; Zverev, I.S. Hydrophysical aspects of oxygen regime formation in a shallow ice-covered lake. *Water Resour.* **2010**, *37*, 662–673. [CrossRef]
7. Petrov, M.P.; Terzhevik, A.Y.; Palshin, N.I.; Zdorovennov, R.E.; Zdorovennova, G.E. Absorption of solar radiation by snow-and-ice cover of lakes. *Water Resour.* **2005**, *32*, 496–504. [CrossRef]
8. Leppäranta, M. *Freezing of Lakes and the Evolution of Their Ice Cover*; Springer: Berlin, Germany, 2015.
9. Pernica, P.; North, R.L.; Baulch, H.M. In the cold light of day: The potential importance of under-ice convective mixed layers to primary producers. *Inland Waters* **2017**, *7*, 138–150. [CrossRef]
10. Palshin, N.; Zdorovennova, G.; Zdorovennov, R.; Efremova, T.; Gavrilenko, G.; Terzhevik, A. Effect of under-ice light intensity and convective mixing on chlorophyll a distribution in a small mesotrophic lake. *Water Resour.* **2019**, *46*, 384–394. [CrossRef]
11. Yang, B.; Wells, M.G.; Li, J.; Young, J. Mixing, stratification, and plankton under lake-ice during winter in a large lake: Implications for spring dissolved oxygen levels. *Limnol. Oceanogr.* **2020**, *65*, 2713–2729. [CrossRef]
12. Huang, W.; Zhang, Z.; Li, Z.; Lepparanta, M.; Arvola, L.; Song, S.; Huotari, J.; Lin, Z. Under-ice dissolved oxygen and metabolism dynamics in a shallow lake: The critical role of ice and snow. *Water Resour. Res.* **2021**, *57*, e2020WR027990. [CrossRef]
13. Cortés, A.; MacIntyre, S. Mixing processes in small arctic lakes during spring. *Limnol. Oceanogr.* **2020**, *65*, 260–288. [CrossRef]
14. Davis, M.N.; McMahon, T.E.; Cutting, K.A.; Jaeger, M.E. Environmental and climatic factors affecting winter hypoxia in a freshwater lake: Evidence for a hypoxia refuge and for re-oxygenation prior to spring ice loss. *Hydrobiologia* **2020**, *847*, 3983–3997. [CrossRef]
15. Boylen, C.; Brock, T. Bacterial decomposition processes in Lake Wingra sediments during winter. *Limnol. Oceanogr.* **1973**, *18*, 628–634. [CrossRef]
16. Deshpande, B.N.; MacIntyre, S.; Matveev, A.; Vincent, W.F. Oxygen dynamics in permafrost thaw lakes: Anaerobic bioreactors in the Canadian subarctic. *Limnol. Oceanogr.* **2015**, *60*, 1656–1670. [CrossRef]
17. Powers, S.M.; Baulch, H.M.; Hampton, S.E.; Labou, S.G.; Lottig, N.R.; Stanley, E.H. Nitrification contributes to winter oxygen depletion in seasonally frozen forested lakes. *Biogeochemistry* **2017**, *136*, 119–129. [CrossRef]
18. Bengtsson, L.; Ali-Maher, O. The dependence of the consumption of dissolved oxygen on lake morphology in ice covered lakes. *Hydrol. Res.* **2020**, *51*, 381–391. [CrossRef]
19. Jansen, J.; MacIntyre, S.; Barrett, D.C.; Chin, Y.-P.; Cortés, A.; Forrest, A.L.; Hrycik, A.R.; Martin, R.; McMeans, B.C.; Rautio, M.; et al. Winter limnology: How do hydrodynamics and biogeochemistry shape ecosystems under ice? *J. Geophys. Res. Biogeosci.* **2021**, *126*, e2020JG006237. [CrossRef]
20. Couture, R.-M.; de Wit, H.A.; Tominaga, K.; Kiuru, P.; Markelov, I. Oxygen dynamics in a boreal lake responds to long-term changes in climate, ice phenology, and DOC inputs. *J. Geophys. Res. Biogeosci.* **2015**, *120*, 2441–2456. [CrossRef]
21. Salmi, P.; Salonen, K. Regular build-up of the spring phytoplankton maximum before ice-break in a boreal lake. *Limnol. Oceanogr.* **2016**, *61*, 240–253. [CrossRef]
22. Obertegger, U.; Obrador, B.; Flaim, G. Dissolved oxygen dynamics under ice: Three winters of high-frequency data from Lake Tovel, Italy. *Water Resour. Res.* **2017**, *53*, 7234–7246. [CrossRef]
23. Tan, Z.; Yao, H.; Zhuang, Q. A small temperate lake in the 21st century: Dynamics of water temperature, ice phenology, dissolved oxygen, and chlorophyll a. *Water Resour. Res.* **2018**, *54*, 4681–4699. [CrossRef]
24. Soja, A.-M.; Kutics, K.; Maracek, K.; Molnár, G.; Soja, G. Changes in ice phenology characteristics of two Central European steppe lakes from 1926 to 2012—Influences of local weather and large scale oscillation patterns. *Clim. Chang.* **2014**, *126*, 119–133. [CrossRef]
25. Hewitt, B.A.; Lopez, L.S.; Gaibisels, K.M.; Murdoch, A.; Higgins, S.N.; Magnuson, J.J.; Paterson, A.M.; Rusak, J.A.; Yao, H.; Sharma, S. Historical trends, drivers, and future projections of ice phenology in small north temperate lakes in the laurentian great lakes region. *Water* **2018**, *10*, 70. [CrossRef]
26. Bengtsson, L.; Svensson, T. Thermal regime of ice covered swedish lakes. *Nord. Hydrol.* **1996**, *27*, 39–56. [CrossRef]
27. Ohata, Y.; Toyota, T.; Fraser, A.D. The role of snow in the thickening processes of lake ice at Lake Abashiri, Hokkaido, Japan. *Tellus A Dyn. Meteorol. Oceanogr.* **2017**, *69*, 1391655. [CrossRef]
28. Kirillin, G.; Leppäranta, M.; Terzhevik, A.; Granin, N.; Bernhardt, J.; Engelhardt, C.; Efremova, T.; Golosov, S.; Palshin, N.; Sherstyankin, P.; et al. Physics of seasonally ice-covered lakes: A review. *Aquat. Sci.* **2012**, *74*, 659–682. [CrossRef]
29. Kelley, D. Convection in ice-covered lakes: Effects on algal suspension. *J. Plankton Res.* **1997**, *19*, 1859–1880. [CrossRef]
30. Vehmaa, A.; Salonen, K. Development of phytoplankton in Lake Pääjärvi (Finland) during under-ice convective mixing period. *Aquat. Ecol.* **2009**, *43*, 693–705. [CrossRef]

31. Bengtsson, L.; Malm, J.; Terzhevik, A.; Petrov, M.; Boyarinov, P.; Glinsky, A.; Palshin, N. Field investigation of winter thermo- and hydrodynamics in a small Karelian lake. *Limnol. Oceanogr.* **1996**, *41*, 1502–1513. [CrossRef]
32. Zdorovennova, G.; Palshin, N.; Efremova, T.; Zdorovennov, R.; Gavrilenko, G.; Volkov, S.; Bogdanov, S.; Terzhevik, A. Albedo of a small ice-covered boreal lake: Daily, meso-scale and interannual variability on the background of regional climate. *Geosciences* **2018**, *8*, 206. [CrossRef]
33. Zdorovennova, G.; Palshin, N.; Zdorovennov, R.; Golosov, S.; Efremova, T.; Gavrilenko, G.; Terzhevik, A. The oxygen regime of shallow lake. *Geogr. Environ. Sustain.* **2016**, *2*, 47–57. [CrossRef]
34. Zdorovennov, R.; Palshin, N.; Zdorovennova, G.; Efremova, T.; Terzhevik, A. Interannual variability of ice and snow cover of a small shallow lake. *Est. J. Earth Sci.* **2013**, *62*, 26–32. [CrossRef]
35. Reliable Prognosis. Available online: https://rp5.ru/Weather_in_the_world (accessed on 25 June 2021).
36. All-Russian Scientific Research Institute of Hydrometeorological Information—World Data Center (VNIIGMI-WDC). Available online: http://meteo.ru/data/162-temperature-precipitation (accessed on 25 June 2021).
37. North Eurasian Climate Center. Available online: http://seakc.meteoinfo.ru/actuals (accessed on 25 June 2021).
38. Mironov, D. *Parameterization of Lakes in Numerical Weather Prediction*; Technical Report 2008; Deutsher WetterDienst: Offenbach am Main, Germany, 2008; p. 44.
39. Golosov, S.; Zverev, I.; Terzhevik, A.; Kirillin, G.; Engelhardt, C. Climate change impact on thermal and oxygen regime of shallow lakes. *Tellus A Dyn. Meteorol. Oceanogr.* **2012**, *64*, 17264. [CrossRef]
40. Climate Data Store. Available online: https://cds.climate.copernicus.eu (accessed on 25 June 2021).
41. Hirsch, R.M.; Alexander, R.B.; Smith, R.A. Selection of methods for the detection and estimation of trends in water quality. *Water Resour. Res.* **1991**, *27*, 803–813. [CrossRef]
42. Guo, R.; Deser, C.; Terray, L.; Lehner, F. Human influence on winter precipitation trends (1921–2015) over North America and Eurasia revealed by dynamical adjustment. *Geophys. Res. Lett.* **2019**, *46*, 3426–3434. [CrossRef]
43. Mikkonen, S.; Laine, M.; Mäkelä, H.M.; Gregow, H.; Tuomenvirta, H.; Lahtinen, M.; Laaksonen, A. Trends in the average temperature in Finland, 1847–2013. *Stoch. Environ. Res. Risk Assess.* **2015**, *29*, 1521–1529. [CrossRef]
44. Zhang, X.; Flato, G.; Kirchmeier-Young, M.; Vincent, L.; Wan, H.; Wang, X.; Rong, R.; Fyfe, J.; Li, G.; Kharin, V.V. Changes in temperature and precipitation across Canada. In *Canada's Changing Climate Report*; Bush, E., Lemmen, D.S., Eds.; Government of Canada: Ottawa, ON, Canada, 2019; Chapter 4, pp. 112–193.
45. Magnuson, J.J.; Robertson, D.M.; Benson, B.J.; Wynne, R.H.; Livingstone, D.M.; Arai, T.; Assel, R.A.; Barry, R.G.; Card, V.; Kuusisto, E.; et al. Historical trends in lake and river ice cover in the Northern Hemisphere. *Science* **2000**, *289*, 1743–1746. [CrossRef] [PubMed]
46. Korhonen, J. Long-term changes in lake ice cover in Finland. *Nord. Hydrol.* **2006**, *37*, 347–363. [CrossRef]
47. Duguay, C.R.; Prowse, T.D.; Bonsal, B.R.; Brown, R.D.; Lacroix, M.P.; Ménard, P. Recent trends in Canadian lake ice cover. *Hydrol. Process. Int. J.* **2006**, *20*, 781–801. [CrossRef]
48. Benson, B.J.; Magnuson, J.J.; Jensen, O.P.; Card, V.M.; Hodgkins, G.; Korhonen, J.; Livingstone, D.M.; Stewart, K.M.; Weyhenmeyer, G.A.; Granin, N.G. Extreme events, trends, and variability in Northern Hemisphere lake-ice phenology (1855–2005). *Clim. Chang.* **2012**, *112*, 299–323. [CrossRef]
49. Efremova, T.; Palshin, N.; Zdorovennov, R. Long-term characteristics of ice phenology in Karelian lakes. *Est. J. Earth Sci.* **2013**, *62*, 33–41. [CrossRef]
50. Sharma, S.; Blagrave, K.; Magnuson, J.J.; O'Reilly, C.M.; Oliver, S.; Batt, R.D.; Magee, M.R.; Straile, D.; Weyhenmeyer, G.A.; Winslow, L.; et al. Widespread loss of lake ice around the Northern Hemisphere in a warming world. *Nat. Clim. Chang.* **2019**, *9*, 227–231. [CrossRef]
51. Solarski, M.; Rzetala, M. Ice regime of the Kozłowa Góra reservoir (Southern Poland) as an indicator of changes of the thermal conditions of ambient air. *Water* **2020**, *12*, 2435. [CrossRef]
52. Magnuson, J.J.; Webster, K.E.; Assel, R.A.; Bowser, C.J.; Dillon, P.J.; Eaton, J.G.; Evans, H.E.; Fee, E.J.; Hall, R.I.; Mortsch, L.R.; et al. Potential effects of climate change on aquatic systems: Laurentian Great Lakes and Precambrian Shield region. *Hydrol. Process.* **1997**, *11*, 825–872. [CrossRef]
53. Bengtsson, L. Ice-covered lakes: Environment and climate-required research. *Hydrol. Process.* **2011**, *25*, 2767–2769. [CrossRef]
54. Baehr, M.M.; Degrandpre, M.D. Under-ice CO_2 and O_2 variability in a freshwater lake. *Biogeochemistry* **2002**, *61*, 95–113. [CrossRef]
55. Golosov, S.; Maher, O.A.; Schipunova, E.; Terzhevik, A.; Zdorovennova, G.; Kirillin, G. Physical background of the development of oxygen depletion in ice-covered lakes. *Oecologia* **2007**, *151*, 331–340. [CrossRef]
56. Bai, Q.; Li, R.; Li, Z.; Lepparanta, M.; Arvola, L.; Li, M. Time-series analyses of water temperature and dissolved oxygen concentration in Lake Valkea-Kotinen (Finland) during ice season. *Ecol. Inform.* **2016**, *36*, 181–189. [CrossRef]
57. Song, S.; Li, C.; Shi, X.; Zhao, S.; Tian, W.; Li, Z.; Cao, X.; Wang, Q.; Bai, Y.; Huotari, J.; et al. Under-ice metabolism in a shallow lake in a cold and arid climate. *Freshw. Biol.* **2019**, *64*, 1710–1720. [CrossRef]
58. Golosov, S.D.; Terzhevik, A.Y.; Zverev, I.S.; Zdorovennov, R.E.; Zdorovennova, G.E.; Bogdanov, S.R.; Volkov, S.Y.; Gavrilenko, G.G.; Efremova, T.V.; Palshin, N.I. Rayleigh-taylor instability as a mechanism of heat and mass exchange in ice-covered lake. *Adv. Curr. Nat. Sci.* **2020**, *11*, 45–51. (In Russian) [CrossRef]

59. Hampton, S.E.; Galloway, A.W.E.; Powers, S.M.; Ozersky, T.; Woo, K.H.; Batt, R.D.; Labou, S.G.; O'Reilly, C.M.; Sharma, S.; Lottig, N.R.; et al. Ecology under lake ice. *Ecol. Lett.* **2017**, *20*, 98–111. [CrossRef] [PubMed]
60. Wen, Z.; Song, K.; Shang, Y.; Lyu, L.; Yang, Q.; Fang, C.; Du, J.; Li, S.; Liu, G.; Zhang, B.; et al. Variability of chlorophyll and the influence factors during winter in seasonally ice-covered lakes. *J. Environ. Manag.* **2020**, *276*, 111338. [CrossRef] [PubMed]

Article

Deriving Six Components of Reynolds Stress Tensor from Single-ADCP Data

Sergey Bogdanov *, Roman Zdorovennov, Nikolay Palshin and Galina Zdorovennova *

Karelian Research Centre, Northern Water Problems Institute, Russian Academy of Sciences (NWPI), 185030 Petrozavodsk, Russia; romga74@gmail.com (R.Z.); npalshin@mail.ru (N.P.)
* Correspondence: sergey.r.bogdanov@mail.ru (S.B.); zdorovennova@gmail.com (G.Z.); Tel.: +7-9116660369 (G.Z.)

Abstract: Acoustic Doppler current profilers (ADCP) are widely used in geophysical studies for mean velocity profiling and calculation of energy dissipation rate. On the other hand, the estimation of turbulent stresses from ADCP data still remains challenging. With the four-beam version of the device, only two shear stresses are derivable; and even for the five-beam version (Janus+), the calculation of the full Reynolds stress tensor is problematic currently. The known attempts to overcome the problem are based on the "coupled ADCP" experimental setup and include some hard restrictions, not to mention the essential complexity of performing experiments. In this paper, a new method is presented which allows to derive the stresses from single-ADCP data. Its essence is that interbeam correlations are taken into account as producing the missing equations for stresses. This method is applicable only for the depth range, for which the distance between the beams is comparable to the scales, where the turbulence is locally isotropic and homogeneous. The validation of this method was carried out for convectively-mixed layer in a boreal ice-covered lake. The results of computations turned out to be physically sustainable in the sense that realizability conditions were basically fulfilled. The additional verification was carried out by comparing the results, obtained by the new method and "coupled ADCPs" one.

Keywords: full set of turbulent stresses; Acoustic Doppler current profilers; interbeam velocity correlations; ice-covered lakes; convectively-mixed layer; anisotropic turbulence

1. Introduction

Acoustic Doppler current profilers (ADCP) are currently viewed as one of the most powerful tools for geophysical flows studies. The product family of these devices includes a lot of versions, which differ from each other by the number of beams, transducer head design, carrier acoustic frequency, cell sizes and time measurements settings. This variety of the device parameters provides flexibility in deployment and makes it possible to adjust the measurements to the broad range of research needs in meteorological (e.g., [1]), oceanological (e.g., [2–4]) and limnological (e.g., [5,6]) studies.

In particular, under the requirement of flow horizontal homogeneity, ADCPs are widely used for mean velocity profiling, as was originally designed. The later instrumental development (e.g., "burst" time settings, velocity measurements extending to 'pulse coherent mode') makes it possible to achieve higher resolution and better accuracy of the measurements, thus triggering the use of ADCP for enhanced studies of turbulence parameters. Within this new domain of ADCP applications, meaningful results have been obtained in fine-scale studies, including the estimations of dissipation rates ε [7–9].

At the same time some special methods have been developed for deriving the parameters of large-scale turbulence, with the special attention to the components $\langle u'_i u'_j \rangle$ of Reynolds stress tensor (u'_i pulsation velocity components in orthogonal frame) (e.g., [3,4,10–12]). The equations for these components are derived from the directly available intensities $\langle b'^2_i \rangle$

of the "beam" velocities pulsations b'_i. However, on the whole the problem remains challenging. Firstly, the number of ADCP beams usually varies from three to four (Janus configuration) and five (Janus+), and the system of equations is not complete. Secondly, in general case by calculating $\langle b'^2_i \rangle$ one can obtain only the relationships between the different required components $\langle u'_i u'_j \rangle$, but not the explicit relations for each of them. For example, with three-beam ADCP, no explicit relations are available. As for the four-beam Janus configuration, after aligning the device axis with the mean velocity, by applying the so-called "variance method" one can derive two explicit relations for off-diagonal stresses (shear stresses), but that is all.

One of the ways to overcome the problem was presented in [13,14]. In both papers the main idea was based on ADCP coupling, when the experimental setup includes two rigidly connected ADCPs. In such a special setup, the design and implementation of the experiment become more complicated. Besides, in both cases the stresses derivation was conjugated with additional restrictions. The method suggested in [13] gave acceptable results only for the case when the axis of the second device was sufficiently (>20°) tilted to the vertical. However, this requirement is not recommended by the device manufacturers; besides, the tilting makes the horizontal homogeneity requirement tougher. In the method, presented in [14], the axes of both devices are vertical, and one pair of beams have the intersection point at some depth. However, in turn, this method also possesses restrictions: it is applicable only to a small range of depths, close to the depth of the intersection point.

In this paper, an alternative method for derivation of full Reynolds stresses from single-ADCP data is presented. As compared to the coupled-ADCP method, presented in [14], this new method is not restricted to stress computations for the special depth, and so gives the opportunity for stress profiling for a range of depths. The missing information is derived not from the additional beam data, but by taking into account the interbeam correlations of the velocity. It is worthy of note that usually these correlations are neglected, by suggesting the statistical independence of velocities at different beams. Meanwhile, in geophysical flows the integral scales of turbulence often are so large that the size of energy-containing eddies at some depths occurs commensurable with the distance between the beams. This is just the case, when beam velocities are correlated, and their covariance includes some "hidden" information, necessary for closing the equation system for $\langle u'_i u'_j \rangle$.

The new method of turbulent stress derivation was applied to the case of the convectively mixed layer (CML) which develops in lakes during under-ice inhomogeneous solar heating of the water column.

2. Method Description
General Framework

In what follows below the simplest ADCP version with three azimuthally symmetric beams is considered (Figure 1). For the standard configuration the angle α_0 between the beam and vertical is 25°. As for the angle 2α between any pair of beams, its value may be determined by the following expression, derived from pure geometrical analysis:

$$sin\alpha = \frac{\sqrt{3}}{2} sin\alpha_0$$

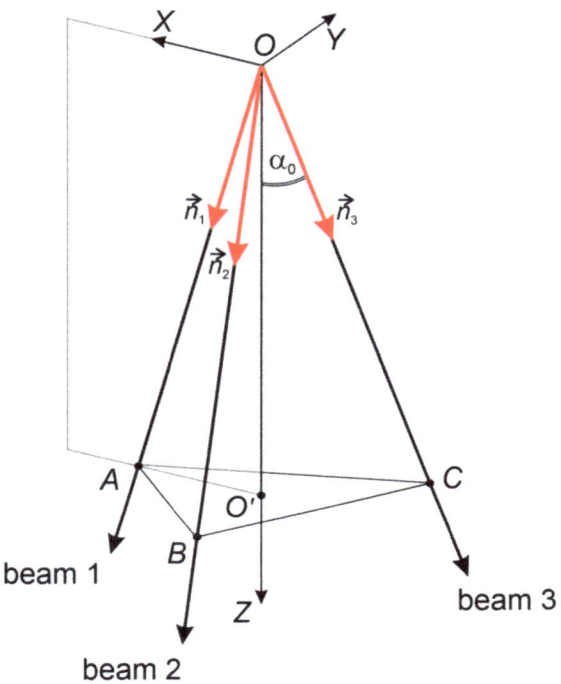

Figure 1. The coordinate system and main notations for three-beam ADCP configuration. Point O corresponds to the device head. Axis X lies in the plane AOO'.

In the orthogonal coordinate system XYZ, which is rigidly connected to the device, axis X is chosen in such a way that it lies on the same plane as beam 1 (Figure 1). In this frame of reference, the unit vectors $\vec{n}_1$, which identify the beams directions, have the following direction cosines:

$$\begin{cases} \vec{n}_1 = (sin\alpha_0, {}^\circ 0, {}^\circ cos\alpha_0), \\ \vec{n}_2 = \left(-\frac{1}{2}sin\alpha_0, {}^\circ -\frac{\sqrt{3}}{2}sin\alpha_0, {}^\circ cos\alpha_0\right), \\ \vec{n}_3 = \left(-\frac{1}{2}sin\alpha_0, {}^\circ \frac{\sqrt{3}}{2}sin\alpha_0, {}^\circ cos\alpha_0\right). \end{cases}$$

The trigonometric coefficients in the above expressions depend only on the basic angle α_0. Later on the correspondent matrix is denoted by **N**.

Each of the beam velocities $\{b_i\}$, measured directly at points A, B, C, is connected to the orthogonal projections u_1, u_2, u_3 of the velocity $\vec{u}$ at the same points by the linear relations, e.g., $b_1(A) = \left(\vec{n}_1 \vec{u}(A)\right)$, or, equivalently (note the summation over the repeated indexes):

$$b_1(A) = N_{1j} u_j(A) \qquad (1)$$

For the flows, which are homogeneous in the horizontal plane, $\langle u_i(A) \rangle = \langle u_i(B) \rangle = \langle u_i(C) \rangle \equiv \langle u_i \rangle$ so the mean values three equations of the type (1) take the form: $\langle b_i \rangle = N_{ij} \langle u_j \rangle$. As the result, the mean velocity components at the given depth are obtained directly as the convolution of the so-called transformation matrix $\hat{T} = \hat{A}^{-1}$ with the "vector" $(\langle b_1(A) \rangle, \langle b_2(B) \rangle, \langle b_3(C) \rangle)$ of mean beam velocities.

The same homogeneity assumption makes it possible to express the beam velocity pulsations intensities $\langle b_i'^2 \rangle$ through the turbulent stresses, e.g.,:

$$\langle b_1'^2 \rangle = \langle u_i'^2 \rangle \sin^2\alpha_0 + \langle u_3'^2 \rangle \cos^2\alpha_0 + 2\langle u_1' u_3' \rangle \sin\alpha_0 \cos\alpha_0 \qquad (2)$$

Expression (2) and two similar ones (for beams 2 and 3) represent three linear equations for six target components $\langle u_i' u_j' \rangle$, so the system is incomplete. Moreover, the explicit relation for any stress through beam pulsation intensities is not available, as was mentioned in the introduction.

In the general case, the problem of yielding the turbulent stresses from three-beam ADCP data is highly problematic. For some flows, not only the external length scales, but also integral scales of turbulence are large enough as compared to the distance between beams. For such cases the correlations between beam velocities are not vanishing, and taking the values of $\langle b_i b_j \rangle$ into account gives the opportunity to overcome the problem of missing equations.

To implement this opportunity, the structural function (SF) of the general type $\tilde{D}_{12}$ should be introduced into consideration. For beams 1 and 2, for example, this SF is defined as:

$$\tilde{D}_{12} = \langle (b_1(A) - b_1(B))(b_2(A) - b_2(B)) \rangle \qquad (3)$$

Under the assumption of local isotropy and homogeneity, the function $\tilde{D}_{12}$ is presented through the longitudinal SF D_{LL} (see Appendix A): $\tilde{D}_{12} = \left(\frac{4}{3}\cos^2\alpha - \sin^2\alpha\right) D_{LL}$. So, after opening the brackets, and taking $\langle b_1(A) b_2(A) \rangle = \langle b_1(B) b_2(B) \rangle$ into account (horizontal homogeneity) Expression (3) takes the form:

$$\left(\frac{4}{3}\cos^2\alpha - \sin^2\alpha\right) D_{LL} = 2\langle b_1 b_2 \rangle - <b_1(A) b_2(B)> - <b_1(B) b_2(A)> \qquad (4)$$

Here $\langle b_1 b_2 \rangle = <b_1(A) b_2(A)>$.

Longitudinal SF D_{LL} is derived directly by calculating along-beam velocity correlations. The second term in the r.h.s. of Equation (4) is also available directly from experimental data. As for the last term in the r.h.s., after taking into account both the horizontal homogeneity and refection invariance, the following expression is obtained in [15]: $<b_1(A) b_2(B)> = <b_1(B) b_2(A)>$. So, all except $\langle b_1 b_2 \rangle$ terms in the Equation (4) are available from the experiment, and one may regard Equation (4) as the explicit expression for $\langle b_1 b_2 \rangle$. With regard to the presentation $\langle b_1 b_2 \rangle = \langle b_1' b_2' \rangle + \langle b_1 \rangle \langle b_2 \rangle$ this expression may be also written as:

$$\langle b_1' b_2' \rangle = \left(\frac{2}{3}\cos^2\alpha - \frac{1}{2}\sin^2\alpha\right) D_{LL} + \langle b_1(A) b_2(B) \rangle - \langle b_1 \rangle \langle b_2 \rangle \qquad (5)$$

The similar presentations are valid for two remaining pairs of beams (13 and 23).

On the other hand, with presentation (1) in mind, three "beam stresses" $\langle b_i' b_j' \rangle (i \neq j)$ may be presented as the linear combination $N_{il} N_{jm} \langle u_l' u_m' \rangle$ of the Reynolds tensor components $\langle u_l' u_m' \rangle$ in the same way as Expression (2) for $\langle b_i'^2 \rangle$ were derived. For example:

$$\langle b_1' b_2' \rangle = \sin^2\alpha_0 (-\langle u_1'^2 \rangle/2 - \sqrt{3}\langle u_1' u_2' \rangle/2 + \langle u_1' u_3' \rangle (\cot\alpha_0)/2 - \sqrt{3}\langle u_2' u_3' \rangle (\cot\alpha_0)/2 + \langle u_3'^2 \rangle \cot^2\alpha_0) \qquad (6)$$

Expression (6) and two similar ones (for $\langle b_1' b_3' \rangle$ and $\langle b_2' b_3' \rangle$) one may regard as three missing equations for $\langle u_i' u_j' \rangle$. Together with Equation (2) (and two similar expressions —for $\langle b_2'^2 \rangle$ and $\langle b_3'^2 \rangle$) they form the closed system of linear inhomogeneous equations.

To represent this system in the compact form, it is reasonable to introduce into consideration the following "vectors":

$$B_i = \left(\langle b_1'^2 \rangle, \langle b_2'^2 \rangle, \langle b_3'^2 \rangle, \langle b_1' b_2' \rangle, \langle b_1' b_3' \rangle, \langle b_2' b_3' \rangle\right),$$

$$R_i = \left(\langle u_1'^2 \rangle, \langle u_1' u_2' \rangle, \langle u_1' u_3' \rangle, \langle u_2'^2 \rangle, \langle u_2' u_3' \rangle, \langle u_3'^2 \rangle\right).$$

With these notations the system of equations becomes:

$$B_i = M_{ij} R_j, \ i,j = 1\ldots 6 \tag{7}$$

The coefficient matrix **M** is derived directly from Equations (2) and (6) and similar ones:

$$\mathbf{M} = \sin^2\alpha_0 \begin{pmatrix} 1 & 0 & 2\cot\alpha_0 & 0 & 0 & \cot^2\alpha_0 \\ 1/4 & \sqrt{3}/2 & -\cot\alpha_0 & 3/4 & -\sqrt{3}\cot\alpha_0 & \cot^2\alpha_0 \\ 1/4 & -\sqrt{3}/2 & -\cot\alpha_0 & 3/4 & \sqrt{3}\cot\alpha_0 & \cot^2\alpha_0 \\ -1/2 & -\sqrt{3}/2 & \cot\alpha_0/2 & 0 & -\sqrt{3}\cot\alpha_0/2 & \cot^2\alpha_0 \\ -1/2 & \sqrt{3}/2 & \cot\alpha_0/2 & 0 & \sqrt{3}\cot\alpha_0/2 & \cot^2\alpha_0 \\ 1/4 & 0 & -\cot\alpha_0 & -3/4 & 0 & \cot^2\alpha_0 \end{pmatrix}.$$

Summing up, it seems reasonable to stress some key points and the step by step procedure. The corresponding algorithm looks as follows:

1. After proper choice of time averaging interval, the mean beam velocities $\langle b_i \rangle$, pulsation intensities $\langle b_i'^2 \rangle$ (Equation (2)) and correlations $\langle b_i' b_j' \rangle$ (Equation (5), $i \neq j$) are calculated directly from experimental data.
2. For each beam the function D_{LL} is calculated. After revealing the inertial interval, its extent is estimated, with the special attention to its upper scale limit l.
3. The range of depths is chosen in such a way that the distance between beams does not exceed the scale l. The maximum depth h is derived from inequality $AB < l$ (see Figure 1): $h < l/\left(\sqrt{3}\tan\alpha_0\right)$.
4. For chosen depths, the turbulent stresses are directly by solving the system (7):

$$R_i = M^{-1}_{ij} B_j, \ i,j = 1\ldots 6 \tag{8}$$

Here the inverse matrix $\mathbf{M}^{-1}$ looks like (here $\tan(\alpha_0)$ is shortly denoted as t):

$$\mathbf{M}^{-1} = \sin^{-2}\alpha_0 \begin{pmatrix} 4/9 & 1/9 & 1/9 & -4/9 & -4/9 & 2/9 \\ 0 & 1/(3\sqrt{3}) & -1/(3\sqrt{3}) & -2/(3\sqrt{3}) & 2/(3\sqrt{3}) & 0 \\ 2t/9 & -t/9 & -t/9 & t/9 & t/9 & -2t/9 \\ 0 & 1/3 & 1/3 & 0 & 0 & -2/3 \\ 0 & -t/(3\sqrt{3}) & t/(3\sqrt{3}) & -t/(3\sqrt{3}) & t/(3\sqrt{3}) & 0 \\ t^2/9 & t^2/9 & t^2/9 & 2t^2/9 & 2t^2/9 & 2t^2/9 \end{pmatrix}.$$

3. Experimental Setup and Results

Method validation was carried out with the velocity data, obtained from the special experiment on the shallow ice-covered lake Vendyurskoe (Karelia, Russia) between 27 March and 6 April 2020. The under-ice convection (most intense during 28–31 March and 4–6 April, when solar radiation was maximal) was clearly observed, with the convectively-mixed layer's (CML) thickness varying from 3 to 6 m. The details of experimental setup are presented in [14].

The measurements were carried out near the northern shore of the lake, the location of the experimental complex is marked by a triangle on Figure 2a; the depth at this location was ~7 m. The instrumental complex included a thermistor chain with 13 temperature sensors (RBR Ltd., accuracy ± 0.002 °C, measurement interval 10 s). The vertical temperature profile is schematically presented in Figure 2b; it clearly demonstrates the splitting of the water body into three sublayers (thin underice gradient sublayer, CML, and the underlying stratified zone), which is typical for developed convection. The CML's lower boundary was determined by thermistor chain data as a depth of the isotherm with a value exceeding the average temperature of the CML by 0.05 °C.

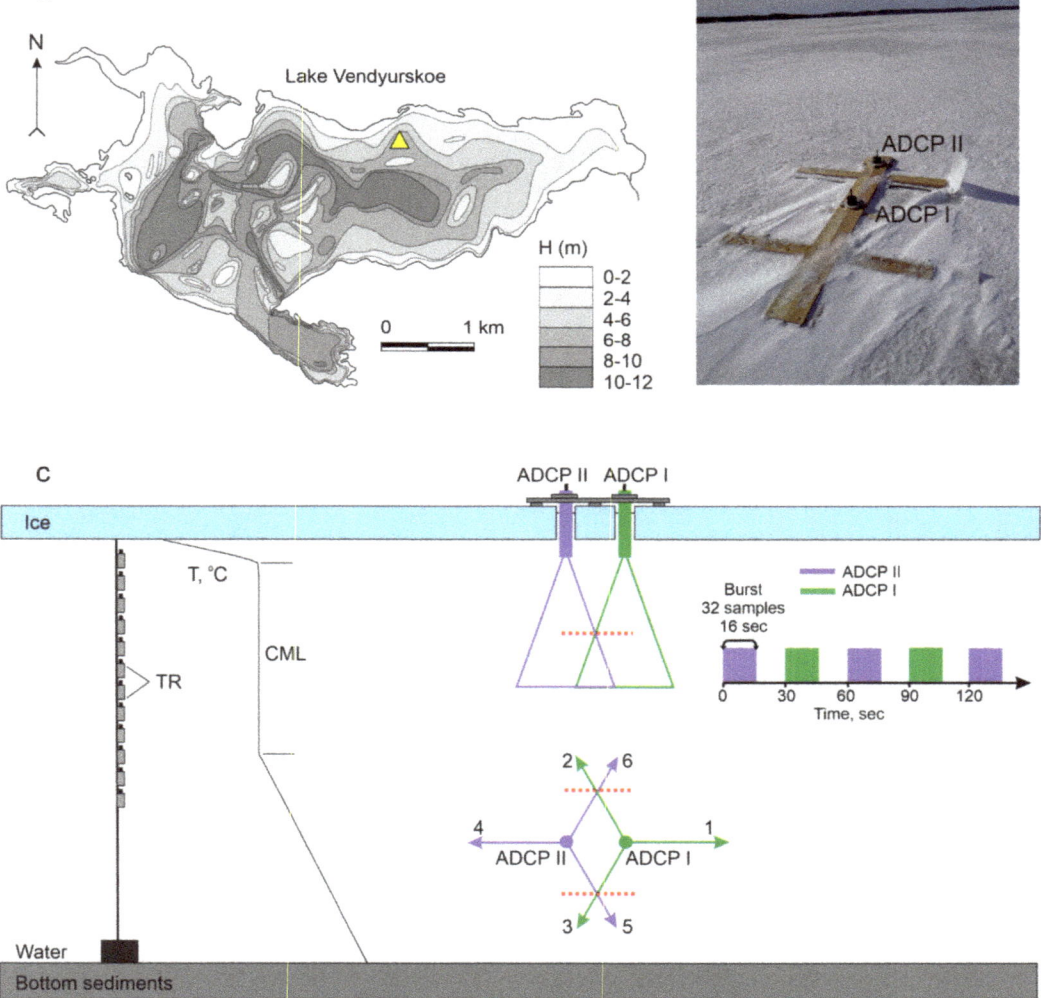

Figure 2. (**a**) Bathymetry of the Lake Vendyurskoe with indication of the measuring complex (yellow triangle). (**b**) Two ADCPs anchored on the ice of Lake Vendyurskoe in Spring 2020. (**c**) Schematic vertical distribution of temperature during springtime underice convection and scheme of the measuring complex. Indexes 1, 2, 3—the beams of the first ADCP, and 4, 5, 6—the second. Red dashed lines serve as the markers of beams' intersection points.

The most essential feature is connected with installing two rigidly connected down-looking ADCPs (2 MHz HR Aquadopp current velocity profiler, Nortek AS, Norway). Both devices were installed on the ice (Figure 2b) with emitters located 3 cm below the lower ice boundary (Figure 2c). The X-axes of both devices were aligned with the separation vector between two emitters, but were oppositely directed (Figures 1 and 2c). Due to the choice of this specific configuration one (X-axes are oriented towards each other) or two (X-axes are oriented away from each other) pairs of beams have intersection points. Figure 2c illustrates the second variant, with the beams in question being 3, 5 and 2, 6. The first variant of the devices' settings was realized from 17:00 on 27 March to 9:30 on 30 March, and the second between 10:00, 30 March and 10:00, 6 April 2020. In both cases the depth of the intersection points was the same (~1.6 m). By fitting the distance between the emitters. The presence of intersection points, as was shown in [14], is a key feature for deriving full stress tensor from coupled-ADCP data.

For both variants of coupled-ADCP setup, the signal discreteness was one minute (32 pulses with a frequency of 2 Hz) and the depth scanning range was 2.875 m (115 cells with a size of 25 mm). To exclude the mutual influence of the two ADCPs, the emitters were set in an asynchronous mode with a 30 s delay (Figure 2c). Then the radial velocities were averaged over 16 s active series; further processing was carried out using these averages, for which the same designations b_i were used. The root-mean-square error of b_i values varied in the range (0.1–0.5) mm/s.

Data processing was carried out in accordance to algorithm presented above. The details of the averaging procedure and the choice of averaging interval (100 min.) are presented in [14]. For specificity, only the results, which correspond to the active solar radiation period of 4–6 April and a depth of 1.6 m (corresponding to the position of beams intersection point) are presented below. For the rest of the data (another dates and depths) the results are similar.

The calculated dynamics of beam velocity intensities $\langle b_i'^2 \rangle$ and interbeam correlations $\langle b_i' b_j' \rangle$ are presented by Figure 3. The daytime maximums of the intensities reached the values (1–2) mm^2/s^2, whereas interbeam correlations, remaining statistically significant, varied roughly from one third to one half of their limits.

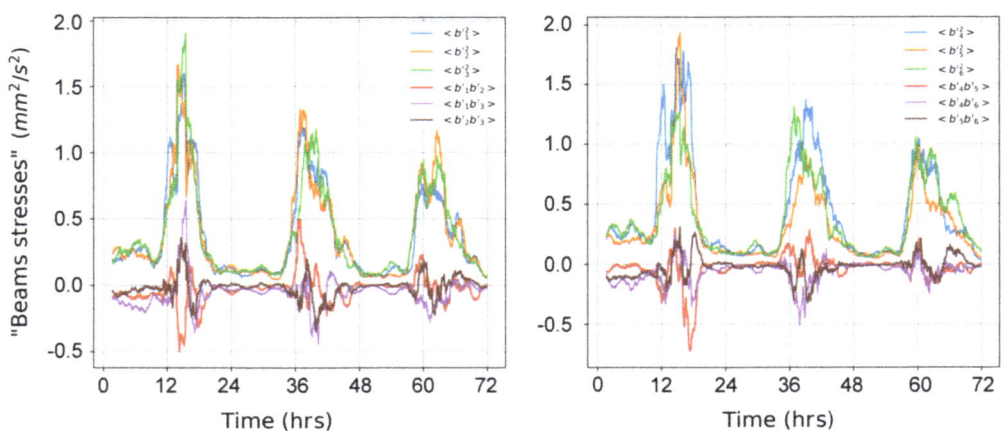

Figure 3. The calculated dynamics of beam velocity intensities and interbeam velocity correlations for both devices. Time readings from 4 April, 00:00.

The next step included the calculations of structure functions D_{LL}. During daylight time the structure function curves clearly demonstrate the presence of the inertial interval for all six beams. For illustration, the daytime sequence of the calculated SF for 4 April is presented by Figure 4 (the curves were averaged by all six beams). The upper bound of the inertial interval reached values up to 1 m, which is not much less as compared to the distance between beams at the depth under consideration. This fact gives grounds (see point 3 of the algorithm above) for involving Equation (6) as a missing equation. So Equation (8) may be used for stress estimations.

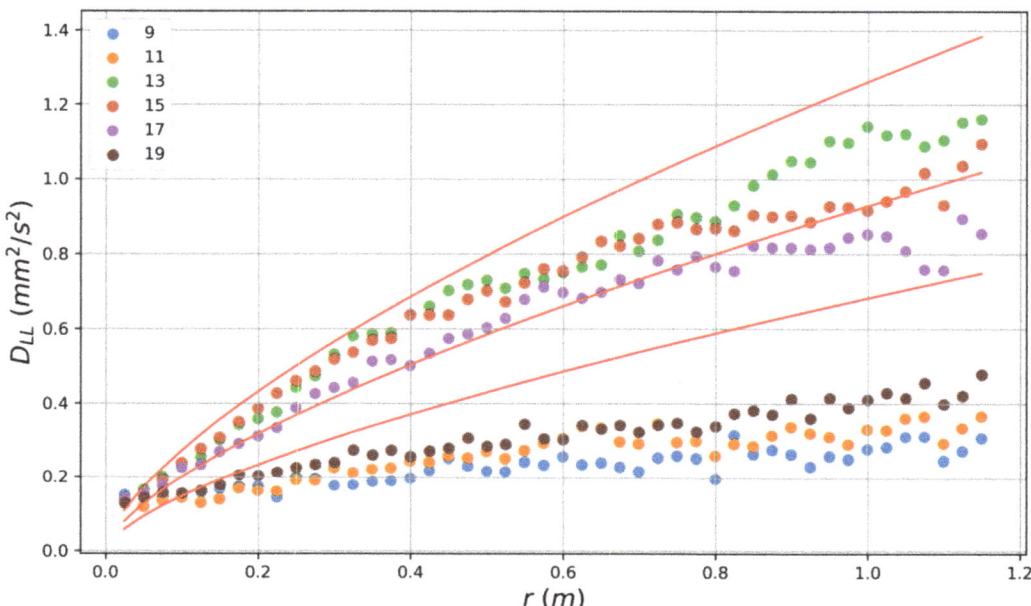

Figure 4. The set of longitudinal SF (averaged over six beams) calculated with a two-hour step for the time interval (09:00–18:00) 4 April 2020. Time averaging over 100 min. Solid lines represent the series of Kolmogorov curves $D_{LL} = C \, \varepsilon^{2/3} \, r^{2/3}$ with ε increasing from $0.2 \cdot 10^{-9}$ to $0.5 \cdot 10^{-9}$. The labels correspond to the measurement time (a.m.).

The calculations of the stresses were carried out by Equation (8) for both devices separately. The results of these independent computations demonstrate qualitative agreement, as Figure 5 illustrates.

Daily maximums of pulsation intensities along axes X, Y, Z achieved the values 7, 4 and 1 mm^2/s^2 correspondently. The value of the anisotropy coefficient $\langle u_3'^2 \rangle / \langle u'^2 \rangle$ was subjected to irregular oscillations within the range (0.05–0.30).

The estimations of standard deviation for stresses were carried out at the same way as presented in [14]. The errors varied from 15% for $\langle u_1'^2 \rangle$ to 25–30% for off-diagonal stresses. It is also worthy to note another criterion of physical sustainability of the results. This criterion—the so-called realizability condition—includes the positive definiteness of pulsations intensities and the Cauchy–Schwarz inequalities $\langle u_i' u_j' \rangle^2 \leq \langle u_i'^2 \rangle \langle u_j'^2 \rangle$ (here no summation over repeated indices). The fulfillment of both restrictions is the crucial point for low-energetic flow computations. In our case, the violations of this criterion were fixed during time intervals (presented by vertical red lines on the middle image of the top panel, Figure 5), which cover less than 5% of the whole observational period. Most of these intervals belong to nighttime, when the turbulence was sufficiently suppressed.

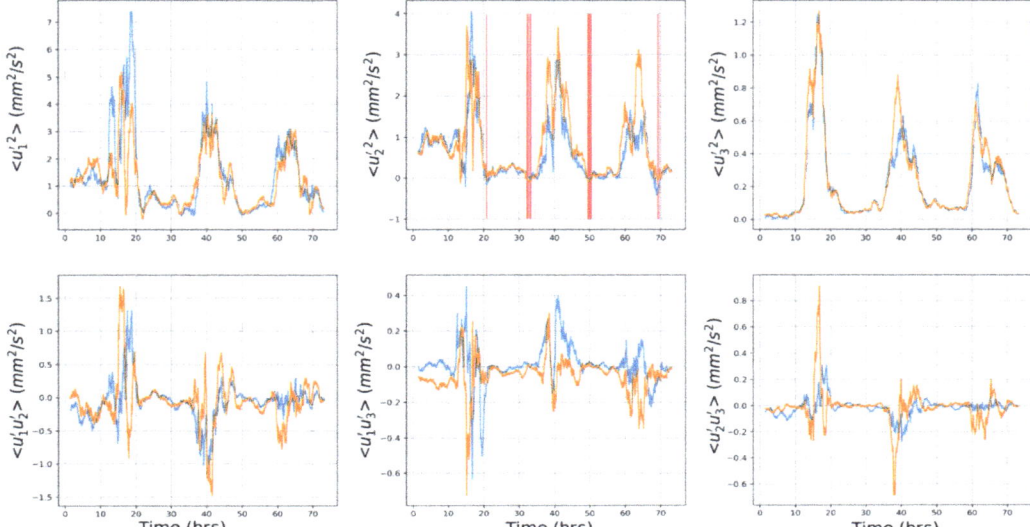

Figure 5. The dynamics of turbulent stresses calculated independently for each device. Top panel—pulsation intensities along axes X, Y, Z. Bottom panel—off-diagonal stresses. The time intervals, when realizability conditions were violated, are marked by the vertical red lines on the middle image of top panel.

As was mentioned in the introduction, the same experimental data were used in [14], but the yielding of stresses was carried out by the coupled-ADCPs method. The comparison of these alternative computations may serve as additional verification of the new method presented in this paper. The results of this comparison are presented in Table 1 and Figure 6. The computations by the new method are presented as the average values of the results, obtained for each device separately.

Table 1. Comparison of stress computations by two independent methods.

Stress Component	Correlation Coefficient, r	Coefficient of Determination, R^2	Linear Regression Coefficient
$\langle u_1'^2 \rangle$	0.96	0.92	1.31
$\langle u_2'^2 \rangle$	0.98	0.96	1.03
$\langle u_3'^2 \rangle$	0.92	0.85	0.61
$\langle u_i'^2 \rangle$	0.99	0.98	1.14

For all three pulsation intensities (diagonal components of the stresses matrix) the correlation coefficient r is higher than 0.9. At the same time, the new single-ADCP method gives the values 1.31 for $\langle u_1'^2 \rangle$ and 0.61 for $\langle u_3'^2 \rangle$, as compared to the coupled-ADCPs method. The values of the component $\langle u_2'^2 \rangle$, calculated by both methods, are practically identical (with deviations within 3%). The best fitting (Figure 7) was observed for turbulent kinetic energy $\langle u_i'^2 \rangle$ (TKE) for which coefficient of determination R^2 achieved the value 0.98 (Table 1).

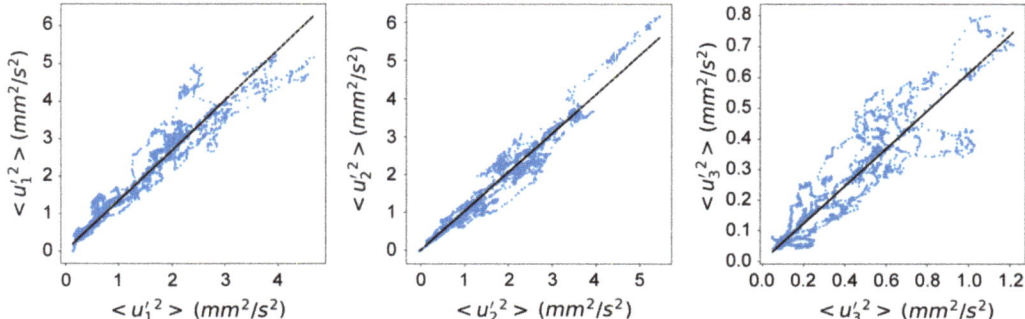

Figure 6. Correlations between the pulsation intensities, derived by two independent methods. Projection of each point on the X and Y axes represent the values, obtained by the coupled-ADCPs and single-ADCP methods correspondently. The linear regression curves are presented by black dashed lines.

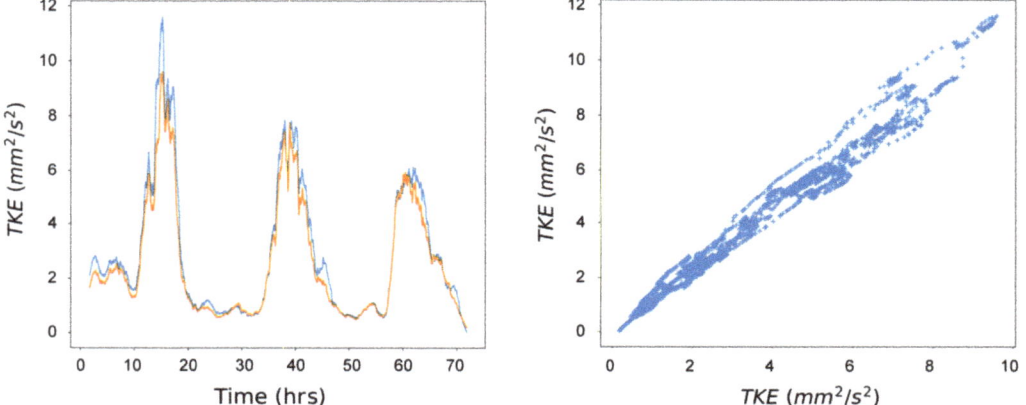

Figure 7. Correlations between the TKE values, derived by two independent methods. Projection of each point on the X and Y axes represent the values, obtained by the coupled-ADCPs and single-ADCP methods respectively.

4. Discussion

The method for yielding turbulent stresses presented in this paper is not restricted to the case of three-beam ADCP. Moreover, with other conditions being equal, four- or five-beam devices, presumably, give some preferences for method implementation. First of all, with such devices, more equations of type (2) and (6) become available. The resulting system turns out to be overdetermined, but can be solved in a least-square sense, as was done in [13] for the case of coupled ADCPs. Though such a solution is only an approximation, one may expect that the procedure does not reduce the accuracy of stress estimation, due to the increase of information involved.

The additional advantage of Janus and Janus+ versions is that the angle 2α between beams is smaller, as compared to the three-beam device. As a result, the distance between beams for given depth becomes smaller too, so the depth range, where the correlations of the beams velocity increments satisfy the local isotropy and homogeneity requirements and Equation (6) are valid, becomes wider.

Author Contributions: Conceptualization, S.B.; methodology, S.B. and R.Z.; software, S.B.; validation, S.B. and N.P.; formal analysis, S.B. and G.Z.; investigation, R.Z., G.Z. and N.P.; resources, G.Z.; data curation, S.B.; writing—original draft preparation, S.B. and G.Z.; writing—review and editing, S.B. and G.Z.; visualization, S.B. and G.Z, supervision, S.B.; project administration, G.Z.; funding acquisition, G.Z. All authors have read and agreed to the published version of the manuscript.

Funding: This research was funded by the Russian Science Foundation project No 21-17-00262 "Mixing in boreal lakes: mechanisms and its efficiency".

Data Availability Statement: All data created or used during this study are available by request to authors.

Acknowledgments: Authors are thankful to G. Kirillin for fruitful discussions.

Conflicts of Interest: The authors declare no conflict of interest.

Appendix A. Derivation of the Relationships between Structure Functions

Consider the plane formed by beams 1 and 2 and introduce the rectangular coordinate system as indicated in Figure A1. In this frame of reference, the beams velocities are presented by:

$$\begin{cases} b_1 = -u_x \sin\alpha + u_z \cos\alpha \\ b_2 = u_x \sin\alpha + u_z \cos\alpha \end{cases}$$

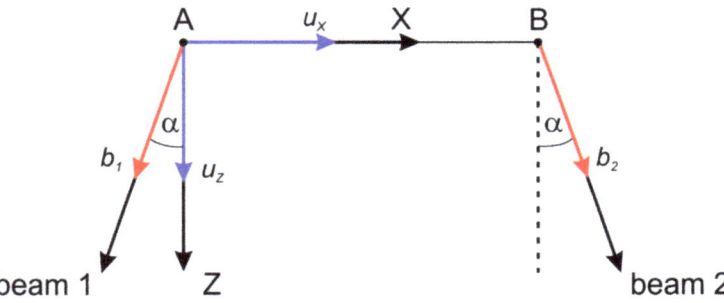

Figure A1. The reference frame for the plane including beams 1 and 2.

Substituting these expressions in (3), one obtains:

$$\tilde{D}_{12} = \langle ((u_z(A) - u_z(B))\cos\alpha - (u_x(A) - u_x(B))\sin\alpha)((u_z(A) - u_z(B))\cos\alpha + (u_x(A) - u_x(B))\sin\alpha) \rangle,$$

or, equivalently:

$$\tilde{D}_{12} = -\langle (u_x(A) - u_x(B))^2 \rangle \sin^2\alpha + \langle (u_z(A) - u_z(B))^2 \rangle \cos^2\alpha - <(u_x(A) - u_x(B))(u_z(A) - u_z(B))\sin\alpha\cos\alpha + (u_z(A) - u_z(B))(u_x(A) - u_x(B))\sin\alpha\cos\alpha \tag{A1}$$

The first term in (A1) includes the so-called longitudinal SF $D_{LL} = <(u_x(A) - u_x(B))^2>$, which is associated with the increments of the velocity components aligned with separation vector $\vec{r} \equiv \vec{AB}$ between points A and B. The second term, in turn, may be presented through the transverse SF $D_{NN} = \langle (u_z(A) - u_z(B))^2 \rangle$, which is defined through the increments of orthogonal velocity components.

The SF of general type is defined as $D_{ij} = \langle (u_i(A) - u_i(B))(u_j(A) - u_j(B)) \rangle$. Here velocity components u_x, u_y, u_z are numbered from 1 to 3. For locally isotropic and homogeneous turbulence, D_{ij} is presented through D_{LL} and D_{NN} by the expression [16]:

$$D_{ij}(\vec{r}) = (D_{LL}(r) - D_{NN}(r))\frac{r_i r_j}{r^2} + D_{NN}(r)\delta_{ij} \tag{A2}$$

The last two terms in (A1) include the cross-correlations of the aligned and orthogonal velocity components, and so are proportional to D_{13}. Due to (A2) both these terms turns to 0, if one takes into account the presentation $(r, 0, 0)$ for vector $\vec{r}$.

Finally the relation (A1) is transformed to:

$$\tilde{D}_{12} = D_{NN}\cos^2\alpha - D_{LL}\sin^2\alpha$$

For locally isotropic and homogeneous turbulence $D_{NN} = 4 D_{LL}/3$, so one obtains the following presentation of $\tilde{D}_{12}$ through the longitudinal SF: $\tilde{D}_{12} = \left(\frac{4}{3}\cos^2\alpha - \sin^2\alpha\right)D_{LL}$.

References

1. Lhermitte, R. Turbulent air motion as observed by Doppler radar. In Proceedings of the 13th Conference on Radar Meteorology, Montreal, QC, Canada, 20–23 August 1968; American Meteorological Society: Boston, MA, USA, 1968; pp. 498–503.
2. Lhermitte, R. Doppler sonar observation of tidal flow. *J. Geophys. Res.* **1983**, *88*, 725–742. [CrossRef]
3. Lohrmann, A.; Hackett, B.; Roed, L. High-resolution measurements of turbulence, velocity, and stress using a pulse-to-pulse coherent sonar. *J. Atmos. Ocean. Technol.* **1990**, *7*, 19–37. [CrossRef]
4. Guerra, M.; Thomson, J. Turbulence Measurements from Five-Beam Acoustic Doppler Current Profilers. *J. Atmos. Ocean. Technol.* **2017**, *34*, 1267–1284. [CrossRef]
5. Bouffard, D.; Zdorovennova, G.; Bogdanov, S.; Efremova, T.; Lavanchy, S.; Palshin, N.; Terzhevik, A.; Vinnå, L.R.; Volkov, S.; Wüest, A.; et al. Under-ice convection dynamics in a boreal lake. *Inland Waters* **2019**, *9*, 142–161. [CrossRef]
6. Bogdanov, S.; Zdorovennova, G.; Volkov, S.; Zdorovennov, R.; Palshin, N.; Efremova, T.; Terzhevik, A.; Bouffard, D. Structure and dynamics of convective mixing in Lake Onego under ice-covered conditions. *Inland Waters* **2019**, *9*, 177–192. [CrossRef]
7. Wiles, P.J.; Rippeth, T.P.; Simpson, J.H.; Hendricks, P.J. A novel technique for measuring the rate of turbulent dissipation in the marine environment. *Geophys. Res. Lett.* **2006**, *33*, L21608. [CrossRef]
8. Lucas, N.S.; Simpson, J.H.; Rippeth, T.P.; Old, C.P. Measuring Turbulent Dissipation Using a Tethered ADCP. *J. Atmos. Ocean. Technol.* **2014**, *31*, 1826–1837. [CrossRef]
9. Volkov, S.; Bogdanov, S.; Zdorovennov, R.; Zdorovennova, G.; Terzhevik, A.; Palshin, N.; Bouffard, D.; Kirillin, G. Fine scale structure of convective mixed layer in ice-covered lake. *Environ. Fluid Mech.* **2019**, *19*, 751–764. [CrossRef]
10. Lu, Y.; Lueck, R.G. Using a broadband ADCP in a tidal channel. Part II: Turbulence. *J. Atmos. Ocean. Technol.* **1999**, *16*, 1568–1579. [CrossRef]
11. Stacey, M.T.; Monismith, S.G.; Burau, J.R. Measurements of Reynolds stress profiles in unstratified tidal flow. *J. Geophys. Res.* **1999**, *104*, 10933–10949. [CrossRef]
12. Rippeth, T.P.; Simpson, J.H.; Williams, E.; Inall, M.E. Measurement of the rates of production and dissipation of turbulent kinetic energy in an energetic tidal flow: Red Wharf Bay revisited. *J. Phys. Oceanogr.* **2003**, *33*, 1889–1901. [CrossRef]
13. Vermeulen, B.; Hoitink, A.J.F.; Sassi, M.G. Coupled ADCPs can yield complete Reynolds stress tensor profiles in geophysical surface flows. *Geophys. Res. Lett.* **2011**, *38*, L06406. [CrossRef]
14. Bogdanov, S.R.; Zdorovennov, R.E.; Palshin, N.I.; Zdorovennova, G.E.; Terzhevik, A.Y.; Gavrilenko, G.G.; Volkov, S.Y.; Efremova, T.V.; Kuldin, N.A.; Kirillin, G.B. Deriving of turbulent stresses in a convectively mixed layer in a shallow lake under ice by coupling two ADCPs. *Fundame. I Prikl. Gidrofiz.* **2021**, *14*, 17–28. [CrossRef]
15. Hinze, J.O. *Turbulence*; McGrawHill: New York, NY, USA, 1975.
16. Monin, A.S.; Yaglom, A.M. *Statistical Fluid Mechanics. Volume II: Mechanics of Turbulence*, 1st ed.; Courier Corporation: Cambridge, UK; The MIT Press: Cambridge, MA, USA, 1971.

Article

Simulating Nutrients and Phytoplankton Dynamics in Lakes: Model Development and Applications

Bushra Tasnim [1], Xing Fang [1,*], Joel S. Hayworth [1] and Di Tian [2]

[1] Department of Civil and Environmental Engineering, Auburn University, Auburn, AL 36849, USA; bzt0022@auburn.edu (B.T.); jsh0024@auburn.edu (J.S.H.)

[2] Department of Crop, Soil, and Environmental Sciences, Auburn University, Auburn, AL 36849, USA; tiandi@auburn.edu

* Correspondence: xing.fang@auburn.edu

Abstract: Due to eutrophication, many lakes require periodic management and restoration, which becomes unpredictable due to internal nutrient loading. To provide better lake management and restoration strategies, a deterministic, one-dimensional water quality model MINLAKE2020 was modified from daily MINLAKE2012 by incorporating chlorophyll-a, nutrients, and biochemical oxygen demand models into the regional year-around temperature and dissolved oxygen (DO) model. MINLAKE2020 was applied to six lakes (varying depth and trophic status) in Minnesota focusing on studying the internal nutrient dynamics. The average root-mean-square errors (RMSEs) of simulated water temperature and DO in six lakes are 1.51 °C and 2.33 mg/L, respectively, when compared with profile data over 2–4 years. The average RMSE of DO simulation decreased by 24.2% when compared to the MINLAKE2012 model. The internal nutrient dynamics was studied by analyzing time series of phosphorus, chlorophyll-a, and DO over several years and by performing a sensitivity analysis of model parameters. A long-term simulation (20 years) of Lake Elmo shows that the simulated phosphorus release from sediment under the anoxic condition results in surface phosphorus increase, which matches with the observed trends. An average internal phosphorus loading increase of 92.3 kg/year increased the average daily phosphorus concentration by 0.0087 mg/L.

Keywords: water quality; chlorophyll-a; phosphorus; phosphorus release; dissolved oxygen

1. Introduction

Eutrophication has been a threat to waterbodies since the beginning of the twentieth century in industrialized countries [1–4]. A large proportion of the anthropogenic increase in nitrogen and phosphorus flux due to industrialization is delivered to ground or surface waters through direct runoff, human and animal wastes, and atmospheric deposition. Over time, excess nutrients are transported to waterbodies [5,6]. When a waterbody undergoes any human-influenced ecosystem changes such as nutrient loading, extreme weather events, or invasive organisms; algal species (cyanobacteria) can form dense overgrowths known as algal blooms. Since these blooms can produce toxins that are harmful to people and wildlife they are often referred to as harmful algal blooms (HAB). HABs cause undesirable changes in aquatic resources such as reduced water clarity, hypoxia, fish kills, loss of biodiversity, and an increase in nuisance species [7,8]. Oxygen is consumed by both living and dead algae which results in low oxygen concentration in lakes typically known as hypoxia (below 2–4 mg/L of dissolved oxygen (DO)). Eutrophication also has a detrimental effect on human health through increased exposure to cyanobacteria toxins [9,10], nitrites, and nitrates [7,8] in drinking water. Since most cities use surface water as the drinking water source, HABs can cause serious problems of off-flavor odor and taste (sometimes described as earthy or musty). In some cases, drinking water no longer remains safe to drink and complete remediation is needed. For example, the state of Ohio committed to spending USD 172 million to clean up Lake Erie as HABs were causing severe drinking water problems [11]. Furthermore, the economic costs of eutrophication, for restoring the ecosystem services (e.g., housing amenity value, recreation opportunities, freshwater provisioning, and food and fiber production) are high [12–14]. Eutrophication can have serious effects on the social health of a community causing decreases in the activity that are dependent on aquatic or seafood harvests or tourism, resulting in disruption of social, and cultural practices.

In North America, more than 41% of lakes are eutrophic; the proportions for the Asia Pacific, Europe, Africa, and South America are 53%, 28%, 48%, and 41%, respectively [15]. Management and restoration solutions to control eutrophication require predicting the lake's nutrient concentration, understanding interactions between nutrients and water quality variables, and quantifying algal growth and decay. Since the 1970s, numerical modeling has shown to be an effective tool to quantify nutrient concentrations [16,17]. Several promising lake models (MINLAKE, PCLake, LAKE2K, CE-QUAL-W2, EFDC, and ELCOM-CAEDYM) have been developed over the past decades. The key state variables of these models are nutrients, principally phosphorus (P), nitrogen (N), and sometimes silica (Si) [18,19] since these nutrients link to primary production.

The Minnesota Lake Water Quality Management Model (MINLAKE) is a one-dimensional (along depth direction), deterministic water quality model with a time step of one day. MINLAKE was developed in 1988 to support lake eutrophication studies and was capable of simulating water temperature, chlorophyll a (Chl$a$), phosphorus, nitrogen, biochemical oxygen demand (BOD), dissolved oxygen (DO) for lakes during the open water season [20]. MINLAKE1988 was further developed to include ice-cover period simulation [21], simplified regional DO model [22], modified year-round nutrient model [23], and hourly water temperature and DO model [24]. MINLAKE nutrient model (MINLAKE98) was applied to three lakes but did not perform well for multiple-year simulation [23]; hence, the model was not used further. The most recent version of MINLAKE with daily simulation, MINLAKE2012 was capable of simulating water temperature and DO in different types of lakes with good agreement with observations [25] but lacks a nutrient model. LAKE2K is also a one-dimensional lake water quality model which simulates carbon, nitrogen, oxygen, phosphorus, silica concentrations, and phytoplankton and zooplankton biomass using water balance, heat balance, and mass balance for the epilimnion, metalimnion, and hypolimnion (three layers) of a lake [26]. A two-dimensional hydrodynamic and water quality model, CE-QUAL-W2 (originally developed in 1990s) can be used in rivers, lakes, reservoirs, estuaries, and even a combination of river segments and multiple reservoirs but is more suitable for relatively long and narrow water bodies [27]. PCLake (1990) is a process-based model to simulate water quality in shallow, non-stratifying lakes in temperate climate zones with a uniform daily time-step for processes. The recent version of PCLake, PCLake+ simulates basic stratification in temperature using mixing depth and two layers: the epilimnion and hypolimnion only [28]. EFDC is a state-of-the-art, versatile model that can simulate one-, two- or three-dimensional flow, transport, and biogeochemical processes in surface water systems such as rivers, lakes, estuaries, and reservoirs. ELCOM is a three-dimensional hydrodynamic model which is often integrated with a water quality model CAEDYM [29] and this coupled model has been used to simulate water quality and algal bloom scenarios [30,31]. Though EFDC and ELCOM-CAEDYM are very flexible and support a variety of conditions, the complexity of these models makes them difficult to apply when data are scarce. Each of these models has some limitations such as modeling for a certain type of waterbody (PCLake), only modeling certain water quality constituents (MINLAKE2012), neglecting some physical processes (CE-QUAL-W2 and Lake2K), or problems due to model complexity (EFDC and ELCOM-CAEDYM). Apart from these models, during the last two decades, new manifestations of eutrophication have emerged [1,32,33]. In order to improve operational control of algal blooms and management applications, models are often integrated into warning systems to predict short-term phytoplankton blooms [34]. For example, the EcoTaihu model has been integrated into a Windows software platform to predict algal blooms in Lake Taihu [35]. Elliott [36] used PROTECH model to study the effect of an increase in water temperature and phosphorus loading on phytoplankton in Lake Windermere.

High phosphorus release from lake sediments is frequently reported as an important mechanism delaying lake recovery after external loading of phosphorus has been reduced [37–39]. A study of 78 stormwater ponds revealed that more than one-third of the sampled ponds may experience internal loading of phosphorus although these ponds are shallow and had short periods of anoxic condition [40]. A long-term survey of 35 lakes in Europe and North America concluded that internal release of phosphorus typically continues for 10–15 years after the external loading reduction [41] but in some lakes, the internal release may last longer than 20 years [38]. In shallow lakes, it is common to observe the negligible change in phosphorus concentrations in lake water even after external load diversion [42]. For example, Lake Trummen in Sweden remained hypereutrophic even after 11 years of sewage (primary source of external loading) diversion. Eventually, internal phosphorus loading was reduced dramatically by removing 1 m of high phosphorus sediment [42].

Since internal loading adds uncertainty to lake restoration processes, a reliable lake water quality tool which (a) focuses on the internal nutrient dynamics, (b) considers all physical processes (ice cover, sediment heat transfer, etc.), (c) can be applied to different types of lakes, and (d) is capable of multiple year simulation, is necessary and very useful. A simple, one-dimensional lake water quality model, which can predict chlorophyll-a, phosphorus, DO in various types of lakes (shallow to deep, small to large surface area, and oligotrophic to eutrophic) for short-term and long-term simulations, is needed for lake management and restoration purpose. Since MINLAKE2012 was capable of simulating water temperature and DO in all lake types with good performance [25], it was selected for further enhancement. The more general daily lake water quality model MINLAKE2020 was developed by including phytoplankton, zooplankton, nutrients, and BOD sub-models into the existing temperature and DO model of MINLAKE2012. To verify the performance of the model, it was applied to six Minnesota lakes having varying depths and trophic status and having 15–70 days of profile data for model calibration/validation over different time spans (2–20 years). The objective of this study was to use MINLAKE2020 to understand the internal nutrient dynamics and explore the interaction/connection among nutrients, phytoplankton dynamics, and stratification/mixing dynamics in lakes. Thus, the inflow sub model developed for MINLAKE2020 was disabled to provide a better understanding of the nutrient dynamics within the lakes.

2. Materials and Methods

Sections 2.1–2.6 provide descriptions and governing equations for MINLAKE2020 to simulate unsteady phytoplankton (quantified using chlorophyll-a concentration), zooplankton, and nutrients for both the open water season and the ice cover periods. Since this study focuses on the nutrients/phytoplankton internal dynamics/cycles, Figures 1–7 for water quality variables modelled by MINLAKE2020 do not include inflow and outflow as a part of mass balance processes.

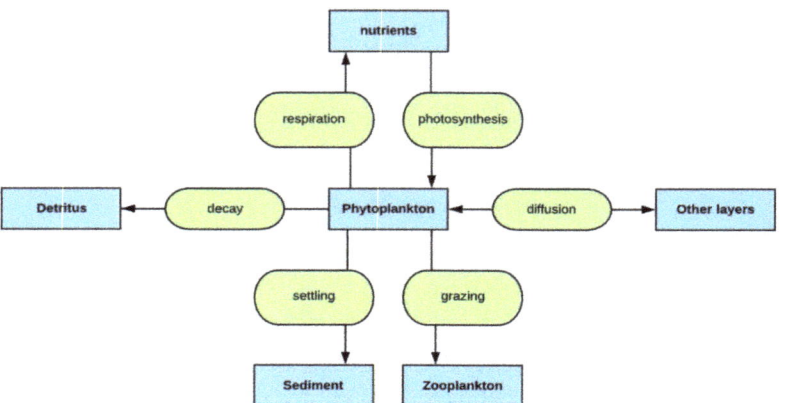

Figure 1. Schematic diagram of phytoplankton (Chlorophyll-a) cycle modeled by MINLAKE2020.

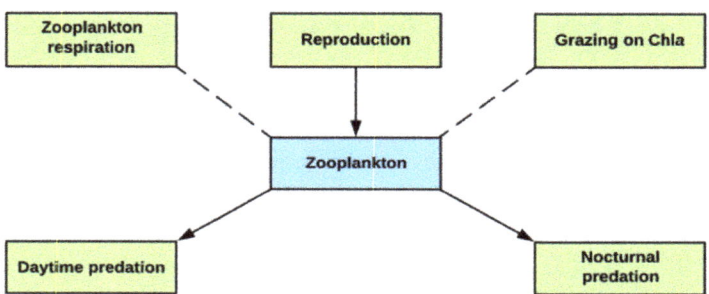

Figure 2. Schematic diagram of zooplankton cycle modeled by MINLAKE2020.

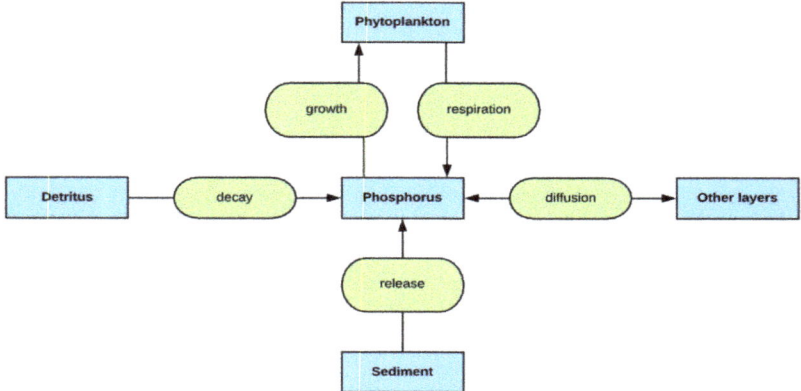

Figure 3. Schematic diagram of phosphorus cycle modeled by MINLAKE2020.

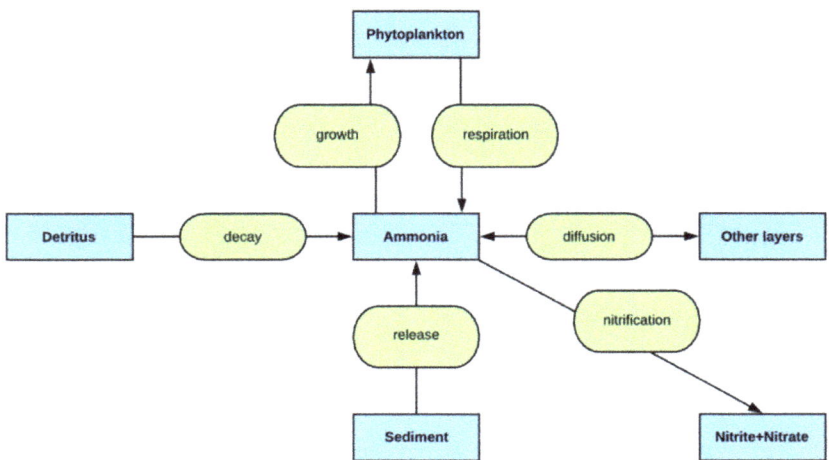

Figure 4. Schematic diagram of ammonia cycle modeled by MINLAKE2020.

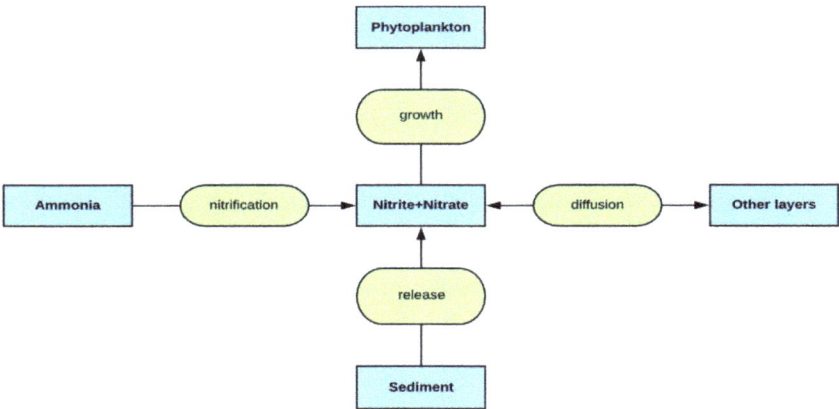

Figure 5. Schematic diagram of nitrate-nitrite cycle modeled by MINLAKE2020.

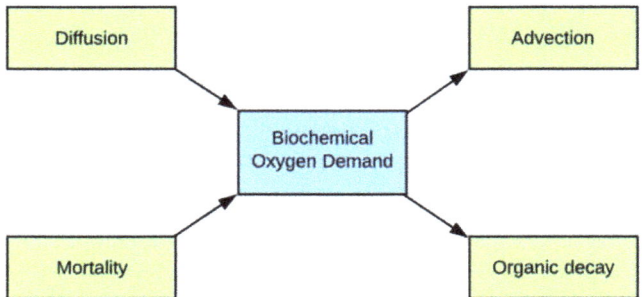

Figure 6. Schematic diagram of BOD cycle modeled by MINLAKE2020.

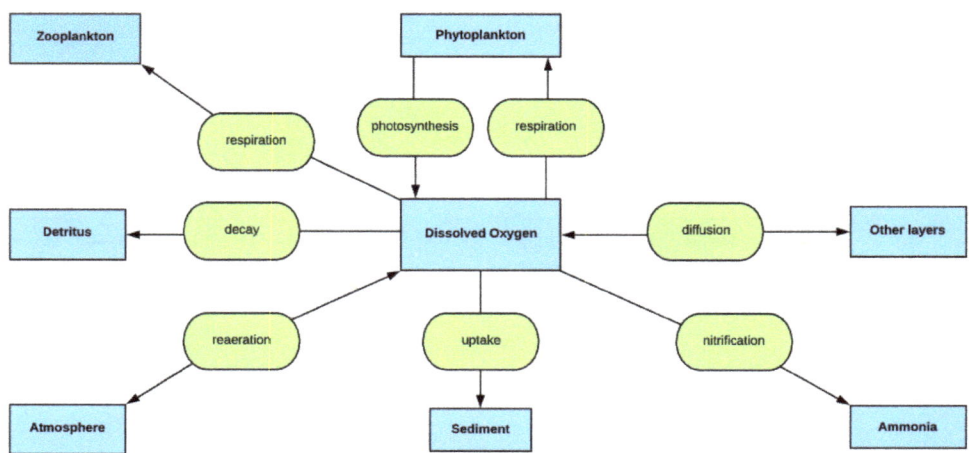

Figure 7. Schematic diagram of DO cycle modeled by MINLAKE2020.

2.1. MINLAKE Overview

MINLAKE models use the basic one-dimensional advection-diffusion equation to simulate the dynamic variations of state variables in horizontal layers of a lake.

$$A\frac{\partial C}{\partial t} + v\frac{\partial(A \times C)}{\partial z} = \frac{\partial}{\partial z}\left(K_z A \frac{\partial C}{\partial z}\right) \pm sources/sinks \quad (1)$$

where C is the concentration of a state variable, v is the vertical settling velocity of the particulate form of some of the state variables (v = 0 for dissolved); z is the vertical coordinate measured positively downward; K_z is the vertical turbulent diffusion coefficient; and A horizontal area of the control volume. The one-dimensional vertical advection-diffusion equation is solved using a series of layers characterized by depth from the water surface, thickness, layer volume, and horizontal areas. The MINLAKE model developed by Riley and Stefan [43] (called MINLAKE88 in this paper) was capable of simulating water temperature and DO profiles with three levels of complexity for phytoplankton. The first and simplest form is a single algal group model with productivity distributed in the mixed layer [43]. The second level of complexity has the algal growth term unique to each layer using Michaelis-Menten equation for a single algal group. The level three considers up to three groups of algae, phosphorus, nitrogen, BOD for lakes during the open water season. MINLAKE88 is not widely used because of its inability to perform multiple-year simulations. Due to a lack of available data, the nutrient model could not be developed further. Gu and Stefan [21] included an ice-cover period simulation in MINLAKE88 to model snow thickness, ice cover thickness, and water temperature but no other state variables. In 1991, Hondzo and Stefan introduced a more general water temperature simulation model for MINLAKE which can be applied to a wide variety of lakes and regions [44]. An important modification of MINLAKE was accomplished in 1994 when Fang [45] developed the regional dissolved oxygen model and combined it with MINLAKE to study the impact of global climate warming on lake water quality and fish habitat in Minnesota lakes. The regional lake model was developed to model different types (categorizing by stratification strength and eutrophication) of lakes by maximum depths (shallow, medium-depth, and deep), surface area (small, medium area, and large), and trophic status (eutrophic, mesotrophic, and oligotrophic). The regional DO model is a simplified version of the DO model that did not simulate daily nutrients and Chl*a* concentrations but used the annual mean Chlorophyll-a concentration with seasonal variation patterns [22] to specify daily Chl*a* for DO simulation. Separate sub-models for winter conditions were developed and integrated with MINLAKE96 to simulate water temperature and DO year-round over many years as long as daily weather data are available [46]. The MINLAKE96 model was further modified and refined as it was used in a study in 2010, to simulate water quality conditions in cisco lakes, which are typically deep mesotrophic or oligotrophic lakes [47]. The MINLAKE96 model was further modified by West-Mack to simulate phosphorus, nitrogen, and chlorophyll-a simulation in 1998 [23,48] and the updated model is called MINLAKE98 here. The mass balance equations for chlorophyll-a, phosphorus, nitrogen, and DO were modified from those of MINLAKE88. The model was tested on three lakes and produced satisfactory results but due to lack of observed data and inability of the model to run for multiple years, the nutrient model was not further developed. The source code of MINLAKE98 was also lost and no longer available for further improvement.

MINLAKE96 was further modified in 2018 to calculate the hourly water temperature and DO using hourly weather conditions [24,49]. MINLAKE2012 is the most recent version of the daily MINLAKE model and can simulate year-round water temperature and DO in various lakes of different regions, using an Excel spreadsheet as the user interface. For all MINLAKE model variants, water temperature is simulated first by solving the following heat transport equation.

$$\frac{\partial T_w}{\partial t} = \frac{1}{A}\frac{\partial}{\partial z}\left(K_z A \frac{\partial T_w}{\partial z}\right) + \frac{H_w}{\rho C_P} \quad (2)$$

where T_w (z, t) is the water temperature in °C, which is a function of depth (z in m) and time (t in d); $A(z)$ (m^2) is the horizontal area for each layer of water as a function of the depth; K_z (m^2/day) is the vertical turbulent heat diffusion coefficient which is a function of depth and time; ρC_p (J/m^3-°C) represents the heat capacity of water per unit volume; H_w (J/m^3-day) is the heat source and/or sink term per unit volume of water. Determination of the turbulent diffusion coefficient is discussed in detail by Fang [45]. In the regional daily MINLAKE model, the vertical heat diffusion coefficient K_z for epilimnion and hypolimnion is calculated using the following equation:

$$K_z = 8.17 \times 10^{-4} \times \frac{A_S^{0.56}}{(N^2)^{0.43}} \quad (3)$$

where K_z is the vertical diffusion coefficient in cm^2/s (1 cm^2/s = 8.64 m^2/day = 0.36 m^2/h), A_s is the surface area of the lake (km^2) and N^2 is the Brunt-Vaisala stability frequency of the stratification (s^{-2}). In the epilimnion, N^2 was set at a minimum value of 0.000075 [42]. Equations (1) and (2) are solved numerically using an implicit finite difference scheme and a Gaussian elimination method with time steps of one day.

2.2. Phytoplankton Simulation

In MINLAKE2020, the chlorophyll-a model was modified to overcome many of the limitations of previous MINLAKE models. The model can simulate up to three algal groups (diatoms, green algae, and blue-green algae) and the shift from diatoms early in the season to green and then blue-green algae during the summer. The algal groups are distinguished by different rates of photosynthesis, respiration, settling, zooplankton grazing, and different nutrient requirements. A schematic diagram of the phytoplankton cycle applicable to all algal groups is presented in Figure 1.

Phytoplankton growth depends on the maximum growth rate of the algae (G_{max}), half-saturation coefficients for nutrients, water temperature, solar irradiance, external nutrient concentrations, and the current Chla concentration. The maximum growth rate of algae varies for different classes of algae. The algal growth limitation by nutrients is modeled using a Michaelis-Menten equation [50]:

$$f(S) = \frac{S}{K_S + S} \quad (4)$$

where $f(S)$ is dimensionless, S is the concentration of the nutrient (P, N, or Si) in water (mg/L), and K_S is the half-saturation constant for the nutrient (mg/L). Algal growth dependence on water temperature is modeled by equations given by Lehman et al. [51]:

$$f(T) = \exp\left(-2.3\left(\frac{T - T_{opt}}{T_{opt} - T_{min}}\right)^2\right) \text{ for } T < T_{opt} \quad (5)$$

$$f(T) = \exp\left(-2.3\left(\frac{T - T_{opt}}{T_{max} - T_{opt}}\right)^2\right) \text{ for } T \geq T_{opt} \quad (6)$$

T is the water temperature (°C). The maximum growth occurs at an optimal temperature, T_{opt} (°C), and the growth rate decreases both above and below T_{opt}. T_{min} (°C) is a low temperature at which phytoplankton growth is reduced to 90% from the optimum. T_{max} (°C) is the high temperature at which growth is reduced by 90%. Phytoplankton growth is usually limited by available light, which is a function of the depth and the light attenuation coefficients due to water (K_w) and algae (K_c). The Haldane equation [52] is used in MINLAKE2020 to calculate the light limitation for algal growth, which was also used in MINLAKE88, MINLAKE98, and the regional DO model:

$$f(L) = \frac{I(z)\left(1 + 2\sqrt{\frac{K_1}{K_2}}\right)}{I(z) + K_1 + \frac{I(z)^2}{K_2}} \quad (7)$$

$$I(z) = \frac{27.25}{TD} RAD \times exp[-(K_w + K_c Chla)z] \quad (8)$$

where $f(L)$ is the light limitation coefficient (dimensionless), $I(z)$ is the photosynthetically active radiation (PAR) as a function of depth, TD is the photoperiod (h) as a function of Julian day, RAD is daily solar radiation, and K_1 and K_2 are the light limitation and inhibition coefficients [53], respectively. The units for $I(z)$, K_1, and K_2 are in µE/m^2/s. In MINLAKE88 and MINLAKE98, light limitation and inhibition coefficients were specified by the user. In MINLAKE2020, light limitation and inhibition coefficients are calculated using the same equations as in the regional DO model [23].

Phytoplankton populations are removed from a water column by four processes: respiration, mortality, settling, and zooplankton grazing (Figure 1). Each phytoplankton population is assigned a fixed or calibrated respiration rate, mortality rate, and settling rate. Respiration affects the available phosphorus and DO immediately whereas mortality contributes with a time lag through detrital decay. In MINLAKE2020, a single class of zooplankton is simulated. To simulate the Chl*a* lost by grazing of zooplankton, the zooplankton population (ZP(t), #/m^3) is simulated in a separate subroutine on daily basis. Grazing is assumed to take place in the evening when zooplankton rises to the upper layers and is dependent on the temperature and Chl*a* concentration. The Michaelis-Menten equation is used to simulate the effect of Chl*a* concentration on grazing. It is assumed that no grazing occurs below a threshold Chl*a* concentration (Chl*a*$_{min}$ in Equation (9)):

$$\underbrace{\frac{\partial Chla}{\partial t} - \frac{1}{A}\frac{\partial}{\partial z}\left(AK_z\frac{\partial Chla}{\partial z}\right)}_{diffusion} + \underbrace{\frac{v_c}{A}\frac{\partial(A\times Chla)}{\partial z}}_{settling} + \underbrace{K_m\theta_m^{T-20}Chla}_{mortality} + \underbrace{K_r\theta_r^{T-20}Chla}_{respiration}$$

$$\underbrace{+GR_{max}\frac{Chla-Chla_{min}}{K_{gchla}+Chla-Chla_{min}}\theta_z^{T-20}ZP\,\Delta T_g\frac{V(IZ)}{V(I)}CF}_{grazing\ by\ zooplankton} \tag{9}$$

$$\underbrace{-G_{max}f(T)\left[f(L):\frac{P}{K_P+P}:\frac{N}{K_N+N}\right]_{min}Chla}_{growth}=0$$

where Chla is the chlorophyll-a concentration (mg/L), v_c is the phytoplankton settling velocity (m/d), K_m, K_r, θ_m, and θ_r are mortality and respiration rates (d^{-1}) and corresponding temperature adjustment coefficients (dimensionless), respectively. G_{max} is the maximum growth rate of phytoplankton (d^{-1}), GR_{max} is maximum grazing rate (mg Chla/individual zooplankton per d), $Chla_{min}$ is minimum chlorophyll concentration for grazing to occur (mg/L), K_{gchla} is half-saturation constant for grazing (mg/L), K_P and K_N are the half-saturation constants for phosphorus and nitrogen (mg/L), respectively; P and N are the available concentration of phosphorus and nitrogen (mg/L), respectively; ZP is the zooplankton density (#/m^3), ΔT_g (<1 d) is time that zooplankton spends in a layer during the night to graze (d), $V(IZ)/V(I)$ is the day depth (layer IZ) and the layer I volume ratio, and CF is the unit conversion from L to m^3 (=0.001).

Zooplankton grazing is only simulated to represent the dynamics of algae. Zooplankton grazing rates vary with different classes of algae. For example, zooplankton is more likely to feed on green algae than blue-green algae [23].

$$GRAZE(k,I) \propto \Delta T_g(I) = (1-\frac{TD}{24})(\frac{\sum_{k=1}^{3}Chla(k,I)}{\sum_{I=1}^{IZ}\sum_{k=1}^{3}Chla(k,I)}) \tag{10}$$

where $\Delta T_g(I)$ is the time of grazing for layer I (d), TD is the photoperiod (h) from sunrise to sunset, $Chla(k, I)$ is the chlorophyll-a concentration of phytoplankton group k in layer I, and IZ is the day-depth layer, where DO ≥ 0.5 mg/L. Zooplankton grazing of phytoplankton (Chla) occurs during nocturnal migration at the day depth. The nocturnal grazing rate is calculated for each layer ($I = 1, \ldots, IZ$) between the day depth and the surface using the volume day depth/layer ratio $V_r = V(IZ)/V(I)$ [43].

2.3. Zooplankton Simulation

MINLAKE2020 includes a zooplankton model to simulate (1) Chla lost by zooplankton grazing and (2) DO consumed by zooplankton respiration. A single class of zooplankton is simulated in the lake environment each day. During the day, zooplankton retreats to deeper water seeking refuge from visual predators. They begin to rise to the surface at dusk while grazing and return to deeper layers at dawn (Figure 2). Zooplankton activity of these two periods is treated separately in the model.

Zooplankton is assumed to have a constant reproduction rate and a time-varying predation rate for determining the zooplankton population $ZP(t)$ as a function of time (t, day). The day depth, light level at the day depth, and predation on zooplankton are calculated as the first step in zooplankton simulation. The day depth of zooplankton is identified as the deepest layer in which the DO concentration is greater than 0.5 mg/L and therefore changes with time depending on DO vertical depth distribution. Below the day depth, the grazing is zero. In MINLAKE2020, dual effects of seasonal predation and light limitation are included to simulate biomanipulation techniques related to methods of increasing the zooplankton population [54]. The dominant zooplankton predators are assumed to be visual predators and zooplankton predation only occurs in the daytime. The light limitation assumes a linear variation of predation between two light levels [55].

$$PD_d = P_d \frac{XI - XI_{min}}{XI_{max} - XI_{min}} \text{ where } 0 \leq \frac{XI - XI_{min}}{XI_{max} - XI_{min}} \leq 1 \qquad (11)$$

where PD_d is the daytime predation rate (d^{-1}) and P_d is the daily predation rate (d^{-1}) calculated using Equation (12). XI is the light intensity at the day depth (µE/m^2/s), XI_{min} is the light intensity at which no predation occurs (µE/m^2/s) and XI_{max} is light intensity above which predation is not light inhibited (µE/m^2/s). When the light intensity XI is less than XI_{min} or larger than XI_{max}, the ratio in Equation (11) is reset to zero or one, respectively.

Both daytime and nocturnal predation are calculated in MINLAKE2020. Daytime predation combines light limitation with a time-varying maximum predation rate. A linear function is used to calculate the time-varying daytime predation rate P_d given in Equation (12) when Julian day DY is between DY_{min} and DY_{max}.

$$P_d = P_{min} + (P_{max} - P_{min}) \left(\frac{DY - DY_{min}}{DY_{max} - DY_{min}} \right) \text{ where } 0 \leq \frac{DY - DY_{min}}{DY_{max} - DY_{min}} \leq 1 \qquad (12)$$

where P_{min} and P_{max} are minimum and maximum predation rates (d^{-1}), respectively. DY_{min} is Julian day of the last day of minimum predation rate and DY_{max} is Julian day of the beginning of maximum predation rate: $P_d = P_{min}$ when $DY < DY_{min}$ and $P_d = P_{max}$ when $DY > DY_{max}$. For example, for Elmo Lake, West and Stefan [23] set P_{min}, P_{max}, DY_{min}, and DY_{max} as 0.05 d^{-1}, 0.7 d^{-1}, 110 (20 April), and 140 (20 May); it means P_{min} occur on and before 20 April and P_{max} occur on and after 20 May.

Zooplankton density in the daytime is determined using Equation (13) including first-order reproduction and daytime predation:

$$ZP(t) = ZP(t-1) + ZP(t-1) \times Repro - PD_d(ZP(t-1) - ZP_{min})\frac{TD}{24} \qquad (13)$$

where *Repro* is the reproduction rate (dimensionless) and $ZP(t-1)$ is the zooplankton density in the previous day.

Nocturnal predation occurs during nocturnal migration at the day depth. The nocturnal predation rate is calculated for each layer between the day depth and the surface.

$$PD(t,I) = PD_n \left(ZP(t)\frac{V(IZ)}{V(I)} - ZP_{min} \right) \Delta T_g(I) \qquad (14)$$

where $PD(t, I)$ is the nocturnal predation rate in layer I during migration (#/m^3) and PD_n is the nocturnal predation rate (d^{-1}) as a constant input parameter. $ZP(t)$ is the zooplankton population in the day depth layer calculated using Equation (13), and ZP_{min} is the minimum concentration of zooplankton for predation to occur (#/m^3). $V(IZ)$ and $V(I)$ are the volumes of the day depth layer and layer I (m^3), respectively. $\Delta T_g(I)$ is the time that zooplankton spent in layer I during the night (d). The daytime $ZP(t)$ minus $PD(t, I)$ gives the zooplankton population for the next day.

The second part of zooplankton simulation is the simulation of phytoplankton grazing by zooplankton which begins with a vertical rise in the evening. Temperature and Chl*a* concentration affect grazing in the water layer (see Equation (9)). A Michaelis-Menten ratio is used to express the effect of Chl*a* concentration on grazing with a refugium effect for no grazing below a threshold Chl*a* concentration (Chl*a*$_{min}$ in Equation (9)).

2.4. Phosphorus Simulation

In most cases, phosphorus is known to be the primary nutrient controlling the trophic state of lakes in the Upper Midwest USA and Canada [56]. Phytoplankton can only use the soluble reactive phosphorus (SRP) which is composed of orthophosphate and polyphosphate ions. The model only simulates the readily accessible phosphorus and indirectly models organic phosphorus as detritus. Phytoplankton growth removes SRP from the water. Respiration releases phosphorus into the water column. Mortality does not directly release phosphorus to the water column but contributes to the detrital mass (BOD); phosphorus is released from the detrital mass through decay. Though diffusion of phosphorus occurs between layers, phosphorus is also transported indirectly between layers by phytoplankton and detritus settling. Figure 3 graphically represents/summarizes the SRP fate and transport modeled by MINLAKE2020.

Phosphorus, accumulated from the detrital biomass, sediment release (zero-order kinetics), and respiration, are used by the phytoplankton, in the presence of sunlight, for growth. Algae need both nitrogen and phosphorus for growth. However, phosphorus is particularly important for algal growth as it is usually in short supply compared to other nutrients. If it is assumed that nitrogen is in abundant supply, phosphorus becomes the only limiting nutrient for algal growth (green and blue-green algae), which is modeled in the application of MINLAKE2020. Uptake depends on the maximum growth rate, light limitation, nutrient limitation, Chl*a* concentration at that time, and the yield ratio of P to Chl*a* as shown in Equation (15).

The differential equation representing SRP fate and transport in a layer is given as Equation (15):

$$\underbrace{\frac{\partial P}{\partial t} - \frac{1}{A}\frac{\partial}{\partial z}\left(AK_z\frac{\partial P}{\partial z}\right)}_{\text{diffusion}}$$

$$\underbrace{+ Y_{PChla}\sum_{n=1}^{3} G_{max}f(T)\left[f(L):\frac{P}{K_P+P}:\frac{N}{K_N+N}\right]_{min} Chla}_{\text{growth of phytoplankton}} \quad (15)$$

$$\underbrace{- Y_{PBOD}K_{BOD}\theta_{BOD}^{T-20}BOD}_{\text{detrital decay}} \underbrace{- Y_{PChla}\sum_{n=1}^{3} K_r\theta_r^{T-20}Chla}_{\text{respiration}} \underbrace{- \frac{S_P}{A}\frac{\partial A}{\partial z}}_{\text{sediment release}} = 0$$

where K_{BOD} and θ_r are decay rate (d^{-1}) and corresponding temperature adjustment coefficient, respectively; S_P is the rate of phosphorus released at the water-sediment interface (g P/m^2/d) at anoxic condition and it is calibrated against available phosphorus/Chl*a*/DO profiles, Y_{PChla} is mass yield ratio of phosphorus to chlorophyll, and Y_{PBOD} is mass yield ratio of phosphorus to BOD.

A phosphorus/chlorophyll yield coefficient (Y_{PChla}) is used to determine the amount of phosphorus consumed during photosynthesis as well as the amount of phosphorus released during algal respiration. In MINLAKE98, the value of Y_{PChla} was derived from the mass yield coefficient of phosphorus to BOD divided by the mass yield coefficient of chlorophyll to BOD [23]. This value is 1.1 mg P/mg Chl*a* for Y_{PChla}, which was assumed to be constant in MINLAKE98, and is close to that presented by Thomann and Mueller [57] of

1.0 µg P/µg Chl*a*. In MINLAKE2020, Y_{Pchla} is set to 1.1 mg P/mg Chl*a* for all simulated lakes. However, the phytoplankton biomass does not depend on phosphorus solely, it also depends on the nitrogen concentration.

There are three source terms for phosphorus: detrital decay (death of phytoplankton), sediment release, and phytoplankton respiration. For many lakes which have a history of progressive eutrophication, the lake sediments have now become the primary source of phosphorus to the water. If the sediment-water interface is anoxic, phosphate ions go to the water at an increased rate, depending upon the concentration difference between porewaters and the overlying water [58]. MINLAKE2020 simulates SRP release back to water when the DO concentration becomes zero.

The daily zero-order internal phosphorus release (same as Equation (15)) was included in MINLAKE88 [20] and MINLAKE98 [46] before. CE-QUAL-W2 [59] includes the zero-order and first-order phosphorus release from sediment. EFDC model [60] has a governing equation for total phosphate including a sediment-water exchange flux of phosphate (g $P/m^2/d$) for the control volume at the bottom. ELCOM-CAEDYM [29] simulates phosphorus release from sediment as a function of temperature, DO and pH. There are some empirical models that quantified long-term internal phosphorus release for the whole lake (not for mass balance equation for a water layer or control volume), for example, Nurnberg [61] determined the internal phosphorus release per year after analyzing the data from various lakes. Stigebrandt et al. [62] and Stigebrandt and Andersson [63] developed a two-layer DIP (dissolved inorganic phosphorus) model including two mass balance equations for the Baltic proper and having a time step of one year. Stigebrandt and Andersson [63] used 47 years (1968–2014) of observational data to derive the phosphorus flux from anoxic bottoms, which is about 1.22 tons $P/km^2/year$ (or 1.22 g $P/m^2/year$) for the Baltic proper. Stigebrandt et al. [62] used a load-response model to explain the evolution of TP in the surface (0–60m) and bottom layers (60 m to bottom) from 1980 to 2005 and suggested that the average specific DIP flux from anoxic bottoms in the Baltic proper is about 2.3 g $P/m^2/year$. If the average anoxic period of the Baltic proper is 100–150 days per year, then the phosphorus flux would be 0.01–0.02 g $P/m^2/d$.

2.5. Nitrogen Simulation

When phosphorus is in excess (e.g., $P \gg K_p$), nitrogen can become the limiting nutrient for algae growth, which is not a usual scenario in many lakes. Nitrogen is available in two forms (ammonium NH_4, and nitrite plus nitrate, represented as NO_{2-3} in this paper). MINLAKE2020 models both ammonia and NO_{2-3} separately. Schematic diagrams of the NH_4 and NO_{2-3} sub models are given in Figures 4 and 5, respectively. Nitrogen N in Equations (9), (15), and (19) is the sum of NH_4 and NO_{2-3} concentrations (in mg N/L).

Diffusion of ammonia and NO_{2-3} occurs between layers. Nitrification is the biological oxidation process (assuming DO) of ammonia to nitrite followed by the faster oxidation of the nitrite to nitrate so that nitrite and nitrate are often modeled as one variable. The uptake of ammonia and nitrate due to phytoplankton growth is calculated as a zero-order sink term with a preference in the uptake of ammonium over the uptake of nitrate. It directly links to the growth of phytoplankton. Respiration of phytoplankton releases ammonia (Equation (16)) but not nitrate. Mortality of phytoplankton does not directly release ammonia or nitrate to the water but contributes to the detrital mass (BOD), and then ammonia is released through the decay of the detrital mass, as shown in Equation (16). Therefore, ammonia is also transported indirectly between layers by phytoplankton (Equation (9)) and detritus settling (Equation (18)). Ammonia and nitrite releases from the sediment are also modeled when DO is low (depending on the DO concentration of the overlying water). The model does not consider the atmospheric deposition of ammonia or NO_{2-3} and denitrification (the reduction of nitrate to nitrogen gas, N_2).

$$\underbrace{\frac{\partial NH_4}{\partial t} - \frac{1}{A}\frac{\partial}{\partial z}\left(AK_z\frac{\partial NH_4}{\partial z}\right)}_{diffusion} + \underbrace{K_{NI}\theta_{NI}^{T-20}(NH_4)}_{nitrification} + \underbrace{\sum_{n=1}^{3}G_{max}f(T)\left[f(L):\frac{P}{K_P+P}:\frac{N}{K_N+N}\right]_{min}}_{growth}$$

$$\times \underbrace{\left(\frac{NH_4}{K_{NH(n)}+NH_4}\right)\left(\frac{NH_4+NO_{2-3}}{K_{TN(n)}+NH_4+NO_{2-3}}\right)Y_{NChla}Chla}_{growth\ (contd.)} \quad (16)$$

$$\underbrace{-Y_{NHChla}\sum_{n=1}^{3}K_r\theta_r^{T-20}Chla}_{respiration} \underbrace{-Y_{NHBOD}K_{BOD}\theta_{BOD}^{T-20}BOD}_{detrital\ decay} \underbrace{-\frac{S_{NH}}{A}\frac{\partial A}{\partial z}\theta_{NH}^{T-20}}_{sediment\ release} = 0$$

$$\underbrace{\frac{\partial NO_{2-3}}{\partial t} - \frac{1}{A}\frac{\partial}{\partial z}\left(AK_z\frac{\partial NO_{2-3}}{\partial z}\right)}_{diffusion} \underbrace{- K_{NI}\theta_{NI}^{T-20}(NH_4)}_{nitrification}$$

$$+ \underbrace{\sum_{n=1}^{3}G_{max}f(T)\left[f(L):\frac{P}{K_P+P}:\frac{N}{K_N+N}\right]_{min}}_{growth}$$

$$\times \underbrace{\left[1-\left(\frac{NH_4}{K_{NH(n)}+NH_4}\right)\right]\left(\frac{NH_4+NO_{2-3}}{K_{TN(n)}+NH_4+NO_{2-3}}\right)Y_{NChla}Chla}_{growth\ (contd.)} \quad (17)$$

$$\underbrace{-\frac{S_{NO}}{A}\frac{\partial A}{\partial z}\theta_{NO}^{T-20}}_{sediment\ release} = 0$$

Equations (16) and (17) provide the governing equations to model NH_4 and $NO_{2\text{-}3}$ in each water layer. K_{NI} and θ_{NI} are the first-order nitrification rate (d^{-1}) and the temperature adjustment coefficient, respectively; $K_{NH(n)}$ and $K_{TN(n)}$ are the half-saturation constants for preferential uptake of ammonia over nitrate and nitrogen uptake for each algae class, respectively; S_{NH}, S_{NO}, θ_{NH}, and θ_{NO} are ammonia and nitrite release rates from sediment (g N/m^2/d) and corresponding temperature adjustment coefficients, respectively; Y_{NHBOD}, Y_{NHChla}, and Y_{NOChla} are the mass yield ratios of ammonia to BOD, ammonia to chlorophyll, and nitrate to chlorophyll, respectively; and K_{BOD} and θ_{BOD} are the first-order BOD or detritus decay rate (d^{-1}) and the temperature adjustment coefficient, respectively.

In MINLAKE2020 phytoplankton growth is simulated by external nutrient limitation and the model does not allow for nitrogen fixation (storage of excess nitrogen for later use) as MINLAKE88 did [49]. MINLAKE2020 simulates the release of ammonia from dead algae indirectly through detrital decay. In the normal range of ammonia concentration, when $K_{NH(n)}$ is small, the Michaelis-Menten ratio for ammonia in Equation (16) will be close to 1.0, and in Equation (17) one minus the Michaelis-Menten ratio will be small; therefore, it has preferential uptake of ammonia over nitrate.

2.6. BOD Simulation

BOD is an important parameter in the DO, phosphorus, and nitrogen cycles. The microbial decay or decomposition of organic matter, which is detritus from the mortality of phytoplankton in MINLAKE2020, consumes oxygen, and therefore, the amount of organic matter is represented as BOD, an oxygen equivalent. However, DO directly affects biological decay processes and phosphorus release under anoxic conditions. In the regional DO model [22], a constant rate for BOD was used for each simulation lake depending on the lake's trophic state. For the year-round lake water quality model [46], different constant rates of BOD for the open water seasons and winter ice cover periods were used over multiple years. However, BOD is an important parameter for nutrient cycles and DO cycle and is affected by mortality, organic decay, diffusion, and advection (Figure 6) and simulated separately in MINLAKE2020.

The differential equation representing BOD in a layer is given as Equation (18):

$$\frac{\partial BOD}{\partial t} - \underbrace{\frac{1}{A}\frac{\partial}{\partial z}\left(AK_z\frac{BOD}{z}\right)}_{\text{diffusion}} + \underbrace{\frac{v_B}{A}\frac{\partial (A \times BOD)}{\partial z}}_{\text{advection}}$$

$$-\underbrace{\frac{1}{Y_{CHBOD}}\sum_{n=1}^{3}K_m\theta_m^{T-20}Chla}_{\text{phytoplankton mortality}} + \underbrace{K_{BOD}\theta_{BOD}^{T-20}BOD}_{\text{detrital decay}} = 0 \qquad (18)$$

where v_B and Y_{CHBOD} represent detritus settling velocity (m/d) and the mass yield ratio of Chla to BOD, respectively. In the model, BOD is increased from two sources (Figure 6). First, detritus travels to adjacent layers via diffusion. Secondly, the mortality of the phytoplankton adds to the detritus concentration. BOD has two sink terms (Figure 6): advection (settling in the vertical direction) and organic decay. Detritus falls from the concerning layers and goes to another layer or the sediment. The microbial decay of organic matter is a function of the detrital mass expressed in oxygen equivalents (BOD). The mortality of cells and a fraction of the grazed phytoplankton are converted from Chla concentrations to oxygen equivalents using the constant carbon/Chla ratio and stoichiometric relationships. The result is a one-to-one correspondence between detrital decay and the utilization of oxygen. This cycle is very important for nutrient calculation as it directly adds to the nutrient load through detrital decay.

2.7. DO Simulation

DO is one of the vital parameters of lake water quality simulation. Aquatic organisms and fish depend on the availability of DO in the waterbody [64]. A schematic diagram of the processes contributing to the DO concentration is given in Figure 7. It shows that DO is added to a water layer through diffusion and photosynthesis; and is removed by respiration of algae and zooplankton, detrital decay (BOD), sediment oxygen demand (SOD), and nitrification. The surface reaeration can add or remove DO depending on whether surface DO is less or greater than saturated DO (a function of surface temperature and lake elevation). Phytoplankton (modeled as Chla) growth can add DO to the water layer through photosynthesis to the point where water could be supersaturated with DO in some cases. These dynamic processes can happen over time scales of less than one day (the time step of the MINLAKE2020 simulation). Therefore, the model outputs DO profiles as an integration of different physical (e.g., mixing), chemical, and biological processes over the day. DO removal from the water layer through phytoplankton respiration is simulated to occur at a constant rate throughout the day while photosynthesis occurs only during the hours with solar radiation. In MINLAKE2020, the adjustments for low DO levels on SOD, BOD, and algal respiration follow Edwards and Owens's [65] formula. SOD is calculated for each layer, and it is treated as a sink term in the one-dimensional (vertical) transport equation [22,57]. Oxygen uptake of the sediment depends on the area and composition of bottom materials in contact with the water [22,66].

The differential equation representing DO dynamics in a layer is given as Equation (19):

$$\frac{\partial DO}{\partial t} - \underbrace{\frac{1}{A}\frac{\partial}{\partial z}\left(AK_z\frac{\partial DO}{\partial z}\right)}_{diffusion} + \underbrace{K_{BOD}\theta_{BOD}^{T-20}BOD}_{BOD} + \underbrace{\frac{S_b}{A}\frac{\partial A}{\partial z}\theta_{SOD}^{T-20}}_{sediment\ oxygen\ demand}$$

$$+ \underbrace{\frac{1}{Y_{CHO2}}\sum_{n=1}^{3}\{K_r\theta_r^{T-20} - G_{max}f(T)\left[f(L):\frac{P}{K_P+P}:\frac{N}{K_N+N}:\frac{Si}{K_{Si}+Si}\right]_{min}\}Chla}_{respiration\ and\ photosynthesis}$$

$$+ \underbrace{\frac{1}{Y_{NHO2}}K_{NI}\theta_{NI}^{T-20}NH_4}_{nitrification} - \underbrace{k_e(DO_{sat} - DO)A(1)/V(1)}_{reaeration\ (surface\ only)} + \underbrace{\frac{TD}{24}K_{zr}\theta_{zr}^{T-20}ZP\frac{0.001}{V_{iz}}}_{zooplankton\ respiration} = 0 \quad (19)$$

where K_{zr} and θ_{zr} are the zooplankton respiration rate (d^{-1}) and the temperature adjustment coefficient, respectively; $K_{NH(n)}$ and $K_{TN(n)}$ are the half-saturation constant for preferential uptake of ammonia over nitrate and for nitrogen uptake for each algae class, respectively; S_b and θ_{SOD} are sediment oxygen demand of sediment (g O/m^2/d) and corresponding temperature adjustment coefficient, respectively; k_e is the surface oxygen transfer coefficient (m/d), A(1) and V(1) are horizontal area (m^2) and lake volume (m^3) for the first or surface layer; Y_{NHO2} and Y_{CHO2} are the mass yield ratios of ammonia to oxygen and chlorophyll to oxygen, respectively. Calibration of the sediment oxygen demand is very important for simulating DO in the hypolimnion.

2.8. Interaction and Connection among Modeling Variables

This study is focused on the internal nutrient dynamics and its interaction/connection to the phytoplankton, BOD and DO dynamics in the lakes; therefore, the inflow and outflow sub-model was disabled. The MINLAKE2020 DO model is different from that of MINLAKE2012, where daily Chla and BOD were specified as model input based on lake trophic status. For MINLAKE2020, several modifications were made to simulate phosphorus, nitrogen, Chla, and BOD concentrations and zooplankton activity in the DO simulation. The photosynthetic oxygen production calculation was modified, and the nitrification and zooplankton respiration were added to the DO model while the rest of the terms were the same as in MINLAKE2012. In MINLAKE 2020, photosynthetic oxygen production is dependent on the simulated daily Chla concentration rather than the data driven Chla pattern and annual mean concentration used in MINLAKE2012.

The diffusion of DO occurs between the water layers in the metalimnion and hypolimnion. The spring and fall overturn periods for temperate lakes completely mix water columns and make all modeling variables distributed uniformly along with the depth. Since zooplankton usually spend the largest amount of their time in the day depth layer, zooplankton respiration is simulated at the day depth only. BOD removes oxygen from the water layer through the decay of detritus which is accumulated from phytoplankton mortality and removed by settling (Figure 6). Nitrification removes oxygen from the water layer through the conversion of ammonia to nitrite then to nitrate. Nitrification is applied to the DO model only if nitrogen is simulated.

From Figures 1–7 and from Equations (9)–(19), one can see the complex interactions and connections among modeling variables. For example, when the settling velocity of the phytoplankton group is increased, it reduces Chl*a* concentration and in turn, affects detritus (BOD) and phosphorus, but phosphorus goes back to affect phytoplankton growth. Additionally, DO concentrations in different water layers connect/integrate changes in water temperature, nutrients, and phytoplankton dynamics (Figure 1). The sediment oxygen demand is an important factor for DO mass balance and results in the anoxic condition in the hypolimnion that leads to phosphorus release from sediment. Therefore, exploring/understanding the internal cycles/dynamics of nutrients/phytoplankton and their interactions in different lakes helps us to gain insights of lake ecosystem and then develop appropriate restoration strategies.

2.9. Model Coefficients and Parameters

MINLAKE2020 model was designed to simulate small lakes ($A_s < 25$ km^2) provided that the user specifies the input data and calibration parameters accordingly. Variation in lake characteristics is reflected in model input data/parameters. Lake bathymetry and weather data (depending on lake geographic location) need to be supplied to the model. For many lakes in Minnesota, depth or elevation contour lines can be downloaded from the Minnesota Department of Natural Resources (MN DNR) LakeFinder website (https://www.dnr.state.mn.us/lakefind/index.html, accessed on 5 June 2019), from which horizontal areas at different elevations/depths can be determined. The weather data include daily air temperature (°F), dew point temperature (°F), wind speed (mph), solar radiation (Langley), sunshine percentage, and precipitation including rainfall (cm) and snowfall (mm). To facilitate the comparison of DO simulation results, in MINLAKE2020, the user can run the DO simulation in two ways: (1) using MINLAKE 2012 regional DO model, i.e., nutrients and Chl*a* are not simulated (called the RegDO model); (2) using MINLAKE2020 (called the NCDO model). This model first simulates nutrients and Chl*a* and then DO [67]. Table 1 lists nutrient, Chl*a*, and DO calibration parameters in MINLAKE2020 and includes descriptions and effects on specific model results.

Table 1. Nutrient, Chl*a*, and DO calibration parameters for MINLAKE2020.

Parameters	Unit	Effect on Model Results	Description of the Parameter
K_{BOD}	d^{-1}	P, BOD, and DO profiles	Detrital decay rate
S_b	g/m^2/d	DO profiles	SOD coefficient at 20 °C
EMCOE (2)		DO profiles	Multiplier for SOD below the euphotic zone
K_r [1]	d^{-1}	Chl*a*, P, BOD, DO profiles	Respiration rate of algae
G_{max} [1]	d^{-1}	Chl*a*, P, DO profiles	Maximum growth rate of algae
K_m [1]	d^{-1}	Chl*a* and BOD profiles	Non-predatory mortality rate of algae
K_P [1]	m/d	Chl*a*, P and DO profiles	Half saturation coefficient for phosphorus
S_P	g/m^2/d	P and DO profiles	Sediment phosphorus release rate
v_c [1]	m/d	Chl*a* and DO profiles	Settling velocity of algae
v_B	m/d	Chl*a* and DO profiles	Settling velocity of detritus
T_{opt} [1]	°C	Chl*a*, P and DO profiles	Optimum temperature for growth of algae
T_{max} [1]	°C	Chl*a*, P and DO profiles	Maximum temperature for growth of algae
T_{min} [1]	°C	Chl*a*, P and DO profiles	Minimum temperature for growth of algae
ZP	m^{-3}	Chl*a* and DO profiles	Zooplankton population
ZP_{min}	m^{-3}	Chl*a* and DO profiles	Minimum zooplankton population for predation
$Chla_{min}$ [1]	mg Chl*a*	Chl*a* and P profiles	Minimum chlorophyll-a for grazing

Note: [1]—these parameters are calibrated for each algal class in MINLAKE2020.

The number of algal classes and the light attenuation coefficient are important input parameters for the simulation. Moreover, the snow and ice model require various coefficients (e.g., snow and ice density, specific heat, thermal conductivity, etc.) as input parameters, which has been well tested in the previous studies [68]. The temperature adjustment coefficients for BOD, photosynthesis, respiration, and sediment oxygen demand were set to 1.047 [69], 1.066 [57], 1.047 [20], 1.065 [70], respectively. Mass ratios of Chl*a* to oxygen, phosphorus to oxygen, and phosphorus to chlorophyll-a are 0.0083, 0.0091, and 1.1, respectively. The inclusion of phytoplankton and zooplankton simulation in MINLAKE2020 calls for many additional input parameters or model coefficients. Most of the zooplankton-related coefficients were taken from West and Stefan's study [23]. Zooplankton respiration rate was set to 0.002 d^{-1} and the reproduction rate was 0.02 d^{-1}. Minimum light intensity for zooplankton predation, light intensity for maximum zooplankton predation, Julian day for the end of low predation period, and Julian day for the beginning of maximum predation were set to 0 $\mu E\ (m^2 s)^{-1}$, 0.1 $\mu E\ (m^2 s)^{-1}$, 110 (20 April), and 140 (20 May), respectively. Minimum seasonal day time predation rate, maximum seasonal day time predation rate, and overnight predation rate were set to 0.05 d^{-1}, 0.7 d^{-1}, and 0.03 d^{-1}, respectively. For any lake, these parameters were kept constant whereas some of the model parameters were updated/calibrated based on the comparison of simulation results with observed data. The first four parameters were used in the regional DO model, but the respiration rate K_r was not calibrated (constant).

3. Lakes Simulated

In this study, six lakes (Table 2) were selected for the model calibration, the sensitivity analysis of model parameters/coefficients, and understanding the interaction/connection among seasonal variations of nutrients and chlorophyll-a: two shallow lakes (Pearl and Carrie), two medium-depth lakes (Riley and Thrush), and two deep lakes (Carlos and Elmo). The maximum depths range from 5.6 to 50.0 m. All six lakes are located in northeastern Minnesota since they have the necessary data for the study. The nearest weather station to the lake was selected for providing weather data: St. Cloud Regional Airport for Pearl, Carrie, and Carlos lakes; Duluth International Airport for Thrush and Riley lakes; and Minneapolis/St. Paul International Airport for Elmo Lake. The geometry ratio ($GR = A_s^{0.25}/H_{max}$, A_s in m^2, and H_{max} in m being the surface area and the maximum depth of the lake) is a very important characteristic parameter of a lake that is related to stratification, lake habitat, etc. [71]. The lake geometry ratio is between 0.75 and 7.53. The lower the geometry ratio, the stronger the lake stratification. Two medium-depth (6 m < H_{max} ≤ 20 m, [72]) and two deep lakes (H_{max} > 20 m, [72]) selected for the study are strongly stratified (geometry ratio less than 3), one medium-depth and one shallow lake are weakly stratified (geometry ratio between 3 and 10). Based on observed Chl*a* concentration, Pearl, Carrie, and Riley lakes are eutrophic (mean Chl*a* > 10 µg/L [73]), Elmo and Carlos are mesotrophic lakes (mean Chl*a* between 4 and 10 µg/L [73]), and Thrush is an oligotrophic lake. The nutrient model was calibrated and validated based on available measured water temperature, chlorophyll-a, phosphorus and DO profile data on particular days, downloaded from the LakeFinder website.

Table 2. Characteristics of six study lakes in Minnesota.

Lake	Surface Area, As, (km²)	Max. Depth H_{max}, (m)	Geometry Ratio (m)$^{0.5}$	Mean Chl*a* (µg/L)	Trophic Status	Simulation Years	Number of Profile Days (Data Points)
Pearl	3.05	5.55	7.53	16.91	Eutrophic	2010–2012	15 (134)
Carrie	0.37	7.90	3.12	6.71	Mesotrophic	2007–2010	36 (342)
Riley	1.19	14.9	2.22	24.00	Eutrophic	1985–1987	16 (148)
Thrush	0.048	14.63	1.01	1.71	Oligotrophic	2008–2015	18 (100)
Elmo	1.039	42.63	0.75	4.45	Mesotrophic	1989–2009	70 (864)
Carlos	10.54	50.00	1.15	3.84	Oligotrophic	2008–2015	54 (308)

4. Result and Discussion

4.1. Model Calibration

For MINLAKE 2020 model application to the six study lakes, the temperature model was calibrated first and then the nutrient model was calibrated. Temperature model calibration ensured that thermal and mixing dynamics were modeled accurately because water temperature and mixing dynamics directly affect nutrients, Chl$a$, and zooplankton processes (Equations (9)–(19)). The wind sheltering coefficient and the multiplier for diffusion coefficient in metalimnion are the main calibration parameters for temperature modeling. Although MINLAKE2020 has an integrated nitrogen model, for this study, only phosphorus was simulated since phosphorus is the limiting nutrient in these six lakes. Green algae and blue-green algae were simulated separately and then combined to represent the total chlorophyll-a concentration. The MINLAKE2020 development also included the inflow and outflow subroutines from MINLAKE88, which were tested/verified to ensure they function properly; however, the inflow/outflow function was not activated for the simulation of the six study lakes (Table 2). Certain inflow and outflow were reported for Carrie, Pearl, and Carlos Lake [74–76] whereas inflows in the other three lakes were minor, and the inflow quality (nutrients and phytoplankton) data were scarce. These approximations are appropriate since the study objective is to examine/understand the internal dynamics and cycles of nutrients over multiple years in six lakes with different stratification and trophic characteristics.

Figure 8 shows an example of the calibration results of water temperature and DO time series at two depths (1 and 7 m) at Lake Carrie including measured data. During 2008–2010, Lake Carrie had measured water temperature and DO profile data for 36 days or 342 data points in total. MINLAKE2020 simulated water temperature and DO with a root mean square error of 1.75 °C and 1.95 mg/L, respectively. Corresponding regression coefficients of measured versus simulated (R^2) are 0.99 and 0.93, respectively. The statistical results summarized in Table 3 show that for the six lakes, MINLAKE2020 model performed better than MINLAKE2012, especially for DO simulations, when simulated profiles were compared with observed profiles. The main reason for this improvement is the simulation of Chl$a$ concentration on daily time step rather than using the specified pattern of observed data. Table 3 shows the model performance improved significantly with the NCDO model in Carlos and Thrush lakes. The average root-mean-square error (RMSE) of DO simulations in six lakes from MINLAKE2020 decreased by 24.2%, and average Nash-Sutcliffe efficiency (NSE) [77] also increased with respect to MINLAKE2012. Chlorophyll concentration affects the solar radiation attenuation in the water column (Equation (8)) and then affects water temperature simulation as shown in Table 3 even though the average RMSE, NSE, and R^2 for temperature (regression coefficient of measured versus simulated) from the two models are almost the same.

For Chl$a$ simulation of six lakes, RMSE ranges from 0.0006 to 0.0276 mg/L. Figure 9 shows an example comparison between simulated Chl$a$ of MINLAKE2020 (NCDO model) and specified Chl$a$ from MINLAKE2012 (RegDO model) for Lake Elmo. From 1980 to 2018, there were 74 days in 10 years with measured temperature and DO profiles (1506 data points) for model calibration but no profile data in 2007 and 2009. The average chlorophyll concentration was 0.0075 mg/L in 74 days but 0.0036 mg/L over 10 days in 2008. Even though there were no profile data in 2009, we identified some Chl$a$ data in 2009 as shown in Figure 9. MINLAKE2012 uses the annual mean Chl$a$ concentration and seasonal variation patterns [22] (depending on trophic status) to specify daily Chl$a$ for DO simulation. Therefore, the RegDO model had higher Chl$a$ in 2007 and 2009 due to the lack of available profile data in these two years whereas the NCDO model predicts the Chl$a$ reasonably well in 2008 and 2009 when comparing with data.

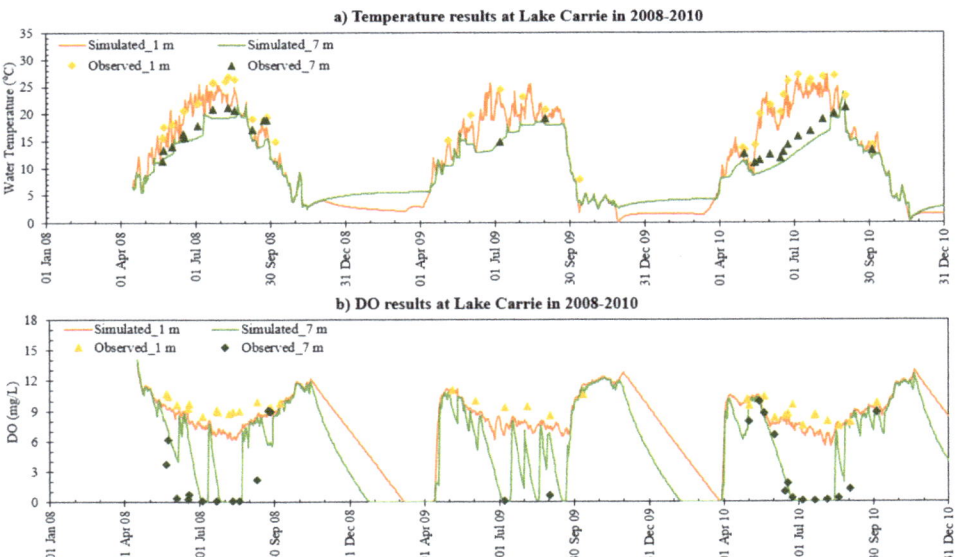

Figure 8. Time series of simulated (**a**) water temperature and (**b**) DO at two depths in Lake Carrie in 2008–2010.

Table 3. Statistical parameters for six lakes when simulated profiles were compared with observed.

Lake Name	NCDO Model (MINLAKE2020)					
	Water Temperature			Dissolved Oxygen		
	RMSE [a] (°C)	NSE [b]	R^2 [c]	RMSE (mg/L)	NSE	R^2
Elmo Lake	0.98	0.98	0.99	2.02	0.70	0.92
Carlos Lake	1.66	0.83	0.97	2.39	0.61	0.90
Riley Lake	1.50	0.50	0.98	1.79	0.80	0.93
Thrush Lake	1.88	0.70	0.95	2.43	0.40	0.92
Carrie Lake	1.75	0.68	0.99	1.95	0.70	0.93
Pearl Lake	1.30	0.87	0.97	3.42	−0.12	0.87
Average ± STD [d]	1.51 ± 0.33	0.76 ± 0.17	0.97 ± 0.01	2.33 ± 0.59	0.52 ± 0.34	0.91 ± 0.02
Lake Name	RegDO Model (MINLAKE2012)					
Elmo Lake	1.03	0.98	0.99	1.89	0.70	0.92
Carlos Lake	1.52	0.85	0.98	4.15	−0.19	0.85
Riley Lake	1.55	0.5	0.98	2.61	0.55	0.91
Thrush Lake	2.01	0.69	0.95	2.91	0.1	0.92
Carrie Lake	1.76	0.64	0.99	1.98	0.69	0.94
Pearl Lake	1.04	0.97	0.98	3.30	0.01	0.89
Average ± STD	1.49 ± 0.39	0.77 ± 0.19	0.98 ± 0.02	3.08 ± 1.11	0.11 ± 0.69	0.90 ± 0.03

Note: [a]—RMSE stands for Root Mean Square Error between simulated and observed, [b]—NSE for Nash-Sutcliffe Efficiency [77], [c]—R^2 stands for regression coefficient of measured versus simulated, [d]—STD for Standard Deviation.

Some lakes, such as Lake Elmo and Carrie Lake (Table 3), do not exhibit a noticeable change in simulated DO concentrations based on the model used for simulation. Some lakes, e.g., Pearl Lake, exhibit a noticeable change in simulated DO concentration depending on the model (RegDO or NCDO model). Figure 10a,b show the time series of DO concentrations simulated by the NCDO model and RegDO model including observed DO at the surface (1 m) and near the bottom (5 m) of Pearl Lake in 2010–2012, respectively. The RegDO model somewhat overpredicts surface DO concentration but the NCDO model underpredicts DO. For the bottom DO, the NCDO model does a better job while the RegDO model overpredicts DO. When BOD is simulated, the winter DO decreases, predicted by MINLAKE2020, are smaller than those predicted by MINLAKE2012 when BOD is specified as a part of the model inputs. There are very limited data for P comparison between simulated and observed, and RMSE ranges from 0.005 to 0.036 mg/L. DO simulation is an overall model performance indication (Table 3).

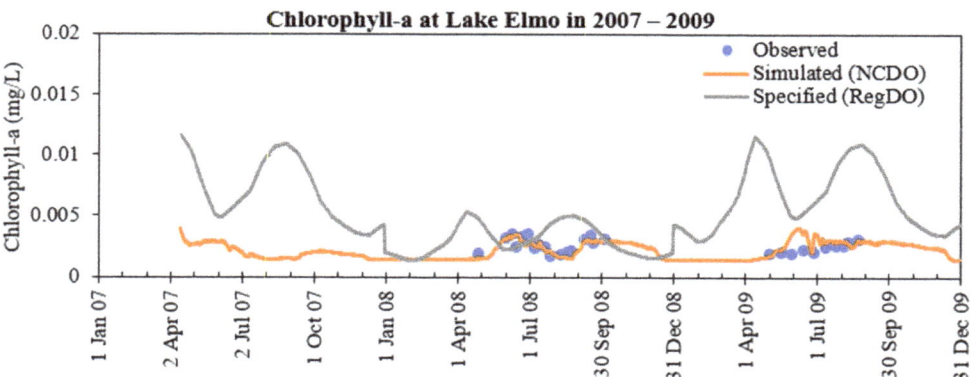

Figure 9. Simulated (NCDO model), specified (RegDO model), and observed chlorophyll-a near the surface in Lake Elmo in 2007–2009.

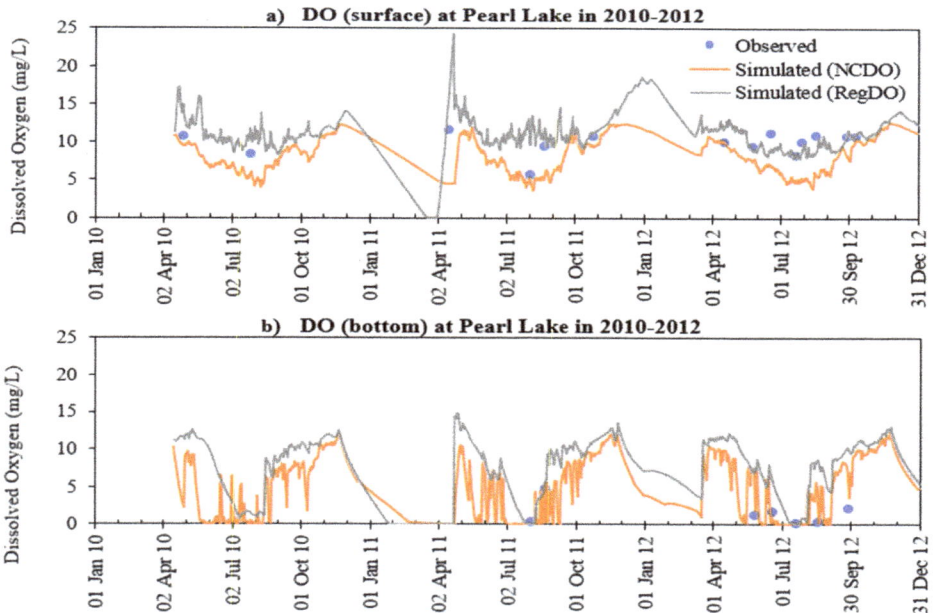

Figure 10. Simulated and observed dissolved oxygen near (**a**) the surface (1 m) and (**b**) bottom (5 m) in Pearl Lake in 2010–2012, respectively.

4.2. Chla and Phosphorus Profiles

Lake Elmo was extensively monitored in 1988 by the Metropolitan Council [65] and had measured phosphorus concentration data at five depths (0, 8, 16, 24, and 32 m) and DO data at 31 depths for open water season. The comparison between simulated and observed concentrations for Chla, phosphorus, and DO on three days in 1988 is presented in Figure 11. Since Elmo is a deep lake and solar radiation cannot penetrate below the euphotic zone, the Chla concentration becomes zero in the deep layers. On 11 April 1988, the lake is more or less well mixed and phosphorus concentration did not vary much throughout the depth but DO concentration gradually declined along with depth due to the contribution of more sink terms (Equation (18)) but for the profile plot, the slope was not steep. The Chla concentration is highest at the surface and did not vary much throughout the depth. On 18 May 1988, the stratification increases and the simulated DO at the bottom is near zero. The Chla concentration is not maximum at the surface, but at 8 m depth from the surface. The phosphorus concentration is higher at the deeper layers because of detrital decay and phytoplankton respiration. On 19 October 1988, the phosphorus near the surface layer is being used by the phytoplankton for growth, the lake became strongly stratified and the bottom layers of the lake become anoxic so that the phosphorus release from sediment contributed to the higher phosphorus concentration at the deeper depths, which increased along with depth from metalimnion and hypolimnion.

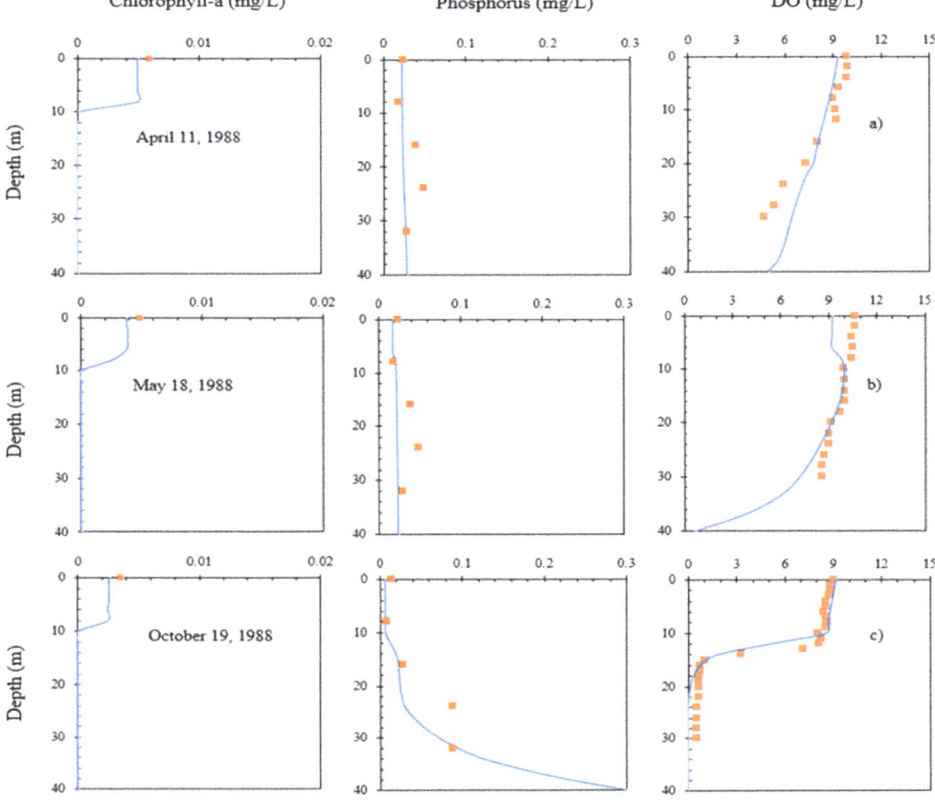

Figure 11. Simulated (**line**) and observed (**dots**) chlorophyll-a, phosphorus, and dissolved oxygen profiles for Lake Elmo on (**a**) 11 April, (**b**) 18 May, and (**c**) 19 October 1988.

Thrush Lake has measured Chl*a* data at both the epilimnion and hypolimnion. The comparison between simulated and observed concentrations for Chl*a* and DO on three days in 1986 is plotted in Figure 12. Thrush is an oligotrophic lake with lower oxygen demands (low BOD and SOD); therefore, the simulated and observed bottom DO is greater than zero, and there is no phosphorus release from sediment in all three days (Figure 12). Since no phosphorus data were available, only simulated phosphorus was plotted and has no increase in the hypolimnion. Due to lower light attenuation coefficients, solar radiation can penetrate through the deepest layers of the lake. The calculated euphotic depth (at 1% surface solar radiation) is equal to or greater than the maximum depth (14.6 m) for all of the three selected dates; therefore, the simulated Chl*a* is not zero near the bottom but increases below the mixed layer, especially on 17 June 1986. On 13 May 1986, the Chl*a* and phosphorus concentration did not vary much throughout the depth. The DO concentration gradually decreased below the mixed layer on 13 May and 17 June 1996, due to the contribution of more sink terms. On 14 October 1986, the lake became well mixed due to the fall overturn and had uniform distribution for simulated Chl*a*, phosphorus, and DO. The model underpredicts the mixed layer depth on 13 May and then overpredicts DO at the hypolimnion.

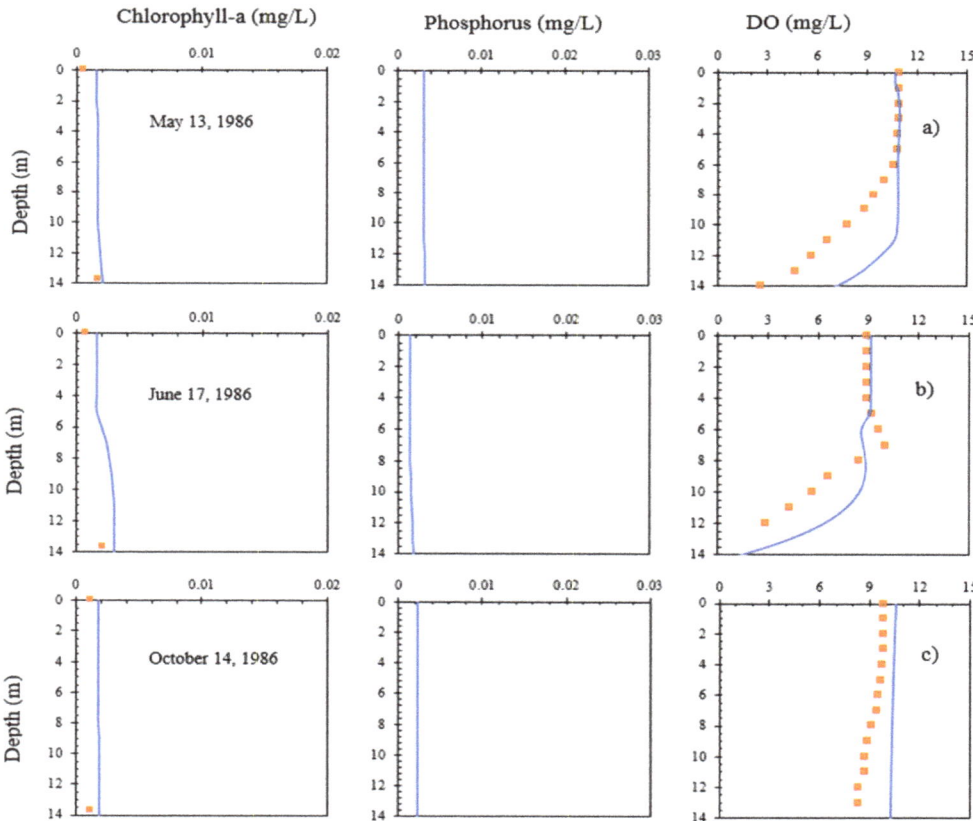

Figure 12. Simulated (**line**) and observed (**dots**) chlorophyll-a, phosphorus, and dissolved oxygen profiles for Thrush Lake on (**a**) 13 May, (**b**) 17 June, and (**c**) 14 October 1986.

4.3. Chlorophyll-a and Phosphorus Interaction

Figures 13 and 14 show examples of simulated time series of P, Chl*a*, and DO with observed data for a deep lake (Elmo) and a shallow lake (Carrie), respectively. Figure 13 shows simulated and observed Chl*a* and phosphorus (P) at three depths (at 1 m, 20 m, 40 m from the surface) of Elmo Lake from 16 April 2007 to 31 December 2009. This is a continuous year-round simulation including two open-water seasons and two ice cover periods, which was not done before using MINLAKE models for Chl*a* and P simulation. The first year of simulation is considered as model warm-up period and the results may have more uncertainties due to the assumed initial conditions, for example, phosphorus concentration from the water surface to 20 m was low during the open water season (Figure 13a,b). The simulated ice cover was from 5 December 2007 to 16 April 2008, 7 December 2008 to 10 April 2009 for Elmo Lake, which is marked by blue shaded regions.

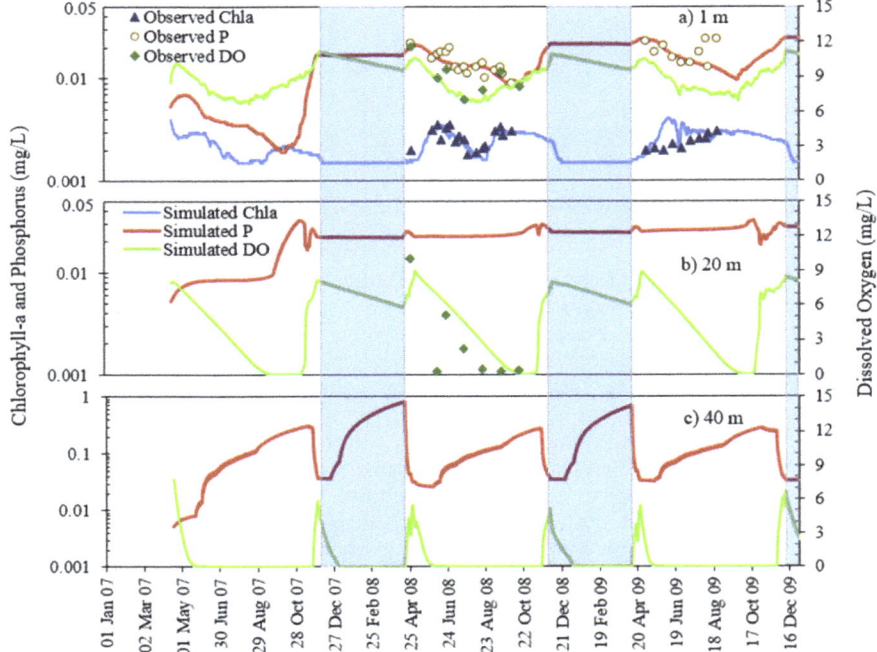

Figure 13. Simulated and observed chlorophyll-a, phosphorus, and dissolved oxygen concentration at (**a**) 1 m, (**b**) 20 m, (**c**) 40 m depth from the surface at Elmo Lake in 2007–2009 (blue shaded area represents the ice cover period). The scale for the major y axis in (**c**) is larger than one in (**a**,**b**).

Near the water surface (1 m), DO concentrations are near saturation as a function of temperature (lowest DO in the middle of summer) and range from 6.69–11.13 mg/L. From late October to late November, before the ice starts to form at the surface of the lake, phosphorus at the surface and other surface layers start to increase due to more mixing and fall overturns. This increase is more evident in 2007 when phosphorus was low in the summer. During the ice cover period, the phosphorus concentration becomes stable at 1 m and 20 m as the organic processes (photosynthesis) become slow due to the near-zero water temperature and/or low light (attenuated by snow cover). After the ice melts out in late spring, the simulated phosphorus concentrations at 1 m increase for a brief period due to spring overturn and then, start to decrease as a result of the phosphorus uptake by phytoplankton. During the early summer (May) of 2008 and 2009, simulated Chl*a* concentration increases gradually from 0.0016 mg/L to a maximum of 0.0033 mg/L (observed on 15 June).

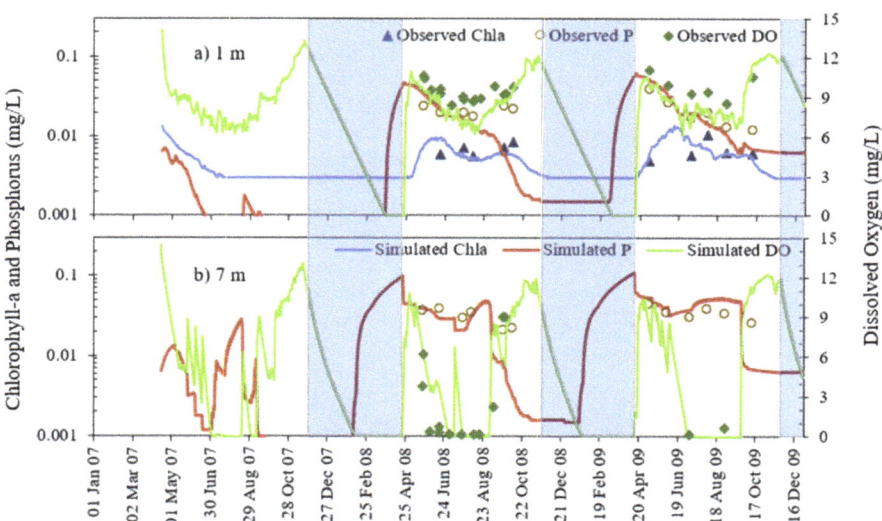

Figure 14. Simulated and observed chlorophyll-a, phosphorus, and dissolved oxygen concentration at (**a**) 1 m and (**b**) 7 m depth from the surface at Carrie Lake in 2007–2009 (blue shaded area represents the ice cover period).

At the deep layers (e.g., 20 and 40 m), DO becomes anoxic during the summer at deep hypolimnion since Elmo Lake is strongly stratified. Anoxic periods at 20 m and 40 m were on average 12 days and 210 days per year from 2007 to 2009, respectively. In the deeper layers, in addition to detrital decay and phytoplankton respiration, sediment release could add up to the available phosphorus. Since there is a long period of anoxic condition at 40 m during the summer, early fall, and some part of the ice cover period, phosphorus release from sediment contributes to a major portion of phosphorus increase. Phosphorus peaks in deepest layers were simulated to occur just after the anoxic condition ends and before the fall or spring mixing/overturns. These overturns sharply reduce phosphorus in very deep layers (e.g., 40 m) and increases phosphorus in other shallower layers. The calculated euphotic depth is 9 m for the simulation period; therefore, there is no photosynthesis below this depth and simulated Chl*a* is zero at 20 m and 40 m. Chl*a* is typically measured near the surface; therefore, there is no measured Chl*a* at the deep depths to compare with simulated values.

Figure 14 shows simulated and observed P, Chl*a*, and DO at two depths (at 1 m, 7 m from the surface) of Carrie Lake from 16 April 2007 to 31 December 2009. The calculated euphotic depth ranges from 3.75 m to 4.23 m from the surface; therefore, the Chl*a* is zero at the bottom layers (e.g., 7 m). Phosphorus release from sediments due to anoxic conditions under the ice cover periods for both 2008 and 2009 winter was not only at 7 m but also at 1 m (does not happen in deep lakes such as Elmo Lake) and triggered algal bloom at 1 m in early summer of 2008 and 2009 (after ice melting). During the summer of 2008 and 2009, the concentration of DO gradually increases at the surface, and sediment phosphorus release decreases which results in a gradual decrease in phosphorus concentration. The DO concentration decreases with depth because of no photosynthesis in the deeper layers (below the euphotic zone) plus sedimentary oxygen demands. Simulated DO at 7 m has some fluctuations in the summer of 2008 and 2009 due to short period strong mixing and results in anoxic conditions only in a few days. The simulated high phosphorus concentrations directly correspond to the simulated anoxic DO conditions in the 2008 and 2009 winters (Figure 14b).

4.4. Sensitivity Analysis

MINLAKE2020 model is sensitive to several calibration parameters. The model was first calibrated with a regression coefficient (R^2 for measured versus modeled profile data) of 0.8972 for DO simulation; then, only one model parameter was changed at a time and the regression coefficient for each new run of DO simulation was determined. Table 4 lists the calibration parameters with the calibrated and uncalibrated values and the regression coefficients of DO simulations with uncalibrated values for Lake Elmo in 2007–2009. In Figures 15 and 16, sensitivity analysis graphs are plotted for five parameters: maximum photosynthesis rate G_{max} (used in Equations (9), (15)–(17) and (19)), sediment phosphorus release rate S_P (used in Equation (9)), half-saturation coefficient of phosphorus K_P (used in Equations (9), (15)–(17) and (19)), minimum Chl$a$ for grazing $Chla_{min}$ (used in Equation (9)), and settling velocities of algae (v_c) and BOD (v_B) (used in Equations (9) and (18)) for Elmo and Pearl Lake, respectively. Units of these model parameters are given in Table 4, and the below corresponding equations.

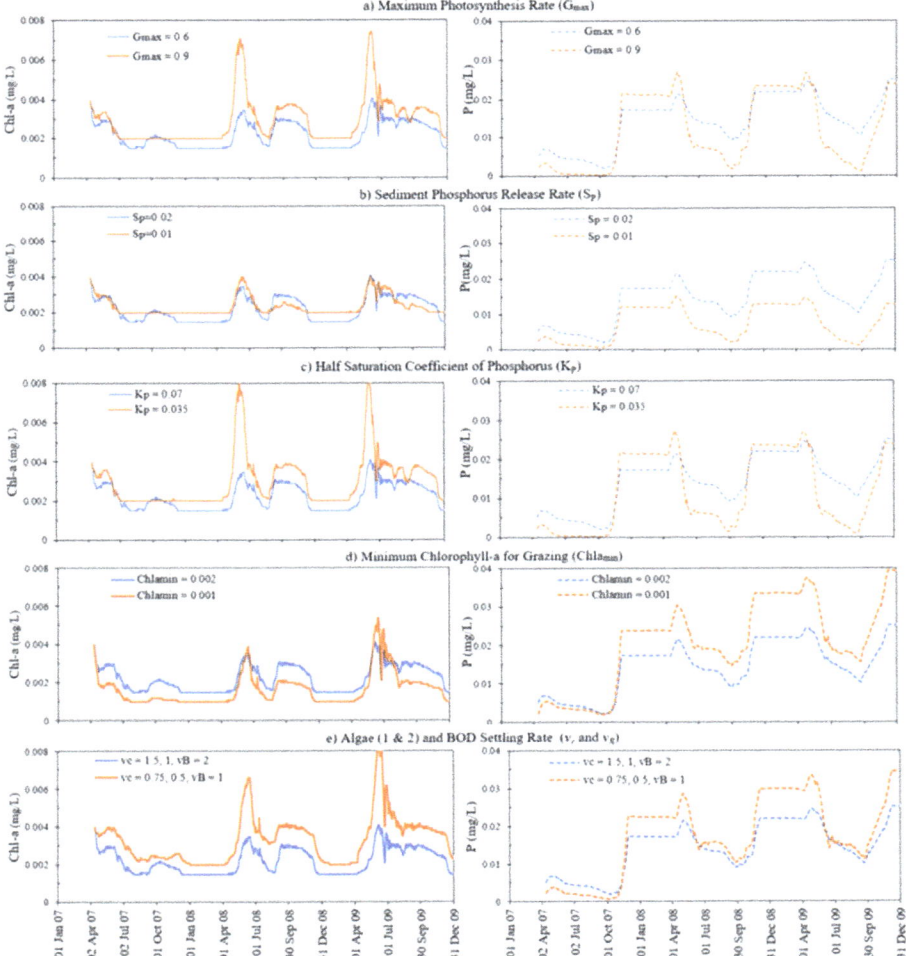

Figure 15. Sensitivity analysis of Chl$a$ (**left panels**) and phosphorus (**right panels**) on (**a**) maximum photosynthesis rate G_{max}, (**b**) sediment phosphorus release rate S_P, (**c**) half saturation coefficient of phosphorus K_P, (**d**) minimum Chl$a$ for grazing $Chla_{min}$, and (**e**) algae and BOD settling rates v_c, and v_B for simulation of Elmo Lake in 2007–2009.

Table 4. Calibration parameters used in MINLAKE2020 for Lake Elmo with regression coefficients of the DO simulation.

Sensitive Parameters	Description of Parameter	Uncalibrated Value	Calibrated Value	Regression Coefficient
K_{BOD}	Detrital decay rate (d^{-1})	0.5	0.05	0.8955
S_b	SOD coefficient at 20 °C (g O/m^2/d)	0.5	1.7	0.7449
EMCOE (2)	Multiplier of SOD below euphotic zone [-]	3	1	0.8931
K_r	Algal respiration rate (d^{-1})	0.03, 0.03	0.06, 0.06	0.8924
G_{max}	Maximum growth rate of algae (d^{-1})	0.9, 0.9	0.6, 0.6	0.8966
K_m	Algal mortality rate (d^{-1})	0.015, 0.015	0.03, 0.03	0.8944
K_P	Half saturation coefficient of phosphorus (mg/L)	0.035, 0.035	0.07, 0.07	0.895
S_P	Benthic phosphorus release rate (g P/m^2/d)	0.01	0.02	0.8931
v_c	Settling velocity for algae (m/d)	0.05, 0.05	0.15, 0.1	0.8931
v_B	Settling velocity for detritus (m/d)	0.05	0.15	0.8931
T_{opt}	Optimum temperature for algal growth (°C)	27, 27	20, 20	0.8931
T_{max}	Maximum temperature for algal growth (°C)	42, 42	25, 25	0.8951
T_{min}	Minimum temperature for algal growth (°C)	0, 0	3, 10	0.8931
ZP	Zooplankton population (m^{-3})	200	100	0.8966
ZP_{min}	Minimum zooplankton for grazing (m^{-3})	50	10	0.8966
$Chla_{min}$	Minimum chlorophyll-a for grazing (mg/L)	0.001, 0.001	0.002, 0.002	0.8924

In Figures 15 and 16, the blue line corresponds to simulation results using the calibrated value of the parameter which has produced satisfactory results. The calibrated maximum photosynthesis rate G_{max} is 0.6 for Lake Elmo. When G_{max} is increased to 0.9, the Chla concentration increases throughout the simulation period on average by 30 percent; the phosphorus concentration increases in the ice cover period because of the detrital decay of increased algae. Higher G_{max} corresponds to higher phytoplankton growth which uses more nutrients (phosphorus) except the slow growth period under ice cover. This results in an average 32% decrease in phosphorus concentration in open water season.

When the sediment phosphorus release rate S_P was decreased from 0.02 to 0.01, the phosphorus concentration decreased by 55% (on average). The phosphorus starts to decrease during the ice cover period in December 2007 as the sediment phosphorus release start contributing to the phosphorus concentration in anoxic condition. After ice melting, as the algae begin to grow and use up phosphorus, the phosphorus concentration decreases even further in summer. The half saturation coefficient of phosphorus is decreased from 0.07 mg/L to 0.035 mg/L. Since the surface layer is not limited by light, the decrease in the half saturation coefficient results in greater phosphorus limitation which results in increased algal growth. The phosphorus concentration decreases except for the ice cover period which is governed by the sediment phosphorus release in anoxic condition. The increase in the minimum chlorophyll-a threshold for grazing results in less grazing of zooplankton and a subsequent increase in Chla. The settling velocity of algae (both green algae and blue-green algae) is a very sensitive parameter for phytoplankton and BOD simulation. In Figure 15e, as the settling velocity of green algae, blue-green algae, and detritus was decreased to 0.75, 0.5, and 1.0, respectively, the phytoplankton concentration increased due to less loss of phytoplankton from the simulating layer from settling. The phosphorus concentration increases due to the release of phosphorus through respiration of the increased algal population.

For a shallow lake such as Pearl, the effect of maximum photosynthesis rate (Figure 16a) is not as straightforward as for a deep stratified lake such as Elmo Lake (Figure 15a). In addition to the predicted increase, the Chla concentration decreases in spring of 2011 and 2012, early summer and fall of 2012 as a result of the spring and fall overturn when G_{max} is increased from 0.25 to 0.4. The Chla is not very sensitive to the half-saturation coefficient (K_P) and the minimum Chla threshold for zooplankton grazing ($Chla_{min}$). However, the phosphorus concentration decreased by an average of 60% when K_P is decreased from 0.03 to 0.005 mg/L (Figure 16c). As K_P decreases, the Chla increases throughout the time period except for a portion of summer and fall of 2012. This happens due to the increased respiration of algae which also causes a greater reduction is phosphorus. The zooplankton grazing, being negligible for Pearl Lake, does not affect the phosphorus concentration much. Hence, phosphorus is not sensitive of zooplankton grazing in Pearl Lake.

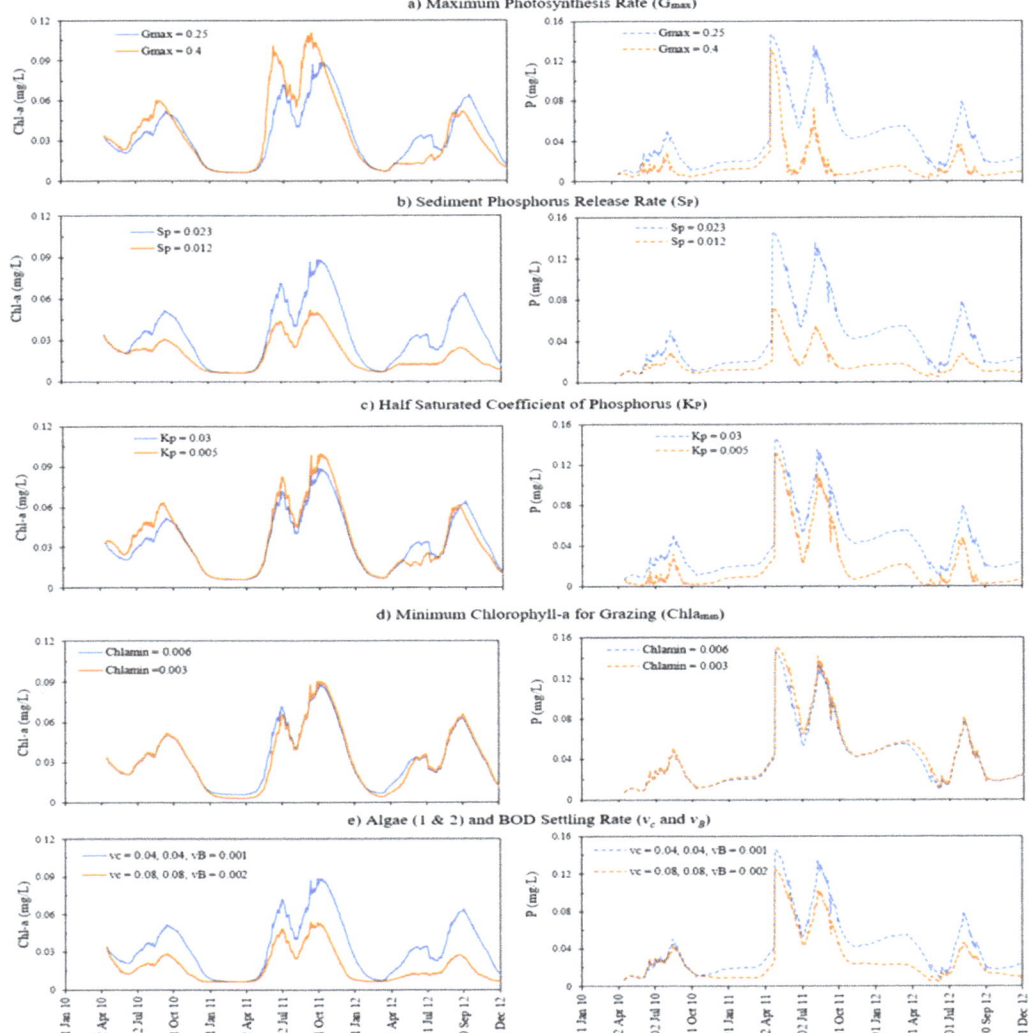

Figure 16. Sensitivity analysis of Chl*a* (**left panels**) and phosphorus (**right panels**) on (**a**) maximum photosynthesis rate G_{max}, (**b**) sediment phosphorus release rate S_P, (**c**) half saturation coefficient of phosphorus K_P, (**d**) minimum Chl*a* for grazing $Chla_{min}$, and (**e**) algae and BOD settling rates v_c, and v_B for simulation of Pearl Lake in 2010–2012.

4.5. Long-Term Simulations Using MINLAKE2020

West and Stefan [23] performed a multiple-year simulation (same calibration parameters) using MINLAKE98 for Lake Riley and Elmo. For Lake Riley, a different set of calibration parameters was needed for different years whereas, for Lake Elmo, the model could simulate successfully for 1985–1990 with the regression coefficient for temperature and DO as 0.91 and 0.79, respectively. For a simulation of 1985–1990 using the MIN-

LAKE2020 NCDO model, the regression coefficient for temperature and DO are 0.9944 and 0.9715 against 146 profile data points, respectively. Simulated phosphorus and Chl*a* concentrations at different depths are satisfactory as well. MINLAKE2020 performed well for multiple-year simulation allowing the user to simulate 20 consecutive years with the same calibration parameters (Figure 17). For a 20-year simulation (1989–2009) using the MINLAKE2020 NCDO model, the regression coefficients for temperature and DO are 0.9888 and 0.9419 against 864 profile data points, respectively. The simulated Chl*a* and phosphorus match reasonably well with observed values with the same trend. Moreover, the phosphorus and Chl*a* concentration at five simulation depths (1 m, 8 m, 12 m, 20 m, and 30 m) match well with the available observed data [54].

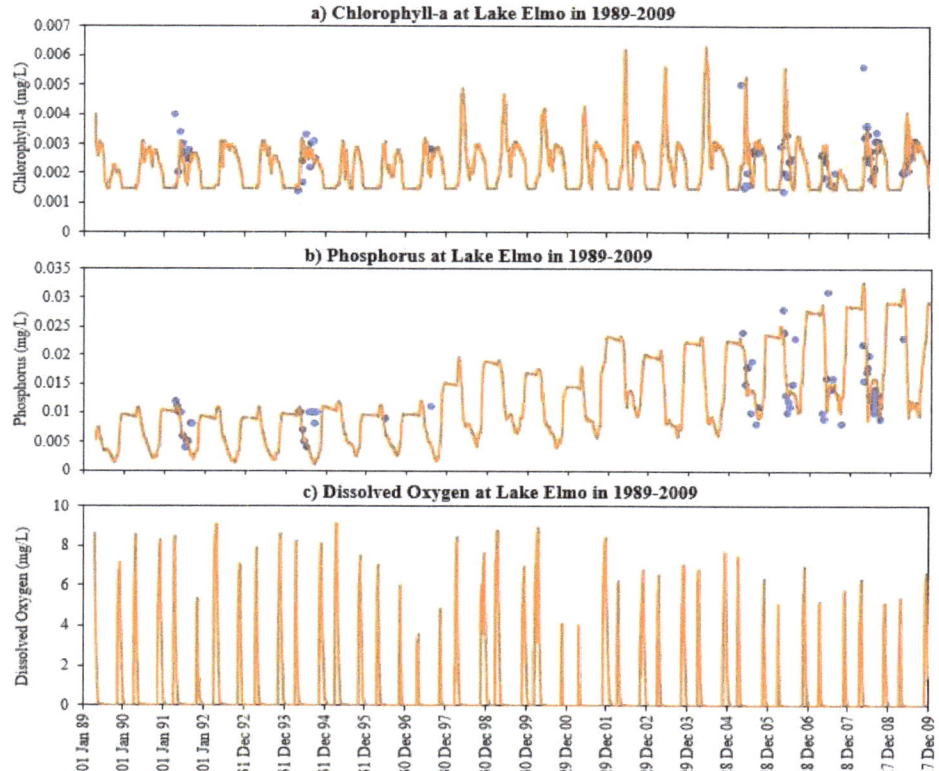

Figure 17. Simulated (**line**) and observed (**dots**) surface (1 m) (**a**) chlorophyll-a, (**b**) phosphorus, and (**c**) dissolved oxygen for long-term simulation of Lake Elmo from 1989 to 2009.

From 1989 to 1996, both phosphorus and chlorophyll-a seasonal variations were reasonably stable. Phosphorus started to increase from 1997 and matched with observed data from 2004 to 2007. The average daily phosphorus was 0.0071 mg/L from 1990 to 1996 and 0.0158 mg/L from 1997 to 2009. Comparing these two periods, the average daily phosphorus increased by 0.0087 mg/L. As phosphorus increased, it resulted in some higher peaks in spring algal blooms as shown in Figure 17a. The phosphorus increase trend is caused by the increase in phosphorus release from the lake sediment which is related to the anoxic condition in lakes. Therefore, the phosphorus release for each layer (the last term in Equation (15) times the layer volume) and then each day (sum for all layers) was outputted and added together for the annual phosphorus release amount. From 1990 (excluding 1989 for the initial condition effect) to 1996, the average yearly sediment phosphorus release is 151.8 kg (21.27 kg of standard deviation) but from 1997 to 2009 average yearly sediment phosphorus release is 244.1 kg (53.28 kg of standard deviation). The average annual phosphorus release increases by 60.8%. Figure 17c shows the time series of the simulated DO at 41 m (1 m above the deepest lake bottom) from 1989 to 2009 and clearly shows many anoxic days in the open water seasons and the ice cover periods, which result in phosphorus release from sediment. From 1990 to 1996, the average anoxic days is 228 but from 1997 to 2009 average anoxic days is 253 (34 days of standard deviation). An average of 25 days more of the anoxic condition is one of the causes of the phosphorus increase trend. From DO simulation it was observed that the anoxia started at a lower depth in the water column (16 m) from 1999 to 2009 compared to earlier simulation years. As a result, the anoxia had a greater horizontal-area coverage. Since the sediment release is a function of the bottom area (Equation (15)), the increased bottom area resulted in the increased sediment phosphorus release.

5. Conclusions

A one-dimensional daily water quality model MINLAKE2020 was developed from the daily temperature and DO MINLAKE2012 model by incorporating phytoplankton, zooplankton, nutrient, and BOD simulation into the model. The inflow-outflow submodel of MINLAKE2020 was disabled for this study to focus on the internal nutrient dynamics inside the lake. The model was applied to six Minnesota lakes with varying characteristics in terms of depth (two shallow lakes, two medium-depth lakes, and two deep lakes) and trophic status (two eutrophic, two mesotrophic, and two oligotrophic lakes). The simulated water temperature, DO, Chl*a*, and phosphorus time series and profiles were compared with available observed data in 15–36 days for two to four years. The model was also applied to long-term simulation over 20 years (1989–2009) for Lake Elmo. Simulation results from the MINLAKE2020 model provide the following conclusions:

1. MINLAKE2020 was calibrated against measured profiles in six Minnesota lakes (Table 4) for the short term (2–4 years) with an average standard error of 1.51 °C for temperature and 2.33 mg/L for DO. The average standard error for DO simulation of these lakes decreased by 24.2% from the original MINLAKE2012 model, which indicates better model performance. DO results reflect/integrate reasonably simulated phosphorus, Chl*a*, and BOD results at different layers (see Figures 10–14 and 17).
2. The addition of phosphorus and Chl*a* simulation in MINLAKE2020 improved model performance in comparison to MINLAKE2012 where Chl*a* was specified input. It greatly affects the DO concentration in some lakes such as Pearl Lake (Figure 10). Thrush Lake and Carlos Lake also showed significant improvement in DO simulation with MINLAKE2020. The standard error decreased by 2.12 mg/L and 1.76 mg/L for Thrush and Carlos Lake, respectively.
3. The deep lakes exhibit a certain trend for phosphorus and Chl*a* simulation year by year whereas the shallow lakes might show a significant change in phosphorus and Chl*a* concentration year by year due to two overturn periods (complete mixing) and the complex interactions/connections among phosphorus, Chl*a*, and DO (Figures 13 and 14), which are evident through governing Equations (9) and (15)–(19) and processes simulated (Figures 1–7).
4. DO concentration is a primary control of internal loading via anoxic release of phosphorus from the lake sediment. MINLAKE2020 was applied to Lake Elmo for a 20-year (1989 to 2009) continuous simulation with a single set of calibration parameters with regression coefficients of 0.99 and 0.94 for temperature and DO profiles, respectively. An increasing trend of surface phosphorus was simulated from 1997 to 2009 which matches well with the observed condition and is directly related to sediment phosphorus release. The average yearly sediment phosphorus release increased from 151.8 kg during the period 1990–1996 to 244.1 kg during the period 1997–2009. This increase is caused by the average 25-day increase in the anoxic condition at the

bottom depth (41 m) and the increase in the anoxic horizontal area (as a result of anoxia at lower depths) in later years.

MINLAKE2020 explains the internal link between phosphorus, Chl*a*, and DO. This model can help in choosing/testing effective ways of lake restoration and management. For example, since Lake Elmo experiences internal loading of phosphorus, removing the external nutrient source might not be an effective restoration technique for this lake. Though MINLAKE2020 has simulated six lakes with good correlation to the observed data, more lakes need to be simulated to verify the model in different scenarios before using it professionally for evaluating lake restoration measures. Since the nutrient model is complex and requires a number of calibration parameters, incorporating automatic calibration (by programming software) will make the model more efficient and user-friendly in the future.

Author Contributions: B.T. developed the final version of the MINLAKE2020 program, conducted the simulations, analyzed the result, and prepared the manuscript draft and revisions. X.F. supervised model development, simulation runs, data analysis, and revised the manuscript. J.S.H. supervised the writing and revised the manuscript. D.T. secured research funds and revised the manuscript. All authors made contributions to the study and writing the manuscript. All authors have read and agreed to the published version of the manuscript.

Funding: This study is partially supported by funding from Auburn University for the PAIR project "A prototype framework of climate services for decision making." Di Tian is PI, Xing Fang and others are six Co-PIs. This study is also partially supported by funding from the OUC-AU Joint Center for Aquaculture and Environmental Science for the project "*Identifying potential bioaccumulation hotspots of heavy metals in dynamic estuarine systems in Alabama, USA and China.*" Matthew N. Waters is PI, Xing Fang and Joel Hayworth are Co-PIs at Auburn University; Jinfen Pan is PI, Min Wang is Co-PI from OUC (Ocean University of China).

Institutional Review Board Statement: Not applicable.

Informed Consent Statement: Not applicable.

Data Availability Statement: Some or all data, models, or code generated or used during the study are proprietary or confidential and may only be provided with restrictions. The model input and output data are specifically designed for a research numerical model. They are available upon request but are not useful for the general public.

Conflicts of Interest: The authors declare no conflict of interest.

List of Symbols, Corresponding Description and Units

Symbols	Description and Units
A	Area (m^2)
BOD	Biochemical oxygen demand (mg/L)
$Chla$	Chlorophyll concentration (mg/L)
$Chla_{min}$	Minimum chlorophyll concentration for grazing to occur (mg/L)
DO	Dissolved Oxygen concentration (mg/L)
DO_{sat}	Saturated oxygen concentration (mg/L)
$f(L)$	Light limiting growth factor (between 0 and 1)
$f(S)$	Michalis-Menten growth limiting factor [-]
$f(T)$	Temperature function for growth [-]
G_{max}	Maximum growth rate (d^{-1})
GR_{max}	Grazing maximum (mg Chla/ind. zooplankton/d)
K_{gchla}	Half-saturation constant for grazing (mg/L)
$I(z)$	Intensity of photosynthetically active radiation (μE/m^2/h)
K_1	Light limitation coefficient (μE/m^2/h)
K_2	Light inhibition coefficient (μE/m^2/h)
K_{BOD}	Organic decomposition rate (d^{-1})
K_e	Surface oxygen gas exchange (transfer) coefficient (m/d)
K_m	Mortality rate (d^{-1})
K_N	Half-saturation constant for nitrogen (mg/L)

Symbol	Description
K_{NI}	Nitrification rate (d^{-1})
$K_{NH(n)}$	Half-saturation constant for preferential uptake of ammonia over nitrate [-]
$K_{TN(n)}$	Half-saturation constant of nitrogen for each algal class (mg/L)
K_P	Half-saturation constant for phosphorus (mg/L)
K_r	Respiration rate (d^{-1})
K_{zr}	Zooplankton respiration rate (d^{-1})
K_z	Eddy diffusivity (cm^2/s)
N	Nitrogen concentration
NH_4	Ammonia concentration (mg N/L)
NO_3	Nitrate concentration (mg N/L)
NO_2	Nitrite concentration (mg N/L)
$NO_{2\text{-}3}$	Nitrite + Nitrate concentration (mg N/L)
n	Phytoplankton species [-]
P	SRP or soluble reactive phosphorus concentration (mg/L)
θ_{BOD}	Temperature adjustment coefficient for BOD [-]
θ_{SOD}	Temperature adjustment coefficient for SOD [-]
θ_m	Temperature adjustment coefficient for mortality [-]
θ_r	Temperature adjustment coefficient for respiration [-]
θ_{zr}	Temperature adjustment coefficient for zooplankton grazing [-]
θ_{NH}	Temperature adjustment coefficient for ammonia [-]
θ_{NO}	Temperature adjustment coefficient for nitrate [-]
θ_{NI}	Temperature adjustment coefficient for nitrification [-]
RAD	Solar radiation (cal/cm^2/d)
S_b	Sedimentary oxygen demand rate at 20°C (g O/m^2/d)
S_{NH}	Ammonia release rate from sediment (g N/m^2/d)
S_{NO}	Nitrite release rate from sediment (g N/m^2/d)
S_P	Phosphorus release rate from sediment (g P/m^2/d)
t	Time (d)
T	Air temperature (°C)
TD	Length of day light (h)
T_{max}	Maximum temperature at which phytoplankton growth is reduced 90 percent (°C)
T_{min}	Minimum temperature at which phytoplankton growth is reduced 90 percent (°C)
T_{opt}	Optimal temperature at which maximum phytoplankton growth occurs (°C)
TP	Total phosphorus (mg/L)
v_c	Phytoplankton settling velocity (m/d)
v_B	Detritus settling velocity (m/d)
V_{iz}	Volume of day depth (m^3)
V_r	Volume day depth/layer ratio [-]
Y_{CHO2}	Mass yield ratio of chlorophyll to oxygen
Y_{NChla}	Mass yield ratio of Nitrogen to Chlorophyll-a
Y_{NHBOD}	Mass yield ratio of ammonia to BOD [-]
Y_{PBOD}	Mass yield ratio of ammonia to BOD [-]
Y_{PCHLA}	Mass yield ratio of ammonia to BOD [-]
z	Depth (m)
ZP	Zooplankton concentration (#/m^3)
ZP_{min}	Minimum concentration of zooplankton for predation to occur (#/m^3)
PD_d	Daytime predation rate (d^{-1})
Pd	Daily predation rate (d^{-1})
XI	Light intensity at the day depth (μE/m^2/s)
XI_{max}	Light intensity at which no predation occurs (μE/m^2/s)
XI_{min}	Light intensity at which no predation occurs (μE/m^2/s)
P_{max}	Maximum predation rate (d^{-1})
DY	Julian day of last day of minimum predation rate [-]
DY_{min}	Julian day of last day of minimum predation rate [-]
DY_{max}	Julian day of beginning of maximum predation rate [-]
$PD(t,I)$	Nocturnal predation rate in layer I (#/m^3)
PD_n	Nocturnal predation rate (d^{-1})
$\Delta T_g(I)$	Time that zooplankton spent in a layer I during the night (d)

References

1. Le Moal, M.; Gascuel-Odoux, C.; Menesguen, A.; Souchon, Y.; Etrillard, C.; Levain, A.; Moatar, F.; Pannard, A.; Souchu, P.; Lefebvre, A.; et al. Eutrophication: A new wine in an old bottle? *Sci. Total Environ.* **2019**, *651*, 1–11. [CrossRef] [PubMed]
2. Moss, B. Cogs in the endless machine: Lakes, climate change and nutrient cycles: A review. *Sci. Total Environ.* **2011**, *434*, 130–142. [CrossRef] [PubMed]
3. Takolander, A.; Cabeza, M.; Leskinen, E. Climate change can cause complex responses in Baltic Sea macroalgae: A systematic review. *J. Sea Res.* **2017**, *123*, 16–29. [CrossRef]
4. Yao, X.; Zhang, Y.; Zhang, L.; Zhou, Y. A bibliometric review of nitrogen research in eutrophic lakes and reservoirs. *J. Environ. Sci.* **2017**, *66*, 274–285. [CrossRef]
5. Liu, Y.; Villalba, G.; Ayres, R.U.; Schroder, H. Global phosphorus flows and environmental impacts from a consumption perspective. *J. Ind. Ecol.* **2008**, *12*, 229–247. [CrossRef]
6. Turner, R.E. Linking landscape and water quality in the Mississippi river basin for 200 years. *Bioscience* **2003**, *53*, 563–572. [CrossRef]
7. Wolfe, A.H.; Patz, J.A. Reactive nitrogen and human health: Acute and longterm implications. *AMBIO J. Hum. Environ.* **2002**, *31*, 120–125. [CrossRef]
8. Townsend, A.R.; Howarth, R.W.; Bazzaz, F.A.; Booth, M.S.; Cleveland, C.C.; Collinge, S.K.; Dobson, A.P.; Epstein, P.R.; Holland, E.A.; Keeney, D.R.; et al. Human health effects of a changing global nitrogen cycle. *Front. Ecol. Environ.* **2003**, *1*, 240–246. [CrossRef]
9. Hudnell, H.K.; Dortch, Q. *Cyanobacterial Harmful Algal Blooms: Chapter 2: A Synopsis of Research Needs Identified at the Interagency, International Symposium on Cyanobacterial Harmful Algal Blooms (ISOC-HAB)*; US Environmental Protection Agency: Washington, DC, USA, 2008; Volume 38, pp. 17–43.
10. Hudnell, H.K. The state of U.S. freshwater harmful algal blooms assessments, policy and legislation. *Toxicon* **2010**, *55*, 1024–1034. [CrossRef] [PubMed]
11. Seewer, J. Ohio wants to put Lake Erie on a new, strict pollution diet. *AP News*, 15 February 2020.
12. Moomaw, W.R.; Birch, M.B. Cascading costs: An economic nitrogen cycle. *Sci. China Ser. C Life Sci.* **2005**, *48*, 678–696.
13. Pretty, J.N.; Mason, C.F.; Nedwell, D.B.; Hine, R.E.; Leaf, S.; Dils, R. Environmental Costs of Freshwater Eutrophication in England and Wales. *Environ. Sci. Technol. Libr.* **2003**, *37*, 201–208. [CrossRef]
14. Dodds, W.K.; Bouska, W.W.; Eitzmann, J.L.; Pilger, T.J.; Pitts, K.L.; Riley, A.J.; Schloesser, J.T.; Thornbrugh, D.J. Eutrophication of U.S. freshwaters: Analysis of potential economic damages. *Environ. Sci. Technol.* **2008**, *43*, 12–19. [CrossRef] [PubMed]
15. Bartram, J.; Chorus, I. *Toxic Cyanobacteria in Water: A Guide to Their Public Health Consequences, Monitoring and Management*; World Health Organization: London, UK, 1999; p. 400.
16. Imboden, D.M. Phosphorus model of lake eutrophication. *Limnol. Oceanogr.* **1974**, *19*, 297–304. [CrossRef]
17. Vollenweider, R.A.; Kerekes, J. *Eutrophication of Waters. Monitoring, Assessment and Control*; OECD Cooperative Programme on Monitoring of Inland Waters; Organization for Economic Co-Operation and Development (OECD): Paris, France, 1982; Volume 156.
18. De Senerpont Domis, L.N.; Van de Waal, D.B.; Helmsing, N.R.; Donk, E.V.; Mooij, W.M. Community stoichiometry in a changing world: Combined effects of warming and eutrophication on phytoplankton dynamics. *Ecology* **2014**, *95*, 1485–1495. [CrossRef] [PubMed]
19. Reynolds, C.S.; Irish, A.E.; Elliott, J.A. The ecological basis for simulating phytoplankton responses to environmental change (PROTECH). *Ecol. Model.* **2001**, *140*, 271–291. [CrossRef]
20. Riley, M.J.; Stefan, H.G. MINLAKE: A dynamic lake water quality simulation model. *Ecol. Model.* **1988**, *43*, 155–182. [CrossRef]
21. Gu, R.; Stefan, H.G. Year-round temperature simulation of cold climate lakes. *Cold Reg. Sci. Technol.* **1990**, *18*, 147–160. [CrossRef]
22. Fang, X.; Stefan, H.G. *Modeling of Dissolved Oxygen Stratification Dynamics in Minnesota Lakes under Different Climate Scenarios*; St Anthony Falls Hydraulic Laboratory, University of Minnesota: Minneapolis, MN, USA, 1994; Volume 55414, p. 260.
23. West, D.; Stefan, H.G. Simulation of Lake Water Quality Using a One-Dimensional Model with Watershed Input. In *Model Description and Application to Lake Riley and Lake Elmo*; Project Report No.430; St. Anthony Falls Laboratory, University of Minnesota: Minneapolis, MN, USA, 1998.
24. Jamily, J.A. Developing an Hourly Water Quality Model to Simulate Diurnal Water Temperature and Dissolved Oxygen Variations in Shallow Lakes. Master's Thesis, Auburn University, Auburn, AL, USA, 2018.
25. Batick, B.M. *Modeling Temperature and Dissolved Oxygen in the Cheatham Reservoir with CE-QUAL-W2*; Vanderbilt University: Nashville, TN, USA, 2011.
26. Chapra, S.C.; Martin, J.L. *LAKE2K, a Modeling Framework for Simulating Lake Water Quality (Version 1.2): Documentation and users Manual*; Civil and Environmental Engineering Department, Tufts University: Medford, MA, USA, 2004.
27. Cole, T.M.; Buchak, E.M. *CE-QUAL-W2: A Two-Dimensional, Laterally Averaged, Hydrodynamic and Water Quality Model, Version 2.0. User Manual*; DTIC Document; DTIC: Fort Belvoir, VA, USA, 1995.
28. Janssen, A.B.G.; Teurlincx, S.; Beusen, A.H.W.; Huijbregts, M.A.J.; Rost, J.; Schipper, A.M.; Seelen, L.M.S.; Mooij, W.M.; Janse, J.H. PCLake+: A process-based ecological model to assess the trophic state of stratified and non-stratified freshwater lakes worldwide. *Ecol. Model.* **2019**, *396*, 23–32. [CrossRef]

29. Hipsey, M.R.; Romero, J.R.; Antenucci, J.P.; Hamilton, D. *Computational Aquatic Ecosystem Dynamics Model: CAEDYM v2*; Contract Research Group, Centre for Water Research, University of Western Australia: Crawley, WA, Australia, 2005; p. 102.
30. Hannoun, I.; List, E.J.; Kavanagh, K.B.; Chiang, W.-L.; Ding, L.; Preston, A.; Karafa, D.; Rachkley, I. *Use of ELCOM and CAEDYM for Water Quality Simulation in Boulder Basin*; Water Environment Federation: Alexandria, VA, USA, 2006.
31. Carraro, E.; Guyennon, N.; Hamilton, D.; Valsecchi, L.; Manfredi, E.C.; Viviano, G.; Salerno, F.; Tratari, G.; Copetti, D. Coupling high-resolution measurements to a three dimensional lake model to assess the spatial and temporal dynamics of the cyanobacterium Planktothrix rubescens in a medium-sized lake. *Hydrobiologia* **2012**, *698*, 77–95. [CrossRef]
32. Pomati, F.; Matthews, B.; Seehausen, O.; Ibelings, B.W. Eutrophication and climate warming alter spatial (depth) co-occurrence patterns of lake phytoplankton assemblages. *Hydrobiologia* **2017**, *787*, 375–385. [CrossRef]
33. Anderson, D.M.; Cembella, A.D.; Hallegraeff, G.M. Progress in Understanding Harmful Algal Blooms: Paradigm shifts and new technologies for research, monitoring, and management. *Annu. Rev. Mar. Sci.* **2012**, *4*, 143–176. [CrossRef] [PubMed]
34. Shimoda, Y.; Arhonditsis, G.B. Phytoplankton functional type modelling: Running before we can walk? A critical evaluation of the current state of knowledge. *Ecol. Model.* **2016**, *320*, 29–43. [CrossRef]
35. Zhang, X.; Recknagel, F.; Chen, Q.; Cao, H.; Li, R. Spatially-explicit modelling and forecasting of cyanobacteria growth in Lake Taihu by evolutionary computation. *Ecol. Model.* **2015**, *306*, 216–225. [CrossRef]
36. Elliott, A.H. Predicting the impact of changing nutrient load and temperature on the phytoplankton of England's largest lake, Windermere. *Freshw. Biol.* **2011**, *57*, 400–413. [CrossRef]
37. Marsden, M.W. Lake restoration by reducing external phosphorus loading: The influence of sediment phosphorus release. *Freshw. Biol.* **1989**, *21*, 139–162. [CrossRef]
38. Sondergaard, M.; Jeppesen, E.; Jensen, J.P.; Amsinck, S.L. Water framework directive: Ecological classification of Danish lakes. *J. Appl. Ecol.* **2005**, *42*, 616–629. [CrossRef]
39. Philips, G.; Kelly, A.; Pitt, J.A.; Sanderson, R.; Taylor, E. The recovery of a very shallow eutrophic lake, 20 years after the control of effluent derived phosphorus. *Freshw. Biol.* **2005**, *50*, 1628–1638. [CrossRef]
40. Taguchi, V.J.; Olsen, T.A.; Janke, B.D.; Gulliver, J.S.; Finlay, J.C.; Stefan, H.G. Internal loading in stormwater ponds as a phosphorus source to downstream waters. *Limnol. Oceanogr. Lett.* **2020**, *5*, 322–330. [CrossRef]
41. Jeppesen, E.; Sondergaard, M.; Jensen, J.P.; Havens, K.E.; Anneville, O.; Carvalho, L.; Coveney, M.F.; Deneke, R.; Dokulil, M.T.; Foy, B.; et al. Lake responses to reduced nutrient loading: An analysis of contemporary long-term data from 35 case studies. *Freshw. Biol.* **2005**, *50*, 1747–1771. [CrossRef]
42. Welch, E.B.; Cooke, G.D. Internal phosphorus loading in shallow lakes: Importance and control. *Lake Reserv. Manag.* **2009**, *21*, 209–217. [CrossRef]
43. Riley, M.J.; Stefan, H.G. *Dynamic Lake Water Quality Simulation Model "MINLAKE"*; St. Anthony Falls Hydraulic Laboratory, University of Minnesota: Minneapolis, MN, USA, 1987; Volume 55414, p. 140.
44. Hondzo, M.; Stefan, H.G. Lake water temperature simulation model. *J. Hydraul. Eng.* **1993**, *119*, 1251–1273. [CrossRef]
45. Fang, X.; Stefan, H.G. Modeling of dissolved oxygen stratification dynamics in Minnesota lakes under different climate scenarios. *Ecol. Model.* **1994**, *71*, 37–68.
46. Fang, X.; Stefan, H.G. *Temperature and Dissolved Oxygen Simulations for a Lake with Ice Cover*; Project Report 356; St. Anthony Falls Hydraulic Laboratory, University of Minnesota: Minneapolis, MN, USA, 1994.
47. Fang, X.; Alam, S.R.; Jacobson, P.; Pereira, D.; Stefan, H.G. *Simulations of Water Quality in Cisco Lakes in Minnesota*; St. Anthony Falls Laboratory, University of Minnesota: Minneapolis, MN, USA, 2010.
48. West, D.; Stefan, H.G. *Simulation of Water Quality and Primary Productivity Control Strategies for Lake McCarrons*; Project Report No.426; St. Anthony Falls Laboratory, University of Minnesota: Minneapolis, MN, USA, 2000.
49. Tasnim, B.; Jamily, J.A.; Fang, X.; Zhou, Y.; Hayworth, J.S. Simulating diurnal variations of water temperature and dissolved oxygen in shallow Minnesota lakes. *Water* **2021**, *13*, 1980. [CrossRef]
50. Monod, J. The growth of bacterial cultures. *Annu. Rev. Microbiol.* **1949**, *3*, 371–394. [CrossRef]
51. Lehman, J.T.; Botkins, D.B.; Likens, G.E. The assumptions and rationales of a computer model of phytoplankton population dynamics. *Limnol. Oceanogr.* **1975**, *20*, 343–364. [CrossRef]
52. Megard, R.O.; Tonkyn, D.W.; Senft, W.H. Kinetics of oxygenic photosynthesis in planktonic algae. *J. Plankton Res.* **1984**, *6*, 325–337. [CrossRef]
53. Megard, R.O.; Combs, W.S.; Smith, P.D.; Knoll, A.S. Attenuation of light and daily integral rates of photosynthesis attained by planktonic algae1. *Limnol. Oceanogr.* **1979**, *24*, 1038–1050. [CrossRef]
54. Shapiro, J.; Forsberg, B.; Lamarra, V.; Lindmark, G.; Lynch, M.; Smeltzer, E.; Zoto, G. *Experiments and Experiences in Biomanipulation*; Interim Report No. 19; Limnological Research Center, University of Minnesota: Minneapolis, MN, USA, 1982.
55. Wright, D.; O'Brien, W.J.; Vinyard, G.L. Adaptive value of vertical migration: A simulation model argument for the predation hypothesis, Evolution and Ecology of Zooplankton Communities. In *Evolution and Ecology of Zooplankton Communities*; Kerfoot, C.W., Ed.; University Press of New England: Lebanon, NH, USA, 1980; pp. 138–147.
56. Dillon, P.J.; Rigler, F.H. The phosphorus-chlorophyll relationship in lakes: Phosphorus-chlorophyll relationship. *Limnol. Oceanogr.* **1974**, *19*, 767–773. [CrossRef]
57. Thomann, R.V.; Mueller, J.A. *Principles of Surface Water Quality Modeling and Control*; Harper Collins Publishers Inc: New York, NY, USA, 1987; p. 644.

58. Goldman, C.R.; Horne, A.J. *Limnology*; McGraw-Hill: New York, NY, USA, 1983.
59. Cole, T.M.; Wells, S.A. *CE-QUAL-W2: A Two-Dimensional, Laterally Averaged, Hydrodynamic and Water Quality Model, Version 3.72 User Manual*; Department of Civil and Environmental Engineering, Portland State University: Portland, OR, USA, 2015.
60. Hamrick, M.J. *The Environmental Fluid Dynamics Code Theory and Computation Volume 3: Water Quality Module*; Tetra Tech, Inc.: Fairfax, VA, USA, 2007; p. 90.
61. Nürnberg, G.K. The prediction of internal phosphorus load in lakes with anoxic hypolimnia. *Limnol. Oceanogr.* **1984**, *29*, 111–124. [CrossRef]
62. Stigebrandt, A.; Rahm, L.; Viktorsson, L.; Ödalen, M.; Hall, P.O.J.; Liljebladh, B. A new phosphorus paradigm for the Baltic proper. *Ambio* **2014**, *43*, 634–643. [CrossRef]
63. Stigebrandt, A.; Andersson, A. The eutrophication of the Baltic Sea has been boosted and perpetuated by a major internal phosphorus source. *Front. Mar. Sci.* **2020**, *7*, 572994. [CrossRef]
64. Jiang, L.; Fang, X. Simulations and validation of cisco lethal conditions in Minnesota lakes under past and future climate scenarios using constant survival limits. *Water* **2016**, *8*, 279. [CrossRef]
65. Edwards, R.; Owens, M. The oxygen balance of streams. *Ecol. Ind. Soc.* **1965**, *6*, 149–172.
66. Henderson-Sellers, B. *Engineering Limnology*; Pitman Advanced Pub. Program: Boston, MA, USA, 1984; p. 356.
67. Tasnim, B. Enhancement and Redevelopment of the Regional Lake Water Quality Model with Applications. Master's Thesis, Auburn University, Auburn, AL, USA, 2020.
68. Fang, X.; Stefan, H.G. Long-term lake water temperature and ice cover simulations/measurements. *Cold Reg. Sci. Technol.* **1996**, *24*, 289–304. [CrossRef]
69. Brown, L.C.; Barnwell, T.O. *The Enhanced Stream Water Quality Models QUAL2E and QUAL2W-UNCAS: Documentation and User Manual*; U.S. Environmental Protection Agency: Athens, GA, USA, 1987.
70. Zison, S.W.; Mills, W.B.; Diemer, D.; Chen, C.W. *Rates, Constants and Kinetic Formulations in Surface Water Quality Modeling*; Tetra Tech, Inc., for USEPA, ORD: Athens, GA, USA, 1978; p. 317.
71. Stefan, H.G.; Hondzo, M.; Fang, X.; Eaton, J.G.; McCormick, J.H. Simulated long-term temperature and dissolved oxygen characteristics of lakes in the north-central United States and associated fish habitat limits. *Limnol. Oceanogr.* **1996**, *41*, 1124–1135. [CrossRef]
72. Stefan, H.G.; Fang, X. Dissolved oxygen model for regional lake analysis. *Ecol. Model.* **1994**, *71*, 37–68. [CrossRef]
73. NAS and NAE. *Water Quality Criteria 1972—A Report of the Committee on Water Quality Criteria*; Environmental Protection Agency: Washington, DC, USA, 1973.
74. Barr Engineering. *Pearl Lake and Mill Creek Bacterial Total Maximum Daily Load Report*; Barr Engineering: Minneapolis, MN, USA, 2012.
75. Engel, L.; Valley, R.; Beck, D.; Anderson, J. *Sentinel Lake Assessment Report Lake Carlos (21-0057) Douglas County, Minnesota*; Minnesota Pollution Control Agency & MN Minnesota Department of Natural Resources: St. Paul, MN, USA, 2010.
76. Engel, L.; Heiskary, S.; Valley, R.; Tollefson, D. *Sentinel Lake Assessment Report Carrie Lake (34-0032) Kandiyohi County, Minnesota*; Minnesota Pollution Control Agency & Minnesota Department of Natural Resources: St. Paul, MN, USA, 2012.
77. Nash, J.E.; Sutcliffe, J.V. River flow forecasting through conceptual models part I—A discussion of principles. *J. Hydrol.* **1970**, *10*, 282–290. [CrossRef]

Article

Simulating Diurnal Variations of Water Temperature and Dissolved Oxygen in Shallow Minnesota Lakes

Bushra Tasnim [1], Jalil A. Jamily [2], Xing Fang [1,*], Yangen Zhou [3] and Joel S. Hayworth [1]

[1] Department of Civil and Environmental Engineering, Auburn University, Auburn, AL 36849, USA; bzt0022@auburn.edu (B.T.); jsh0024@auburn.edu (J.S.H.)
[2] Connecticut Department of Transportation, Newington, CT 06111, USA; jzj0065@auburn.edu
[3] Key Laboratory of Mariculture, Ministry of Education, Ocean University of China, Qingdao 266100, China; zhouyg@ouc.edu.cn
* Correspondence: xing.fang@auburn.edu; Tel.: +1-334-844-8778

Abstract: In shallow lakes, water quality is mostly affected by weather conditions and some ecological processes which vary throughout the day. To understand and model diurnal-nocturnal variations, a deterministic, one-dimensional hourly lake water quality model MINLAKE2018 was modified from daily MINLAKE2012, and applied to five shallow lakes in Minnesota to simulate water temperature and dissolved oxygen (DO) over multiple years. A maximum diurnal water temperature variation of 11.40 °C and DO variation of 5.63 mg/L were simulated. The root-mean-square errors (RMSEs) of simulated hourly surface temperatures in five lakes range from 1.19 to 1.95 °C when compared with hourly data over 4–8 years. The RMSEs of temperature and DO simulations from MINLAKE2018 decreased by 17.3% and 18.2%, respectively, and Nash-Sutcliffe efficiency increased by 10.3% and 66.7%, respectively; indicating the hourly model performs better in comparison to daily MINLAKE2012. The hourly model uses variable hourly wind speeds to determine the turbulent diffusion coefficient in the epilimnion and produces more hours of temperature and DO stratification including stratification that lasted several hours on some of the days. The hourly model includes direct solar radiation heating to the bottom sediment that decreases magnitude of heat flux from or to the sediment.

Keywords: diurnal variation; hourly model; water temperature; dissolved oxygen; shallow lakes; and sediment heat flux

Citation: Tasnim, B.; Jamily, J.A.; Fang, X.; Zhou, Y.; Hayworth, J.S. Simulating Diurnal Variations of Water Temperature and Dissolved Oxygen in Shallow Minnesota Lakes. *Water* **2021**, *13*, 1980. https://doi.org/10.3390/w13141980

Academic Editors: Lars Bengtsson and Bahram Gharabaghi

Received: 1 June 2021
Accepted: 15 July 2021
Published: 19 July 2021

Publisher's Note: MDPI stays neutral with regard to jurisdictional claims in published maps and institutional affiliations.

Copyright: © 2021 by the authors. Licensee MDPI, Basel, Switzerland. This article is an open access article distributed under the terms and conditions of the Creative Commons Attribution (CC BY) license (https://creativecommons.org/licenses/by/4.0/).

1. Introduction

Water quality of natural water systems is of great concern in the modern world because of the importance of water in human life and the environment. Water temperature and dissolved oxygen (DO) are the two most important water quality parameters for aquatic systems since temperature affects DO and the availability of DO in lakes affects freshwater fish species and populations [1]. In the later part of the twentieth century, due to the advent of modern computers, various numerical models have been developed to predict water quality parameters in different types of waterbodies such as riverine systems, estuaries, lakes, and reservoirs. Most of the time, lakes are simulated using one-dimensional models that assume well mixed or uniform conditions along horizontal layers and only recognize major variations in water quality along the vertical (depth) direction. The assumption with one-dimensional models is that all inflow quantities and constituents are instantaneously dispersed throughout the horizontal layers [2]. The turbulent diffusion approach and the mixed-layer approach [3] are the two approaches commonly used to model water temperature and DO in a lake. There are quite a number of water quality models for simulating the water quality of a lake. Almost all the models developed previously have one common characteristic: they are all based on a daily time step. In this study, a deterministic one-dimensional year-round daily water quality model—MINLAKE2012 [4] was modified to capture the hourly fluctuations of water quality parameters in shallow lakes (named MINLAKE2018).

The Minnesota Lake Water Quality Management Model (MINLAKE) is a deterministic one-dimensional model with a time step of one day that was developed in the 1980s for lake eutrophication studies and control strategies [5]. When the model was developed, the lack of temporal data did not allow the development of a diurnal-nocturnal model with a shorter time step, e.g., one hour. Since horizontal variations of water temperature and DO in freshwater lakes are relatively small compared to vertical variations, the one-dimensionality of MINLAKE is appropriate for freshwater lakes. The model has been successfully applied to many lakes over several years with satisfactory results [6–8]. To provide decision-makers with a useful tool for lake management and restoration, the MINLAKE model has been frequently reviewed, modified, and updated to improve accuracy and confidence. For example, a regional MINLAKE model was developed in the early 1990s and comprised of two separate sub-models—a regional water temperature model [9] and a regional dissolved oxygen model [10]. Though MINLAKE models were originally developed to simulate water quality during periods of open water it was observed that heat and oxygen transfer processes through the open water surface were substantially altered by winter ice and snow cover [11]. As a result, separate sub-models were developed for winter conditions and integrated with MINLAKE to provide the capability to simulate year-round water temperature and DO, and the revised model is called MINLAKE96. To account for water-sediment heat exchange, a separate sub-model, which calculates temperature profiles in the sediment below the water-sediment interface, was developed [12]. Fang and Stefan [12] found that heat fluxes between lake water and sediment could be substantial. Both simulated and field measurements have shown that shallow lakes can become 1–2 °C warmer under a thick winter ice/snow cover without significant radiation penetration through the snow/ice-covered surface, or significant flow into and out of a lake [13]. The year-round water temperature simulation model has been expanded significantly by simulating the winter ice and snow cover and including heat exchange between water layers and sediment [14]. Moreover, the model was further refined for coefficients used to accurately project water quality in deep cisco lakes, resulting in the MINLAKE2012 model. This is the version of the MINLAKE model used in our study.

Xu and Xu [15] revealed that for water quality parameters, daily models work better for deep dimictic lakes that are stratified for a majority of the year. In shallow lakes, fluctuations of water quality parameters are very dynamic and more diurnal. Also, shallow productive tropical lakes may show less stratification and more extreme diel variations in their physicochemical parameters [16]. Relatively weak temperature stratification occurs during the daylight hours, but it is removed or destroyed by nocturnal cooling and wind mixing. Similarly, vertical variations in DO occur during daytime hours, but they become mixed during the night [16]. Hourly DO has been simulated using Bayesian Model Averaging (BMA) and Adaptive Neuro-Fuzzy Inference System (ANFIS) models [17,18]. Both models are strongly dependent on observed data, which are scarce in most lakes, since monitoring programs to collect the data are time-consuming and costly.

To help water quality management decision-makers understand the dynamic fluctuation of water quality parameters in shallow lakes, an hourly model is necessary. For example, in 2015, the surface water temperature in Pearl Lake in Minnesota varied at a maximum of 6.9 °C during a day, and the maximum fluctuation in surface DO was 6.2 mg/L. Lake dynamics are a very complex process, and some processes happen faster and hence affect changes in water temperature and DO within a day. Moreover, the sediment sub-model of MINLAKE (same as all other lake temperature models) only considers heat conduction from the overlying water to sediments when water is heated by shortwave solar radiation that is attenuated by lake water containing algae, total suspended solids, etc. This modeling approach does not account for all the heat sources of bottom sediment. Solar radiation can penetrate through shallow water and directly heat sediments at the lake bottom. As a result, the sediment heat budget equations were modified to account for direct solar radiation reaching sediment areas in this study. Moreover, the main driving factor is weather conditions that change diurnally.

The main objective of this study was to develop and validate an hourly lake water quality model MINLAKE 2018, which can be used as a decision-making tool for lake management. The MINLAKE2012 model was modified to produce hourly output for water quality and sediment temperature. To simulate hourly water quality parameters, hourly weather data are required [15,19]. After modifications, the model was calibrated for the hourly time step. The model was validated using hourly water temperature data in five Minnesota lakes. Direct solar radiation heating of the sediment was added to both the MINLAKE2012 and MINLAKE2018 models for comparison. It was observed that sediment heat exchange was particularly important for hourly models. Results from the MINLAKE2012 and MINLAKE2018 models were compared to understand the impact of diurnal weather changes in water temperature and dissolved oxygen.

2. Materials and Methods

2.1. Daily Year-Round Water Temperature Model

The MINLAKE2018 model is based on previous studies and efforts made to simulate lake water temperature and dissolved oxygen, and specifically developed from the MINLAKE2012 model (a one-dimensional, deterministic year-round daily model which assumes no inflow or outflow in the lake). MINLAKE solves the one-dimensional, unsteady heat transfer Equation (1) to simulate vertical water temperature profiles in lakes:

$$\frac{\partial T_w}{\partial t} = \frac{1}{A}\frac{\partial}{\partial z}\left(K_z A \frac{\partial T_w}{\partial z}\right) + \frac{H_w}{\rho C_p} \quad (1)$$

where T_w (z, t) is the water temperature in (°C), which is a function of depth (z) and time (t); $A(z)$ (m^2) is the horizontal area for each layer of water as a function of the depth, K_z (m^2/day) is the vertical turbulent heat diffusion coefficient, which is a function of depth and time; ρC_p (J/m^3-°C) represents the heat capacity of water per unit volume, and H_w (J/m^3-day) is the sum of heat source and sink terms per unit volume of water. The determination of turbulent diffusion coefficients for lake temperature modeling has been discussed in detail by Fang [8]. In the regional daily water temperature MINLAKE model, the vertical heat diffusion coefficient K_z (m^2/day) for epilimnion and hypolimnion is calculated based on Equation (2):

$$K_z = 7.06 \times 10^{-3} \times \frac{A_s^{0.56}}{(N^2)^{0.43}} \quad (2)$$

where A_s is the surface area of the lake (km^2), and N^2 is the Brunt-Vaisala stability frequency of the stratification (s^{-2}). In the epilimnion, N^2 was set at a minimum value of 0.000075 [20].

Equation (1) is developed for up to 80 layers (depending on the lake maximum depth and modeler's choice) of water in a lake and solved numerically using an implicit finite difference scheme and a Gaussian elimination method with time steps of one day. Temperature in a waterbody is usually affected by ambient weather conditions such as solar radiation, sky condition, wind speed, wind direction, air temperature, and precipitation. In cold regions, ice and snow thickness impact solar radiation penetration into the lake water. The surface heat flux terms (kCal/m^2-day) through the water surface during the open water seasons can be represented as:

$$\Delta H = H_{sn} + H_a - (H_{br} + H_c + H_e) \quad (3)$$

where H_{sn} is net shortwave solar radiation, H_a is net atmospheric longwave radiation, H_{br} is longwave back radiation from the water to the atmosphere, H_c is heat conduction/convection, and H_e is the evaporation through the surface water.

Solar radiation is the only energy source that can penetrate water layers due to the short wavelengths. Short wavelength radiation is of the highest energy. Consequently, solar radiation is a heat source in more than just the top layer of a lake [7]. Beer's law states that the radiation intensity (heat flux) decreases exponentially with depth [7]. The heat transfer equations of the above-mentioned source and sink terms are given below:

Shortwave Solar Radiation $\quad H_{s(i+1)} = H_{s(i)} \times \exp(-k \times \Delta z)$
Atmospheric Longwave Radiation $\quad H_a = \epsilon_a \sigma T_{aa}^4$
Back Radiation $\quad H_{br} = \epsilon_{ws} \sigma T_{as}^4$
Evaporation $\quad H_e = \left[2.7(T_{swv} - T_{av})^{\frac{1}{3}} + 3.1 W_z\right](e_s - e_{air})$
Conduction $\quad H_c = 0.61\left[2.7(T_{swv} - T_{av})^{\frac{1}{3}} + 3.1 W_z\right](T_{sw} - T_{air})$

In Equation (4), $H_{s(i)}$ and $H_{s(i+1)}$ are shortwave solar radiation reaching the top and bottom of a water layer i (kcal/m²-day), k is an attenuation coefficient of water (1/m), and Δz is the thickness of the water layer (m). Equations (5) and (6) represent atmospheric longwave radiation and back radiation, respectively, where H_a is atmospheric radiation (kcal/m²-day), ϵ_a is the atmospheric emissivity directly related to air temperature and cloud cover, T_{aa} is the atmospheric absolute temperature (°K), σ is the Stefan–Boltzmann constant [11.7×10^{-8} cal/(cm² °K⁴ day)], and H_{br} is back radiation from the water surface. For back radiation estimation, the emissivity of water surface ϵ_{ws} is set constant (0.97), and the water temperature of the top mixed layer is used as T_{as}. Evaporation heat loss is one of the most complicated parts of water temperature calculations. Equations (7) and (8) represent heat transfer through evaporation (H_e) and conduction (H_c) in W/m², where T_{swv} and T_{av} are the virtual temperatures [21] of the water surface and the air, respectively, in °K; e_s and e_{air} are saturated and actual vapor pressure in millibars, and W_z is the wind speed at two meters above the surface (m/s). In Equation (8), the constant coefficient 0.61 is the Bowen ratio, and T_{sw} and T_{air} are surface water temperature and air temperature (°C), respectively. The total heat absorbed in a water layer, $HQ(i)$ (kcal/day) is calculated as:

$$HQ(i) = H_{sn(i)}\left[A_{(i)} - A_{(i+1)} exp(-k \Delta z_i)\right] \quad (9)$$

where, i = the horizontal layer number, ranging from 1 to a user-specified MBOT layer, which divides the lake into MBOT layers from the water surface to the lake bottom corresponding to the maximum depth H_{max}. $H_{sn(i)}$ is the shortwave solar radiation reached the top surface of a water layer, $A_{(i)}$ and $A_{(i+1)}$ are the areas of the top and bottom surfaces of the water layer, respectively, and Δz_i is the depth of the water layer.

To simulate the effect of the snow and ice layers, the snow and ice thickness submodels were integrated with the MINLAKE model [11,22].

2.2. Daily Dissolved Oxygen Model

For better prediction of water quality, the regional DO model [8] was integrated with the year-round water temperature model in MINLAKE2012. The model solves the vertical unsteady mass transport or diffusion Equation (10) to estimate vertical profiles of DO in a lake day by day over many years. The transport equation for dissolved oxygen is given:

$$\frac{\partial C}{\partial t} = \frac{1}{A}\frac{\partial}{\partial z}\left(K_z A \frac{\partial C}{\partial z}\right) + S \quad (10)$$

where $C(z, t)$ is the DO concentration (mg/L), K_z (m²/day) is the vertical turbulent diffusion coefficient for DO as a function of depth and time, and $S(z, t)$ represents the sum of all the source and sink terms of DO (mg/L-day). The DO source and sink equation is given by:

$$S = P - R - S_{SOD} - S_{BOD} + F_s \quad (11)$$

The main source of DO in a lake is oxygen production due to photosynthesis $P(z, t)$; surface reaeration F_s could be a source or sink term, while the main sinks of the DO are the respiration processes in the water body $R(z, t)$, sediment oxygen demand $S_{SOD}(z, t)$, and carbonaceous oxygen demand and nitrogenous oxygen demand represented together as $S_{BOD(z,t)}$:

Photosynthesis
$$P = k_g \times Chla$$
$$k_g = k_{gT} \times k_{gL}$$
$$k_{gT} = Pmax \times 1.036^{(T-20)}$$

Surface Aeration
$$F_s = K_e(C_s - C_1)\frac{A_s}{V(1)}$$

Respiration
$$R = \frac{1}{YCHO2} k_r \theta_r^{T-20} Chl_a$$

Biological Oxygen Demand
$$S_{BOD} = k_b \theta_b^{T-20} BOD$$

Sediment Oxygen Demand
$$S_{SOD} = \frac{S_b}{A}\frac{dA}{dz}$$

Equation (12) represents oxygen production by photosynthesis, where k_g is the photosynthetic oxygen production rate (mg O_2 (mg Chl-a)$^{-1}$ h^{-1}) in the MINLAKE model when biomass in a lake is represented using Chlorophyll-a concentrations (abbreviated as Chl-a). When the conditions of nutrients, sunlight, and water temperature are favorable, algal blooms may occur. With any of these conditions are limiting, algal growth will be restricted. This multiplicative relationship [23] is presented in Equation (13): k_{gT} includes temperature correction on the photosynthetic oxygen production rate, k_{gL} is the light limitation determined using the two-parameter Haldane kinetics equation to describe light-limited growth and inhabitation [24], and the limitation due to nutrients is directly represented or linked to Chl-a concentrations. $Pmax$ in Equation (14) is the maximum photosynthesis rate for oxygen production in mg O_2 (mg Chl-a)$^{-1}$ h^{-1}. Reaeration is simulated based on the gas-transfer theory presented by Holley [25] using k_e, the bulk surface oxygen transfer velocity (m/day) in Equation (15), where A_s is the lake surface area (m^2), C_1 (mg/L) and $V(1)$ are the oxygen concentration and water volume in the surface layer, and C_s is the DO saturation concentration (mg/L) as a function of surface temperature and lake elevation. Equation (16) represents plant respiration as a first-order kinetic process where $YCHO2$ is a yield coefficient representing the ratio of mg chlorophyll-a to mg oxygen (0.008), k_r is the respiration rate coefficient (day^{-1}), and θ_r is the temperature adjustment coefficient. Equation (17) represents biochemical oxygen demand (BOD) as a function of detrital mass expressed in oxygen equivalents. Here k_b is the first order decay coefficient (day^{-1}), θ_b is the temperature adjustment coefficient, and BOD is detritus as oxygen equivalent (mg/L). In the regional DO model and MINLAKE2018, $k_r = k_b = 0.1$ day^{-1} and $\theta_r = \theta_b = 1.047$ are used [8]. Sedimentary oxygen demand (SOD) from the lake bottom sediments is a boundary condition (sediment surface flux), but since it occurs for each layer it is treated as a sink term in the one-dimensional (vertical) transport equation [26]. Sedimentary oxygen flux per control volume ($A*dz$) is given by Equation (18), where S_b is the sedimentary oxygen demand coefficient (g O_2/m^2-day), which is directly related to lake trophic status. There are several factors that are commonly considered responsible for SOD variation in a water body. The most important factor responsible for SOD variation is the temperature near the sediment-water interface [27], the velocity of the water overlying the sediment [28], the organic content of the sediment, and the oxygen concentration of the overlying water [3].

2.3. Sediment Temperature Simulation

One of the important issues in modeling water temperature is heat transfer between water and sediment. However, in shallow lakes, the direction of the heat flux reverses frequently on daily or hourly timescales. Therefore, sediment heat exchange has been included in the year-round daily water temperature model by Fang and Stefan [29] for all layers, from the water surface to the lake bottom. The sediment temperature model simulates sediment temperature up to 10 m below the sediment/water interface (divided into 10 layers) using a heat conduction equation.

$$\frac{\partial T_s}{\partial t} = K_s \left(\frac{\partial^2 T_s}{\partial z^2} \right) \tag{19}$$

where T_s (°C) is the sediment temperature and K_s (m^2/day) is the sediment thermal diffusivity. The boundary conditions for the sediment temperature model are given by the following equations:

$$T_s = T_{w(i)} \text{ at the water} - \text{sediment interface}$$
$$\frac{\partial T_s}{\partial z} = 0 \text{ at 10 m below the lake sediment} \tag{20}$$

where $T_{w(i)}$ is the simulated water temperature in the water layer (i) at the previous time step. The first boundary condition assures the continuity of temperature at the water-sediment interface. The second boundary condition implies an adiabatic boundary (there is no heat transfer) at 10 m below the sediment surface, where seasonal temperature fluctuations are damped out. The initial sediment temperature at the sediment-water interface is set equal to the initial water temperature at the sediment surface. Sediment temperature increases/decreases exponentially with sediment depth until it approaches a constant value at 10 m below the sediment/water interface. Sediment temperatures at 10 m below the sediment-water interface (T_{S10}) at different water depths are very important input data (depending on the geographic location of the lake) that have been studied by Fang and Stefan [30].

2.4. Modifications to the Daily Model

To change the daily MINLAKE2012 model into an hourly model (MINLAKE2018), several modifications were made to the temperature and dissolved oxygen models. The first modification was to change the time step of the model from 1 day to 1 h. A new variable 'dt' was introduced in MINLAKE2018 program: $dt = 1$ for the daily model, $dt = 1/24$ for the hourly model, and dt can be changed to other time steps in the future, which depends on available weather data with shorter time intervals (e.g., 2 or 3 h). Since water quality variables in lakes change in response to the ambient weather conditions, in the daily MINLAKE2012 model, daily weather data (air temperature, dew-point temperature, wind speed, solar radiation, cloud cover ratio or sunshine percentage, and precipitation, specially, snowfall for the winter ice cover simulation) are used as input. For MINLAKE2018, available hourly weather data, especially air temperature and solar radiation, are necessary and have been prepared/used for the hourly model after revising the code to read hourly weather input.

In addition to changing the time step and weather input file, model coefficients that are either physical-based or determined empirically for the daily MINLAKE2012 model were revisited and reevaluated, and they were modified to represent the hourly values by multiplying by dt (for most cases). For the temperature model, the hourly solar radiation is directly used in Equation (4); other heat source and sink terms for the daily model were multiplied by dt and used the simulated hourly water temperature. The same modification approach was applied to coefficients required for estimating dissolved oxygen sink terms (such as BOD, SOD, and plant respiration) and surface reaeration. The maximum or optimal photosynthetic oxygen production rate $Pmax$ (Equation (14)) at 20 °C is 9.6 mg O$_2$ (mg Chl-a)$^{-1}$ h^{-1} [24,31] so that no change is needed. In the daily MINLAKE2012 model, daily solar radiation was redistributed as a sinusoidal function over the photo period and used to compute hourly oxygen production by photosynthesis that was summed as daily oxygen production [8]. In the hourly MINLAKE2018 model, hourly solar radiation from the weather input file was used directly to compute the hourly photosynthetic oxygen production. The vertical heat diffusion coefficient K_z (m^2/h) was calculated using Equation (21) [21,32] in epilimnion and Equation (2) in metalimnion and hypolimnion:

$$K_z = 1.74 \times W \times dt \qquad (21)$$

K_z in Equation (2) for the daily model varies with lake surface area but does not change with time during a simulation, but hourly K_z in Equation (21) is a function of hourly wind speed W in mph (mile/h). Wang et al. [33] reported that the wind speed showed significant correlation to the half-hourly turbulent heat flux and energy budget over a small lake in open water season. The wind speed is the most significant factor governing physical processes (evaporation, heat flux, and energy budget) in lakes for time periods shorter than daily [19,33]. Moreover, the wind speed changes throughout the day and causes the short-term mixing in the lakes. Since the short-term mixing was not of interest in daily model MINLAKE2012, wind speed was not used in turbulent diffusion coefficient calculation but in computing the wind (kinetic) energy applied on the lake surface, which is balanced with the potential energy of lifting colder/denser water at lower depths to determine the mixed layer depth in each day [8]. Equation (21) was compared with Equation (2) and another equation ($K_z = 28 \times dt \times W^{1.3}$) used by Riley and Stefan (1987); Equation (21) seemed to perform the best for hourly model, though additional study on specifying K_z is needed. The unsteady heat transfer Equation (1) and dissolved oxygen transport Equation (10) were then used to simulate hourly temperature and dissolved oxygen through all layers of the lake.

In this study, the sediment temperature model was also modified for shallow lakes. First, the thermal diffusivity K_s in Equation (19) was multiplied by dt to account for the hourly time step. To accurately simulate the sediment temperature gradient, the 10-m sediment below the lake bottom was divided into 20 layers each of 0.5 m thickness, while it was 10 layers for the daily model. In shallow lakes, solar radiation is likely to penetrate the whole water depth and directly heat the bottom sediment below the water. In the daily model, the heat absorbed by any water layer was quantified as a subtraction of the heat reached on the top and bottom surfaces of the layer as shown in Equation (9), which means all the heat reaching the area $A(i) - A(i+1)$ was used to warm up water only (not directly heat the bottom sediments). Therefore, direct solar heating to sediment was ignored even though heat exchange between the water and sediment was considered [12] when sediment temperature is simulated. The heat absorbed by a water layer due to radiation attenuation (k) was modified as:

$$HQ(i) = H_{sn(i)} A_{(i+1)}[1 - \exp(-k\Delta z_i)] + \int_0^{A_{(i)} - A_{(i+1)}} H_{sn(i)}[1 - \exp(-kdz)]dA \qquad (22)$$

The second part of Equation (22) more accurately accounts for absorbed solar radiation by water in the area of $A(i) - A(i+1)$ since some solar radiation heats the bottom sediment (and not the water). Since lake horizontal area $A(i)$ is not uniform across all water depths due to slope gradient, small area dA and depth dz were introduced for the integration. Equation (22) was integrated and further simplified into Equation (23) by assuming each horizontal area is circular:

$$HQ(i) = H_{sn(i)} A_{(i+1)}[1 - \exp(-k\Delta z_i)] + H_{sn(i)}\left(A_{(i)} - A_{(i+1)}\right) - HQsed(i) \qquad (23)$$

$HQsed(i)$ becomes a new heat source term for the first sediment layer at the water layer i to more accurately simulate the sediment temperature profile using Equation (19). Heat reaching the lake sediment is $HQsed(i)$ (kcal/h) and calculated using Equation (24):

$$HQsed(i) = \frac{2\pi H_{sn(i)}}{k \tan \alpha} \left[\left(r_{(i)} - \frac{1}{k \tan \alpha}\right) - \left(r_{(i+1)} - \frac{1}{k \tan \alpha}\right) \exp(-k\Delta z_i) \right] \qquad (24)$$

where $r_{(i)}$ and $r_{(i+1)}$ are the radius of the top and bottom surface areas of the water layer i, respectively; and tanα is equal to $[r_{(i)} - r_{(i+1)}]/\Delta z_i$ and approximates the slope of the lake bottom for layer i. These changes subsequently may change the sediment temperature profile and sediment heat flux of the lake.

2.5. Modeled Shallow Lakes

After making the modifications for the hourly model, five shallow lakes were selected for the current study (Table 1): Carrie Lake, Pearl Lake, Belle Lake, Portage Lake, and Red Sand Lake. The maximum depths of these lakes range from 4.5 to 7.9 m. All of the selected lakes are located in northeastern Minnesota. The nearest weather station to the lake provided weather data: St. Cloud Regional Airport for Belle, Carrie, and Pearl lakes, and Brainerd Lakes Regional Airport for Portage and Red Sand lakes. The geometry ratio ($A_s^{0.25}/H_{max}$, where A_s in m^2 and H_{max} in m are the surface area and the maximum depth of the lake, respectively) is a very important characteristic parameter of a lake that affects stratification, lake habitat, etc. Since all the study lakes have a geometry ratio between 3 and 10, they are weakly stratified. The lower the geometry ratio, the higher the stratification of a lake. From Table 1, Carrie Lake is relatively more stratified (geometry ratio 3.12 m$^{0.5}$) and Red Sand Lake is the least stratified (geometry ratio 8.34 m$^{0.5}$). Based on chlorophyll-a concentration, Belle, Pearl and Portage lakes are eutrophic (mean Chl-a > 10 µg/L, [34]) and Carrie and Red Sand lakes are mesotrophic (mean Chl-a between 4 and 10 µg/L, [34])

Table 1. Characteristics of five study lakes in Minnesota.

Lake	Surface Area, A_s, (km^2)	Max. Depth H_{max}, (m)	Geometry Ratio [1] (m)$^{0.5}$	Mean Chl-a (µg/L)	Trophic Status	Simulation Years	Number of Days with Profile Data
Carrie	0.37	7.90	3.12	6.71	Mesotrophic	2008–2012	50
Belle	3.71	7.60	5.77	27.10	Eutrophic	2008–2012	73
Pearl	3.05	5.55	7.53	16.91	Eutrophic	2008–2012	36
Portage	1.54	4.57	7.71	15.98	Eutrophic	2008–2015	86
Red Sand	2.11	4.57	8.34	4.43	Mesotrophic	2008–2015	87

Note: [1] Geometry ratio of a lake is $A_s^{0.25}/H_{max}$ where A_s in m^2 and H_{max} in m.

3. Modeling Results

3.1. Model Calibration

The hourly model was calibrated and validated based on available measured water temperature and dissolved oxygen profile data at a specific hour in particular days that were downloaded from LakeFinder (https://www.dnr.state.mn.us/lakefind/index.html, accessed on 12 March 2019). The temperature model of MINLAKE2018 was calibrated first and then for the dissolved oxygen model in each of the five lakes. All the calibrated parameters are listed in Table 2. EMCOE is an array of empirical coefficients introduced from the MINLAKE2012 program [35]. The classic approach to modeling is to divide the observed data into two parts and use them for calibration and validation, respectively; due to the scarcity and discontinuity of profile data in the study lakes (Table 1), all profile data were used together for model calibration. Fortunately, all five lakes had 30-min measured near-surface water temperatures over several years, and Pearl Lake had 30-min measured temperature at six depths; these 30-min data were used for validation purposes. A comparison of suggested and calibrated values of model calibration parameters is presented in Table 3. The maximum photosynthesis rate for oxygen production Pmax was not used as a calibration parameter in previous studies and had some variations in these five shallow lakes during the calibration, indicating that further study of Pmax variation is necessary in future studies.

Table 2. Calibration parameters for hourly and daily MINLAKE models [35].

Calibration Parameter	Effect on Model Results	Description of the Parameter
Wstr	Temperature and DO profiles	Wind sheltering coefficient
BOD	DO Profiles	Biochemical oxygen demand depending on lake trophic status
Sb20	DO Profiles	Sediment oxygen demand, lake tropic dependent
EMCOE(1)	Temperature and DO Profiles	Multiplier for diffusion coefficient in the metalimnion
EMCOE(2)	DO Profiles	Multiplier for SOD below the mixed layer
Pmax	DO Profiles	Maximum photosynthesis rate for oxygen production

Table 3. Suggested and calibrated values of model calibration parameters for five study lakes.

Parameter/Lakes	Red Sand Lake	Portage Lake	Carrie Lake	Pearl Lake	Belle Lake
Wstr	0.47 (0.47)	0.37 (0.37)	1 (1.0)	0.6 (0.4)	0.67 (1.0)
BOD	0.75 (0.75)	1.5 (1.5)	1 (0.75)	0.75 (1.5)	1.5 (1.0)
Sb20	0.75 (0.75)	1.5 (1.5)	1.5 (0.75)	0.75 (1.5)	1.5 (1.8)
EMCOE(1)	1 (7)	1 (3)	1 (3)	1 (0.8)	1 (4)
EMCOE(2)	1 (3)	1.1 (1)	0.82 (1.2)	1 (0.7)	1 (0.5)
Pmax	9.6 (16.8)	9.6 (8.5)	9.6 (9.6)	9.6 (8.5)	9.6 (7.7)

Note: the suggested value for each parameter is given followed by calibrated value inside brackets.

Figure 1 shows an example how the calibration affects the water temperature and DO profiles in Belle Lake. Wind sheltering coefficient is an important calibration parameter since wind is very important for mixing in a lake. However, the wind is usually obstructed by tree canopy surrounding a lake. The wind sheltering coefficient in MINLAKE model accounts for the effective portion of wind which takes part in mixing (ranges from 0 to 1). Wind sheltering coefficient of 1 ensures the wind is fully used to mix a lake water without any obstruction; therefore, it should be called the wind utilization coefficient in the future. Figure 1 shows that, on 7 July 2010 in Belle Lake, as we increased the wind sheltering coefficient, there was increased mixing and the water temperature at the hypolimnion matches better with the observed data (compared with wind sheltering coefficient of 0.67). However, on 22 September 2008, the wind sheltering coefficient affects the mixed layer depth. As the wind sheltering coefficient changed from 0.67 to 1, the mixed layer depth increased from 4.5 m to 6.2, complying with the observed data.

The results summarized in Table 4 show that for these shallow lakes, the hourly MINLAKE2018 model performed better than the daily model, especially for DO simulations, when simulated profiles were compared with observed profiles. The main reason for this improvement is the consideration of diurnal changes as illustrated by comparison of the observed and simulated temperature or DO at observed time. Table 4 shows the model performance improved significantly with the hourly model in Portage and Carrie lakes. The average root-mean-square error (RMSE) of temperature simulations in five lakes from hourly MINLAKE2018 decreased by 17.3%, and average Nash-Sutcliffe model efficiency (NSE) [36] increased by 10.3% with respect to daily model MINLAKE2012. Similarly, for the DO hourly model, average RMSE decreased by 18.2% and NSE increased by 66.7%.

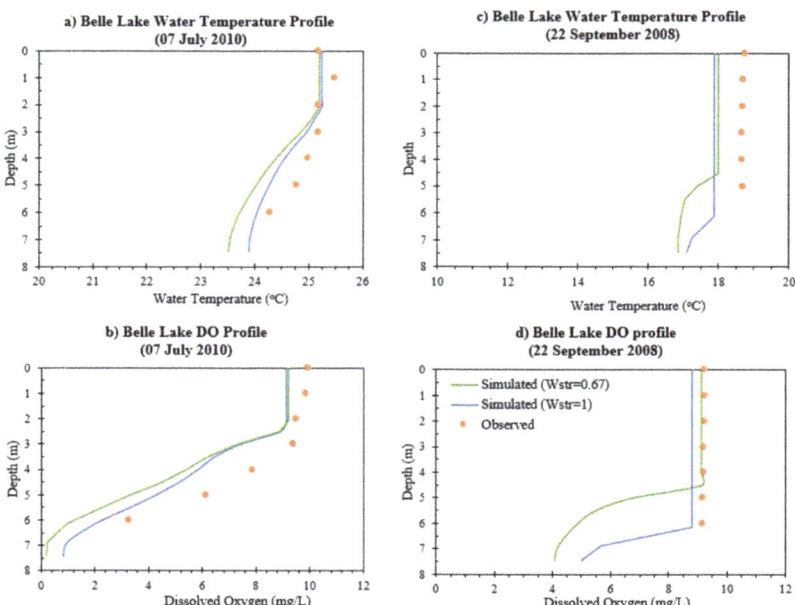

Figure 1. Effect of wind sheltering coefficient (increasing from 0.67 to 1) on simulated (**a**) water temperature and (**b**) dissolved oxygen profiles on 7 July 2010; (**c**) water temperature and (**d**) dissolved oxygen profiles on 22 September 2008 in Belle Lake including observed profiles.

Table 4. Statistical error parameters for the hourly and daily MINLAKE models against observed profile data in five study lakes.

Lake Name	Hourly Model (MINLAKE2018)					
	Water Temperature			Dissolved Oxygen		
	RMSE [1] (°C)	NSE [2]	Slope [3]	RMSE (mg/L)	NSE	Slope
Carrie Lake	2.21	0.85	1.04	1.69	0.78	0.96
Pearl Lake	1.03	0.98	0.98	2.23	0.35	1.00
Belle Lake	1.03	0.96	1.03	1.53	0.69	1.00
Red Sand Lake	1.86	0.94	0.97	2.77	0.36	0.99
Portage Lake	1.41	0.97	0.98	1.91	0.31	0.99
Average ± STD [4]	1.48 ± 0.32	0.96 ± 0.02	0.98 ± 0.01	2.02 ± 0.49	0.50 ± 0.22	0.99 ± 0.02
Lake Name	Daily Model (MINLAKE2012)					
Carrie Lake	2.47	0.77	1.08	2.76	0.42	0.92
Pearl Lake	1.04	0.97	0.98	2.58	0.13	0.98
Belle Lake	1.14	0.96	1.01	2.09	0.43	0.94
Red Sand Lake	2.48	0.79	0.97	2.90	0.29	0.98
Portage Lake	1.82	0.86	1.03	2.03	0.22	0.96
Average ± STD	1.79 ± 0.69	0.87 ± 0.09	1 ± 0.03	2.47 ± 0.39	0.30 ± 0.13	0.96 ± 0.03

Note: [1] RMSE stands for Root Mean Square Error, [2] NSE for Nash-Sutcliffe Efficiency [36], [3] Slope of linear regression between simulated and observed, [4] STD for Standard Deviation.

Figure 2 shows the observed water temperature and DO profile data versus simulated results using hourly and daily models in two lakes. In Figure 2a, the observed water temperatures versus daily simulated have a slope of 1.03 which means the daily model slightly underpredicts the observed water temperatures in Portage Lake. The hourly model slightly overpredicts the observed water temperatures having a slope of 0.98. In Figure 2b, the daily DO model has a slope of 0.96 whereas the hourly model has a slope of 0.99 (near 1) and performs better. Both the models slightly overpredict the observed DO. In Figure 2c, Carrie Lake daily water temperature has a slope of 1.08 whereas the hourly water temperature has a slope of 1.04. Carrie Lake daily DO has a slope of 0.92 whereas the hourly DO has a slope of 0.96. Figure 2 shows that the simulated hourly water temperature and DO match better with observed data compared to the daily model, although DO model performance still has room for improvement.

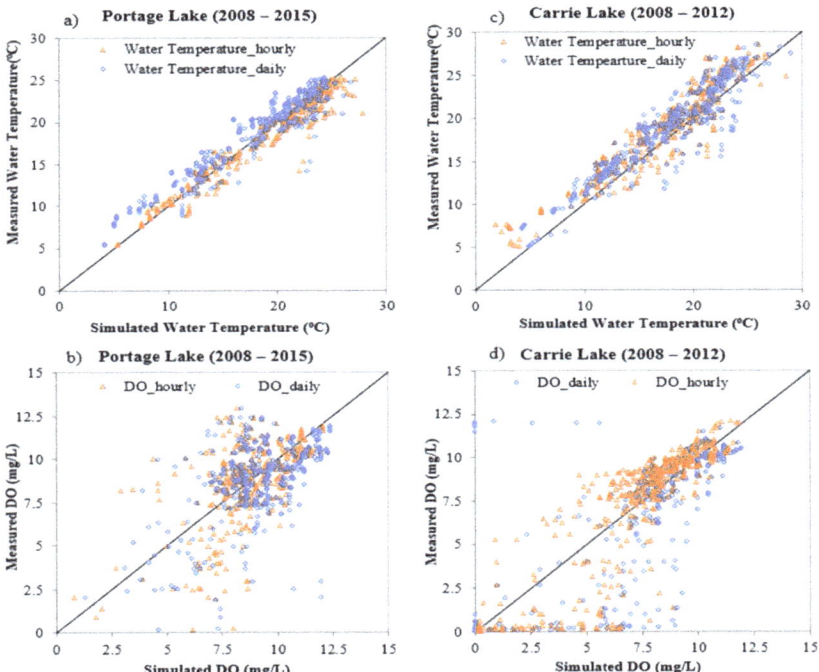

Figure 2. Measured versus simulated water temperature and DO in Portage Lake and Carrie Lake for simulation over several years.

3.2. Diurnal Variations

In this study, the daily MINLAKE2012 model has been successfully revised into an hourly MINLAKE2018 model by making various program changes discussed in Section 2.4. In the daily MINLAKE2012 model, the output is a single profile for the whole day. Simulated water temperatures and DO concentrations from the daily model are typically close to those of later afternoon hours. For the hourly model, 24 profiles of water temperature and DO were simulated for a day giving more detail into diurnal changes and lake processes. Figure 3 shows the simulated and observed maximum and minimum hourly surface water temperature of Portage Lake and Belle Lake during each day in 2009. The Minnesota Department of Natural Resources (MNDNR) had a long-term monitoring program to continuously measure surface water temperatures (~1 m) in a 30-min interval for 25 sentinel lakes including the five study lakes (Table 1), and a few lakes (such as Pearl Lake) had measured time-series at several depths also. Therefore, the maximum and minimum observed hourly temperatures in each day were extracted from the 30-min monitored database (MNDNR, 2018) to compare with simulated hourly results. The simulated maximum and minimum water temperatures (Figure 3) have a root mean square error (RMSE) against observed ones of 1.04 °C and 0.90 °C for Belle and 1.32 °C and 1.44 °C for Portage in 2009, respectively. At Portage Lake, the daily simulated surface temperature is close to the minimum observed surface temperature except for September. At Belle Lake, the daily simulated temperature matches well with the observed minimum hourly temperature until April, then the daily temperature lies between the maximum and minimum hourly temperature; matches with the minimum (observed or simulated) again from July to August and October to December. During September, the daily simulated temperature matches well with the maximum hourly temperature. During the winter, observed minimum and maximum temperatures are lower than simulated ones (Figure 3) because there is a strong temperature increase gradient below the ice-water interface and the water depth of the temperature sensor should reduce the ice thickness (vary with time) from the open-water depth. The diurnal temperature variations can be quantified by differences of the maximum and minimum hourly temperatures in each day and are 0.05–9.47 °C for the simulated results (mean difference 4.03 °C with a standard deviation of 2.48 °C) and 0.05–11.41 °C for the observed data (mean difference 3.69 °C with a standard deviation of 2.14 °C) at Portage Lake in 2009. The maximum water temperature occurs at 5 p.m. for 92% of the days in the open water season and at 2 p.m. for 85% of the days in the ice cover period. The minimum water temperature mostly occurs at 5 a.m. all year round (100% of days in open water season and 95% of days in the ice cover period). The average absolute difference of simulated and observed diurnal temperature variations is 0.96 °C at Portage Lake and 0.97 °C at Belle Lake in 2009. The average diurnal DO variations in each day are 0.60 mg/L (standard deviation of 0.71 mg/L and the maximum variation of 2.41 mg/L) for Belle Lake and 0.49 mg/L (standard deviation of 0.50 mg/L and the maximum variation of 5.63 mg/L) for Portage Lake for simulation results. In 2009, at Portage Lake, the maximum DO occur at 5 p.m. (because of continuous photosynthetic oxygen production in the daytime) for 95% of the days in open water season and 88% of the days in the ice cover period. The minimum DO mostly occurs at 7 a.m. all year round (95% of days in the open water season and 89% of days in the ice cover period) because of no photosynthesis during the night. There are no continuous hourly DO measurements available in the five lakes to compare with hourly simulated DO time series.

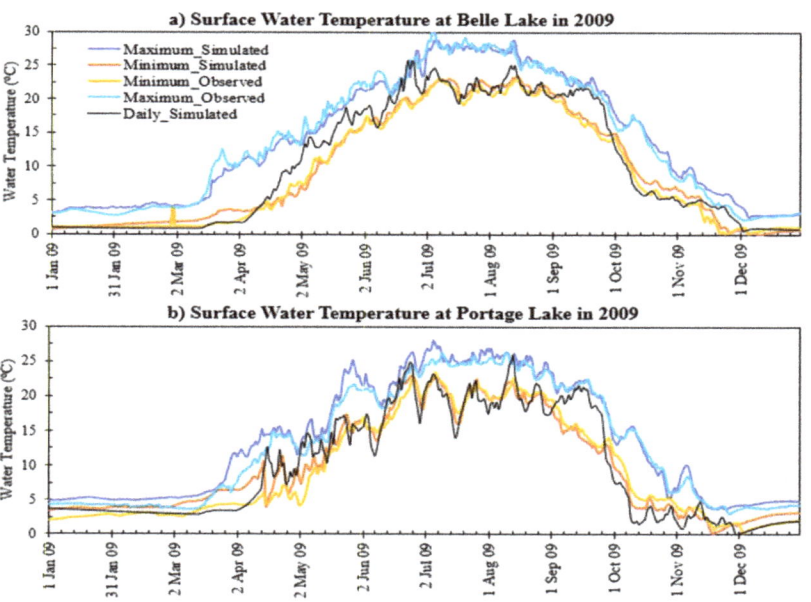

Figure 3. Time series of simulated and observed maximum and minimum hourly surface water temperatures each day at (**a**) Portage Lake and (**b**) Belle Lake in 2009 including simulated daily temperatures.

In Figure 4 the simulated daily temperatures from the daily MINLAKE2012 matched well with the observed and simulated temperatures at 4 p.m. for Belle Lake and Portage Lake over multiple years. In summer months, the simulated daily temperatures were slightly higher than observed ones but in other months slightly lower than observed. The mean difference of daily simulated water temperatures and observed water temperatures at 4 PM is 1.03 °C (standard deviation of 1.44 °C) in Belle Lake and 1.86 °C (standard deviation of 1.05 °C) in Portage Lake.

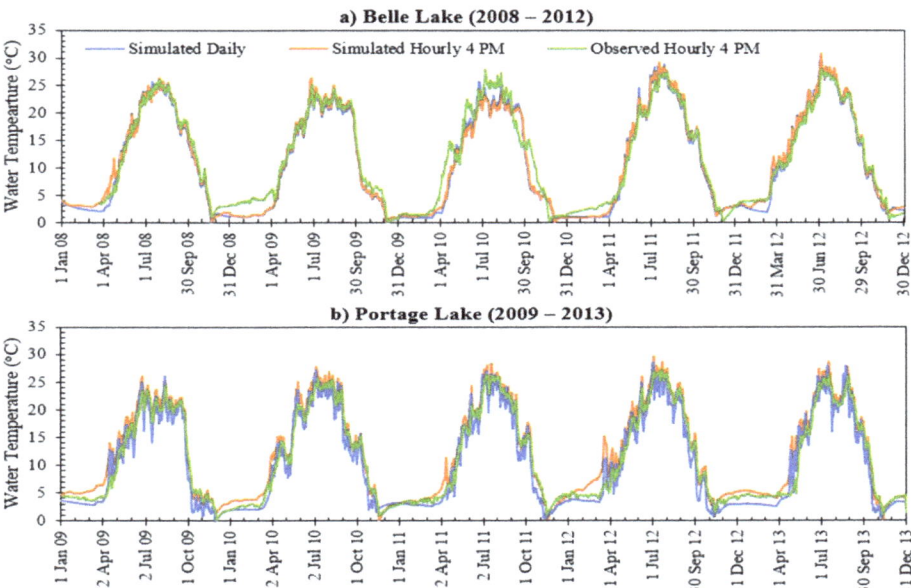

Figure 4. Time series of simulated daily, simulated and observed hourly surface (1 m at Belle and 1.5 m at Portage Lake) temperatures at 4 PM at Belle Lake (2008–2012) and Portage Lake (2009–2013).

The simulated hourly temperatures at 4 PM closely follow with the observed hourly temperatures at 4 PM for 4 years in Belle and Portage Lake. The simulated and observed water temperatures at 4 PM have a mean absolute difference of 1.35 °C (standard deviation of 1.29 °C) in Belle Lake and 1.21 °C (standard deviation of 1.23 °C) in Portage Lake. For Belle Lake, simulated temperatures in 2009 winter and 2010 summer had relatively large differences from observed data; for Portage Lake, simulated temperatures slightly over-predicted in 2009–2013.

3.3. Profile Comparison

Figure 5 shows example of simulated water temperature and DO profiles for five different times throughout the day (6–24 h). Figure 5 also includes daily profiles simulated from daily MINLAKE2012 and the observed profiles in two days (water temperature) in Pearl Lake and two days (DO) in Belle Lake. The observed profiles on both days (5 June and 15 July 2008) were collected at 10:00 a.m. Therefore, the simulated profiles at 10:00 a.m. closely match the observed data. The hourly model outputs show relatively large diurnal variations in epilimnion and metalimnion, and simulated variations of DO in hypolimnion are larger than corresponding variations of temperature because there are additional DO sink terms (Equation (11)). The simulated daily profiles are reasonably close to simulated profiles at 4:00 p.m. (16:00), which is in agreement with the surface water temperature comparison shown in Figure 4. The daily model seems to overpredict stratification (surface and bottom temperature or DO difference) in those days and both lakes. On 15 July 2008, the hourly model predicts the surface DO correctly whereas overpredicts the DO in the hypolimnion at Belle Lake. For the daily model, the measured profile was compared with the simulated daily profile of water temperature or dissolved oxygen, regardless of the hour of measurement. However, for the hourly model, each observed profile can be compared with the simulated hourly profile in the observed hour and day. Simulated surface temperatures on 15 and 31 July 2008 have a diurnal variation of 3.41 and 1.63 °C in Pearl Lake, and simulated surface DO on 12 June and 15 July 2008, have a diurnal variation of 0.85 and 1.69 mg/L in Belle Lake, respectively.

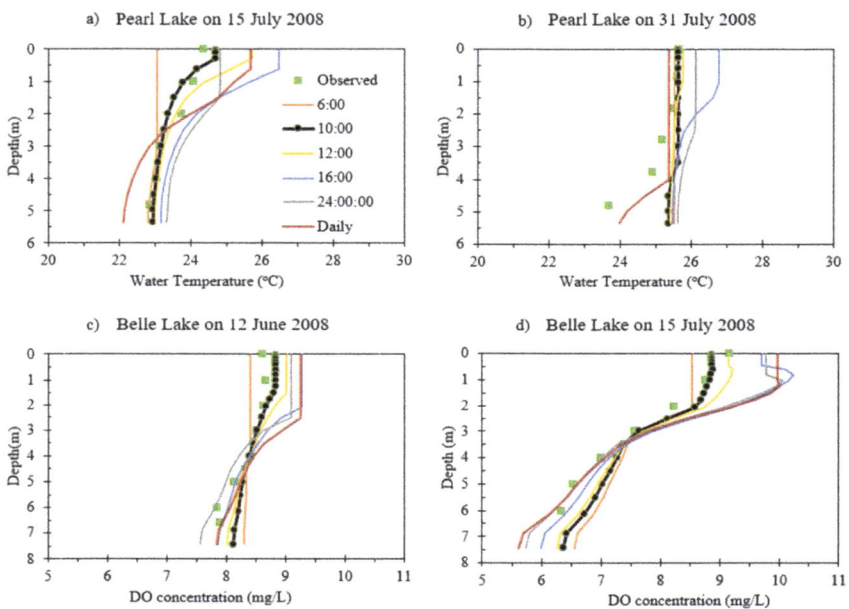

Figure 5. Simulated water temperature profiles in Pearl Lake and DO profiles in Belle Lake at five different hours comparing with observed (green squares) and simulated daily profiles.

3.4. Comparison of Long-Term Surface Temperature Simulation

Table 5 lists the statistical parameters (RMSE, NSE, and slope) of the simulated hourly water temperatures against observed data in the five study lakes. The average RMSE of long-term water temperature simulation in the five lakes was 1.50 °C with a standard deviation of 0.32 °C. The Nash-Sutcliffe model efficiency or NSE ranges from 0.95 to 0.99, which is close to the optimal value of 1.0, and indicates the developed hourly model performs very well in comparison to observed hourly data. The slopes of 0.97–0.99 also show a good match between simulated and observed water temperature. Since Pearl Lake has observed water temperature at six depths, the error parameters were calculated at surface depth (1.2 m) and other five depths (1.7 m, 2.4 m, 3.4 m, 4.4 m, and 5 m). For Pearl Lake, the average RMSE for water temperature simulation at six different depths is 1.30 °C with a standard deviation of 0.15 °C. Figure 6 graphically shows the good performance of the hourly MINLAKE2018 against the available hourly measured data.

Table 5. Statistical error parameters for the hourly MINLAKE model against observed time-series data.

Surface Depths	Carrie Lake	Pearl Lake	Belle Lake	Red Sand Lake	Portage
RMSE	1.82	1.22	1.19	1.95	1.33
NSE	0.95	0.98	0.98	0.94	0.99
Slope	0.98	0.98	0.99	0.97	0.99
Pearl Lake	1.7 m	2.4 m	3.4 m	4.4 m	5.0 m
RMSE	1.08	1.18	1.47	1.47	1.42
NSE	0.98	0.98	0.97	0.96	0.97
Slope	0.98	0.99	0.98	0.97	0.98

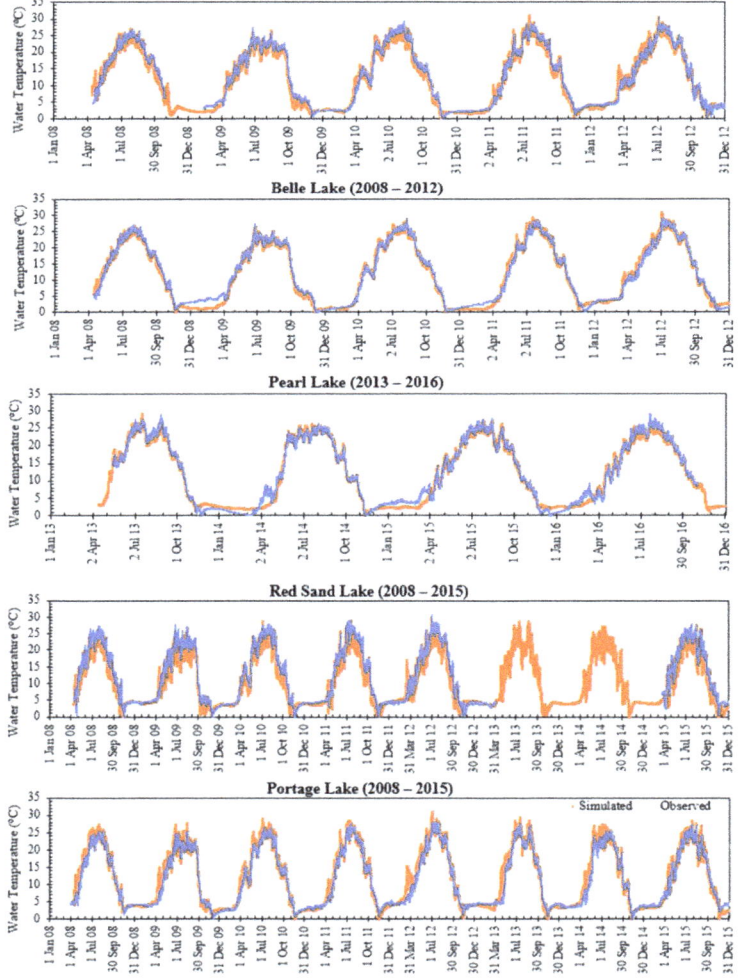

Figure 6. Time series of observed (orange) and simulated (blue) hourly surface water temperatures in five study lakes over 4–8 years of simulation.

3.5. Comparison of Heat Flux, DO Production and Reaeration

Figure 7a plots hourly air temperature and solar radiation variations and Figure 7b shows the time series of calculated hourly and daily heat fluxes that enter and exit the water surface of Carrie Lake in ten days (1 June to 10 June 2009). The daily heat flux was averaged and plotted over 24 h (Figure 7). The hourly flux in (sum of H_{sn} and H_a in Equation (3)) has a clear diurnal variation for most of the days while the hourly flux out (sum of H_{br}, H_c, and H_e in Equation (3)) has almost no diurnal variation but some fluctuations in each day. At night when the solar radiation H_{sn} is absent, water loses heat to the atmosphere as the flux out is greater than the flux in, while during the day, due to the increase of shortwave solar radiation, the water body gains heat to increase the water temperature in epilimnion (Pearl Lake in Figure 5). The daily model has a constant heat flux in and flux out over a 24-h period whereas the hourly model considers the heat flux variations hour by hour (24 values for each day). As a result, the hourly model can represent daily variations more accurately than the daily model, which is evident in Figure 7b. Results from the daily model illustrate that heat could transfer in one direction (cooling or warming) for several consecutive days while the results from the hourly model show that heat transfer from and to the waterbody is a more dynamic process that occurs within the day and depends on the time of day.

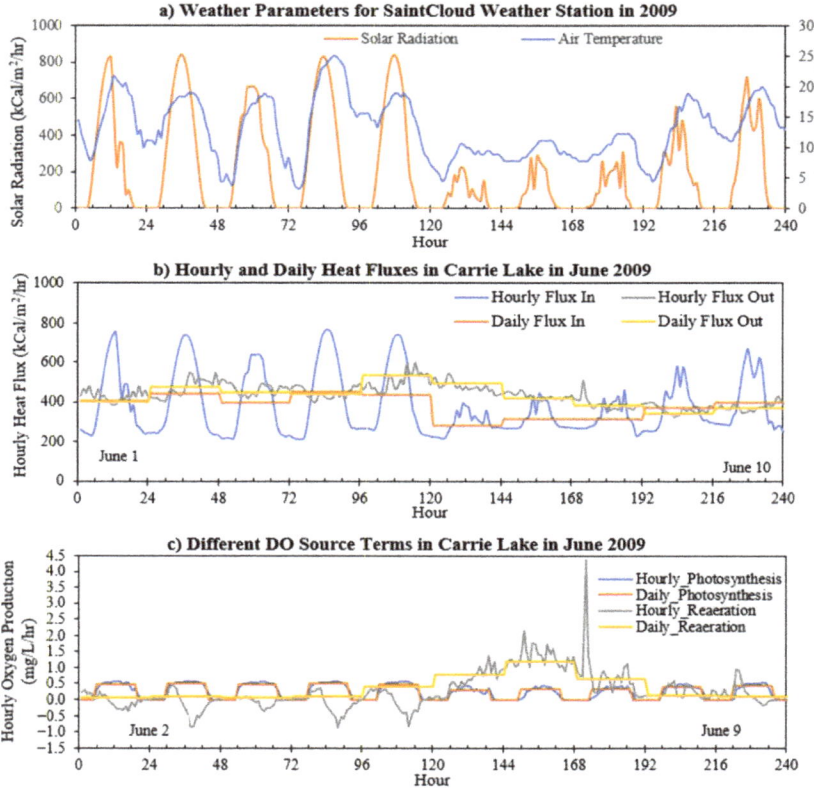

Figure 7. Time series of (**a**) air temperature and solar radiation, (**b**) calculated daily and hourly heat fluxes, and (**c**) two DO source terms through the water surface at Carrie Lake on 1–10 June 2009.

Figure 7c shows a comparison of the time series of dissolved oxygen source terms (photosynthesis and surface reaeration) calculated by the hourly and daily models (expressed as mg/L/h). Hourly reaeration during most of the day is a sink term for the first four days of June when saturated DO (C_s in Equation (15)) is less than the surface DO concentration. On 5 June, the hourly reaeration rate is still negative whereas the daily reaeration rate is positive. From 6 June to 10 June, hourly reaeration becomes positive which means that saturated DO is higher than surface DO. On 8 June after midnight, the hourly reaeration rate suddenly increases and becomes 4.32 mg/L/h at 4 a.m. because of a strong wind speed of 19 mph (the average wind speed was 6.18 mph for the first 10 days of June at Saint Cloud). The daily model cannot account for hourly variations and has a constant reaeration rate of 0.57 mg/L/h. In the daily model, hourly photosynthesis was calculated by redistributing daily solar irradiance as a sine function over the photoperiod (14 h in June) and added together to get a daily oxygen production by photosynthesis for solving the daily DO balance equation [10], while the hourly model uses hourly solar radiation from the weather data file to get the hourly photosynthetic oxygen production and solve the hourly DO balance equation. In Figure 7c daily photosynthesis was averaged over the photoperiod starting from 6:00 a.m. to compare with hourly photosynthesis. In the presence of light, hourly photosynthetic oxygen production increases and then becomes zero during the night when there is no oxygen production. Figure 7c shows that both the hourly and daily models have similar estimates on photosynthetic oxygen production but differ in surface aeration.

3.6. Impact of Direct Solar Radiation Heating on Sediment Bed

In previous MINLAKE models, the direct solar radiation heating of sediments was ignored. In the hourly MINLAKE model, the sediment heating code was modified to account for the solar radiation directly heating the sediment layers. In Figures 8 and 9, a comparison was made between the previous and modified sediment heat transfer models for Carrie Lake in March and June of 2009.

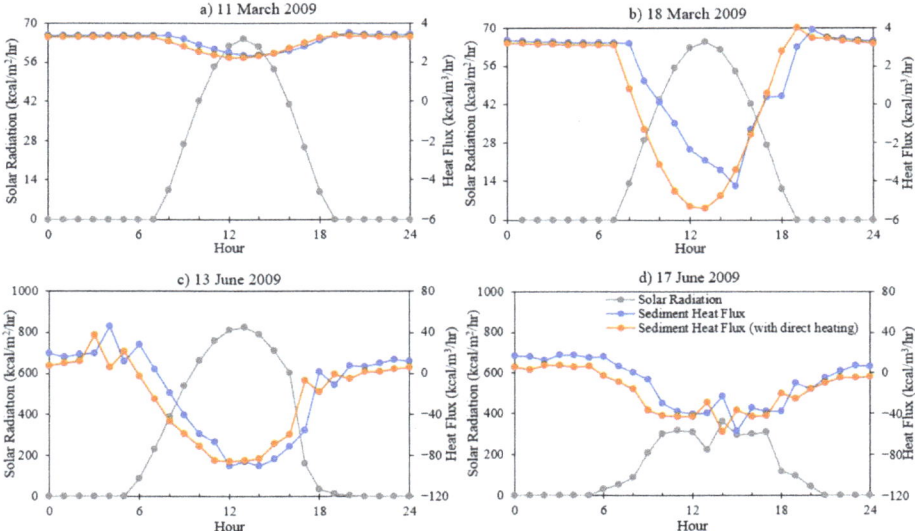

Figure 8. Comparison of heat flux calculated by the modified and previous sediment models on an hourly basis on (**a**) day with ice and snow cover, (**b**) day with ice cover but no snow cover, (**c**) high solar radiation day in June, (**d**) low solar radiation day in June. The y-axis scales on (**a**,**b**) are different from on (**c**,**d**).

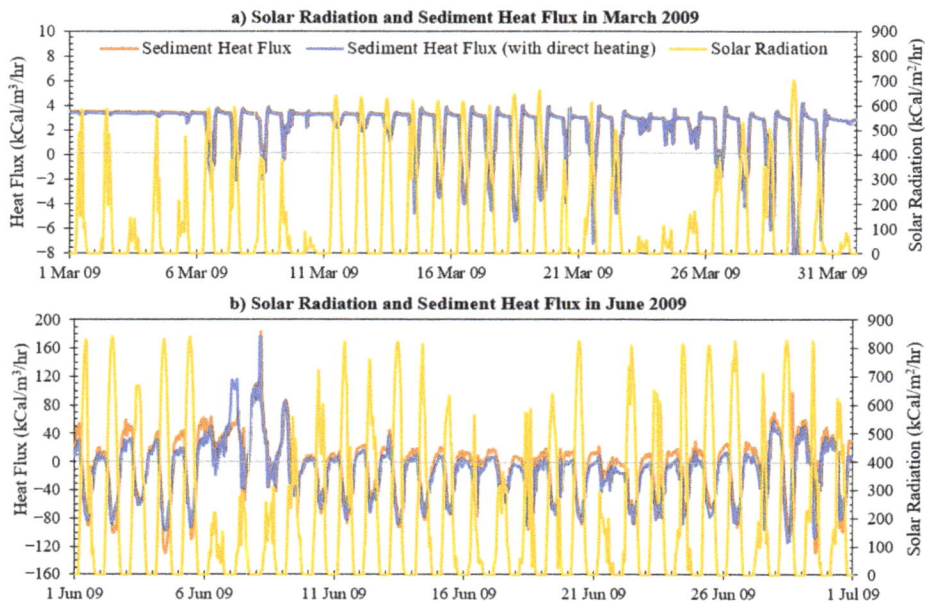

Figure 9. Sediment heat flux distribution in (**a**) March 2009 and (**b**) June 2009 in Carrie Lake.

During the night when there is no solar radiation, heat moves from sediment to water, i.e., positive fluxes/numbers in Figure 8. During the day when there is incoming solar radiation, the pattern of the sediment heat flux (with direct heating) coincides with the timing of solar radiation as we are considering direct heating of sediment. In the previous sediment flux calculation, there was a time lag between the incoming solar radiation and the negative heat flux. Sediment heat flux changes in direction and magnitude depending on the solar radiation which is shown in Figure 8 for four different cases. On days with ice and snow cover, when a thin layer of snow can attenuate most of the solar radiation, heat moves from sediment to water throughout the day. On 18 March 2009, the lake has ice cover but no snow cover. The sediment heat flux starts decreasing as solar radiation heats up the water. At one point, the heat goes from water to sediment and as the water cools down at night, heat flux changes its direction again. Sediment heat flux in summer is much larger than that in winter and depends on the magnitude of solar radiation (Figure 8c for a sunny day and 8d for a cloudy day). High solar radiation results in larger negative heat flux (heat going from water to sediment) as shown in Figure 8c.

Figure 9 shows that sediment heat flux is mainly dependent on solar radiation. This addition of direct solar heating is important for the overall dynamics of shallow water layers in daytime hours. Overall, due to the modification, the modified heat flux magnitude is reduced both in positive and negative heat flux situations. As the sediment is directly heated by solar radiation, the heat flux difference between sediment and water decreases, and this trend continues throughout the entire day (24 h).

4. Discussion

4.1. Short-Term Mixing Prediction

Figure 10 shows the simulated hourly and daily water temperatures near the surface and at bottom depths in June in Belle (H_{max} = 7.6 m), Carrie (H_{max} = 7.9 m), and Portage (H_{max} = 4.6 m) lakes. In 2009 Belle Lake was completely mixed from 6 June to 9 June and on 30 June from both the hourly and daily model simulations. On 28–29 June, the daily model results show that the lake is well mixed whereas the hourly model predicts weak stratification. This occurs due to the sudden increase in daily wind speed on those days. The hourly model simulates water temperature hour by hour using hourly wind speed which increased gradually and hence, no complete mixing was simulated by the hourly model. Carrie Lake is more stratified than Belle Lake (Figure 10b), which is related to a smaller geometry ratio (Table 1). Observed half-hourly water temperature data in these lakes were collected and converted into hourly observed data for comparison. At the surface, the simulated water temperatures match well with the observed data with an RMSE of 1.2 °C, 1.7 °C, and 1.4 °C for Belle (2008–2011), Carrie (2008–2015), and Portage (2008–2015), respectively. Since Portage is a very shallow lake with a maximum depth of 4.3 m, there are small differences between the surface and bottom water temperatures. For most of the days, the daily model predicts essentially the same temperatures at the lake surface and bottom whereas the hourly model predicts the temporal variations of water temperature more correctly. Owing to the lack of necessary details associated with the daily weather data, the daily model underpredicts the water temperature in well-mixed conditions at Portage Lake.

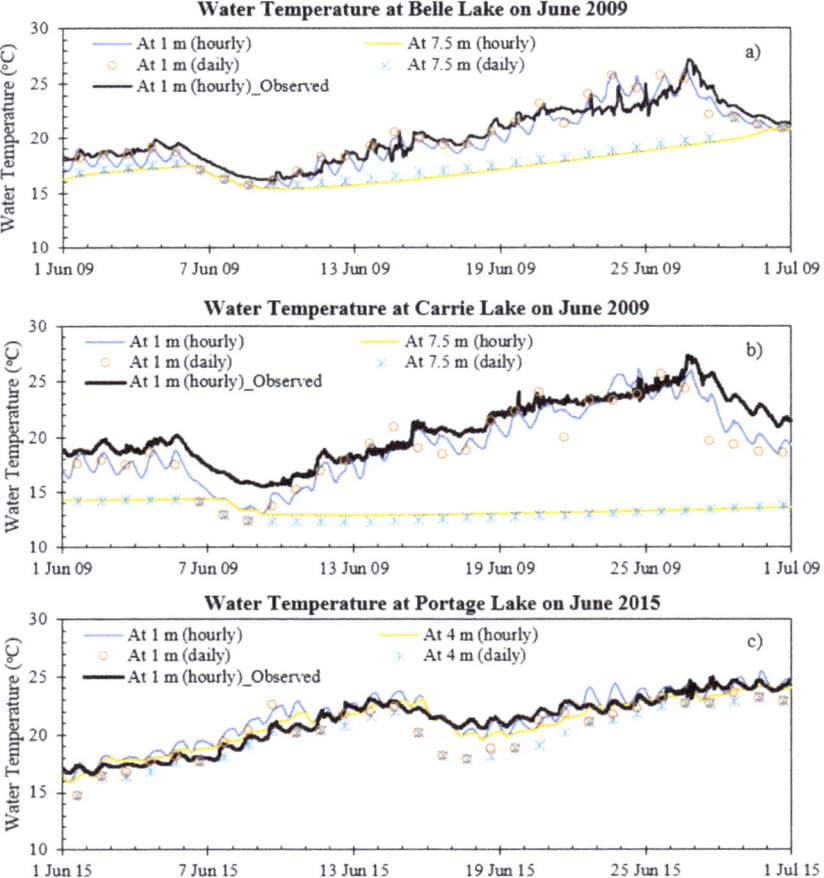

Figure 10. Time series of simulated daily and hourly water temperatures at two depths (1 m, 7.5 or 4 m) at (**a**) Belle Lake in June 2009, (**b**) Carrie Lake in June 2009, and (**c**) Portage Lake in June 2015 including observed hourly surface temperatures. Daily values were plotted at 4:00 p.m. each day.

The predicted mixing from temperature simulation plays an important role in not only water temperature distribution but also dissolved oxygen (Figure 11) and nutrient concentration distributions with depth. At Belle Lake, the daily simulation represents complete mixing on 28–29 June whereas the hourly model shows anoxic condition at the lake bottom (Figure 11a). At Carrie Lake (Figure 11b), the lake becomes well mixed on 7 June whereas the hourly model simulates the low bottom DO condition on 7 June and higher DO concentrations starting from 10 June. Frequent mixing is observed in Portage Lake (Figure 11c): each day has a period of well-mixed conditions following several hours of weak stratification of DO. The daily model only accounts for the diminishing of DO from the top layer to the bottom layer and does not show the occurrence of the mixing within the day or how DO again recovers to be the same as the top layer.

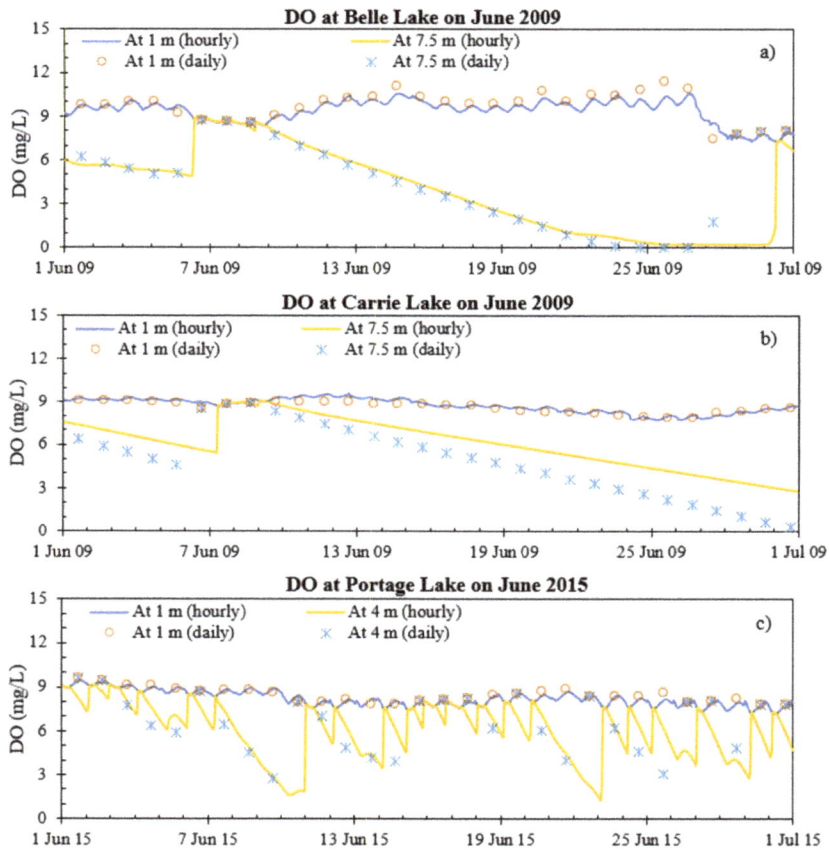

Figure 11. Time series of simulated daily and hourly DO at two depths (1 m, 7.5 or 4 m) at (**a**) Belle Lake in June 2009, (**b**) Carrie Lake in June 2009, and (**c**) Portage Lake in June 2015. Daily values were plotted at 4:00 p.m. each day.

4.2. Stratification Prediction

Carrie Lake has a depth of 7.9 m with a geometry ratio of 3.12. Portage Lake is comparatively shallow (4.57 m) with a geometry ratio of 7.71. From the water temperature graphs of Figure 10, it is evident that Carrie Lake is more stratified compared to Portage Lake. If the difference of water temperature or DO between the surface layer and bottom layer is more than 1 °C or 1 mg/L, the lake is considered stratified. Based on this criterion, for the hourly model, Carrie Lake is stratified 720 h (30 days) out of 720 h in June 2009 for water temperature and dissolved oxygen. However, the evaluation of stratification using the daily model reveals that the lake is stratified on 28 days in June 2009 for temperature and DO. The daily model cannot account for the mixing and stratification happening on time step shorter than a day. As a result, the daily model cannot capture the accurate scenario of stratification. For Red Sand Lake, the lake is stratified for 298 h and 359 h for water temperature and dissolved oxygen, respectively. This shows the dependence of stratification on lake depth and geometry ratio.

Pearl Lake has 30-min observed water temperature data that were measured from 2013 to 2016 at 6 depths (1.2, 1.7, 2.4, 3.4, 4.4, and 5.0 m) by MNDNR. The average chlorophyll-a concentration at Pearl Lake is 16.91 µg/L as a eutrophic lake and the geometry ratio is 7.71 for a weakly stratification or polymictic lake. MINLAKE2018 model was successfully run from January 2013 to December 2016, and time series of simulated temperatures in 2015 summer at 1.2 m and 5 m were compared with measured data in Figure 12.

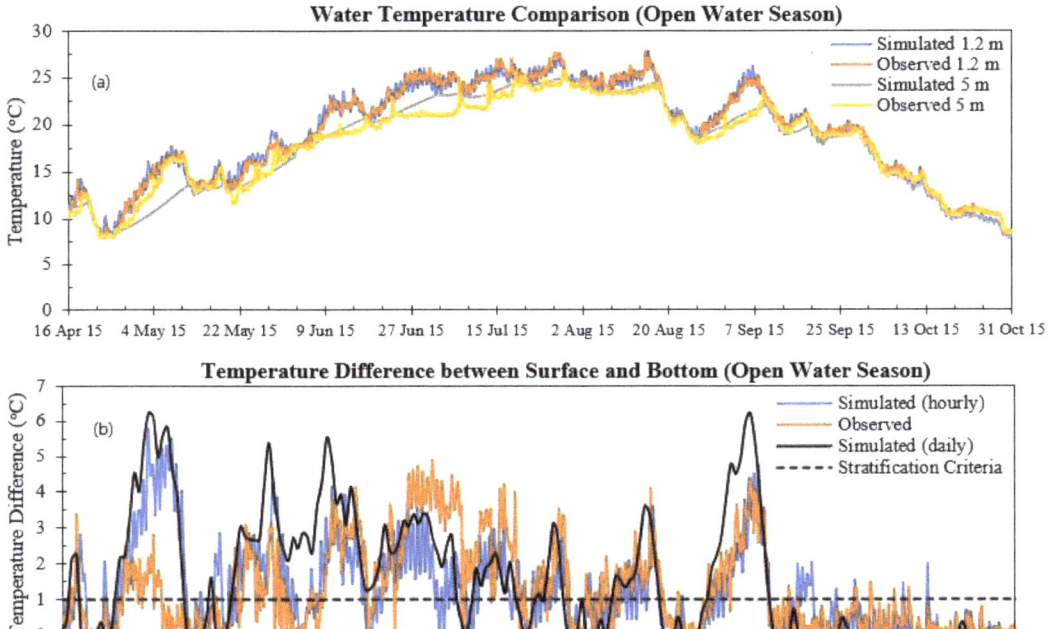

Figure 12. (a) Time series of observed and simulated hourly water temperature at 1.2 m and 5 m in Pearl Lake in open water season in 2015, (b) differences of simulated and observed hourly temperatures between 1.2 m (surface) and 5 m (near the bottom) during open water season in 2015 in Pearl Lake including simulated daily temperature differences and 1 °C difference as stratification criterion.

Figure 12a shows that at 1.2 m depth near the surface, simulated hourly water temperatures match well with observed data in the open water season (1 April–30 September) of 2015. The simulated and observed hourly water temperatures at 5 m were used to verify whether the model can successfully reflect the attenuation of solar radiation and the heat diffusion mechanism inside the lake. At 5 m depth, the model underestimates the water temperature from 27 April to 9 May and overestimates from 27 June to 9 July. Apart from those periods, the model predicts water temperature at 5 m depth correctly. At 1.2 m and 5 m depths, the simulated water temperatures in 2015 have NSE values of 0.99 and 1.00, respectively, when compared with observed data.

In Figure 12b, the stratification of the lake over the same period was quantified. The stratification criterion was set as 1 °C and the observed stratification was compared with the stratification calculated by hourly and daily models. From 27 April to 9 May, the lake is stratified and the difference between the surface and bottom temperature is much larger than observed as shown in Figure 12a. From 27 May to 12 June, the hourly model represented the fluctuation of stratification correctly during this period, whereas the daily model shows continuous stratification over the entire time period. From 27 June to 9 July, for both hourly and daily models, the differences between the surface and bottom water temperature are much smaller than the observed ones as shown in Figure 12a. From 2 August to 7 August, the daily model overestimates the temperature gradient whereas the hourly model captures the temperature gradient perfectly. For the rest of the year 2015, the daily model underestimates the temperature gradient and shows no stratification during this fall season. But in reality and in the hourly model simulation, occasional stratification was observed in fall seasons. From comparing with hourly model results and observations, it reveals that the daily model cannot capture the rapid change of stratification and mixing in shallow lakes, which is one major drawback of the daily model. The daily model fails to predict stratification or mixing fluctuation, which happens for a short period, i.e., 3–6 h on some days. The daily model assumes longer periods of stratification although the shallow lakes may mix several times (each time for a few hours). Moreover, the daily model cannot correctly predict weak stratification over several hours on some days in the fall season.

The stratification statistics for five study lakes during the ice cover periods and open water seasons are given in Table 6 as percentage hours or percentage days for three years. Comparing hourly and daily models (Table 6), the major discrepancy (3–24%, average 12%) in stratification estimates happens in dissolved oxygen stratification during the ice cover period. However, in the ice cover period, for temperature stratification, there is no major discrepancy (average 5%) in the stratification scenarios of the hourly model and daily model since temperature varies only from 0 to slightly larger than 4 °C. In the open water season, some discrepancy (13–16%) for lakes with higher geometric ratios is observed. Overall, the stratification hours of dissolved oxygen increase with the change in model time step from daily to hourly.

Table 6. Water temperature and dissolved oxygen stratification in study lakes presented as % hours of stratification (hourly model) or % days of stratification (daily model) in 2009–2011.

Lake Name	Geometry Ratio (Secchi Depth)	% Hours or Days of Temperature Stratification			
		Ice Cover Period		Open Water Season	
		Hourly Model	Daily Model	Hourly Model	Daily Model
Carrie	3.12 (1.48 m)	89	89	65	64
Belle	5.77 (1.46 m)	86	80	37	35
Pearl	7.53 (1.85 m)	93	81	80	67
Portage	7.71 (2.00 m)	89	83	26	25
Red Sand	8.34 (3.04 m)	90	89	32	16
Lake Name	Geometry Ratio (Secchi Depth)	% Hours or Days of DO Stratification			
		Ice Cover Period		Open Water Season	
		Hourly Model	Daily Model	Hourly Model	Daily Model
Carrie	3.12 (1.48 m)	75	66	58	71
Belle	5.77 (1.46 m)	88	76	56	52
Pearl	7.53 (1.85 m)	93	79	67	47
Portage	7.71 (2.00 m)	89	67	48	38
Red Sand	8.34 (3.04 m)	90	87	37	42

4.3. Application in Lake Management

DO is considered to be the most important water quality parameter of a lake. With the increasing anthropogenic nutrient loading, stratification becomes increasingly important in the consumption of DO and the formation of hypoxia [37–39]. Hypoxia in the waterbody influence biogeochemical cycles of nutrients and exert severe negative impacts on aquatic ecosystems, such as mortality of benthic fauna, fish kills, habitat loss, and physiological stress [40–42]. Given the significance and the recent increase in hypoxic events in lakes, it has become necessary to enhance our understanding of the natural and anthropogenic drivers of hypoxia and the internal feedback mechanisms. Large diurnal fluctuations of oxygen between nighttime hypoxia and daytime supersaturation have been observed in shallow tidal creeks, lagoons, and estuaries [43]. It was observed that high primary production during daytime results in supersaturated DO levels, while at night respiration overwhelms the DO supply, often leading to hypoxia [44]. Variations in the extent and duration of low oxygen events can lead to substantial ecological and economic impacts. Moreover, sometimes the mixing of the inflow is very dynamic and an hourly model might be appropriate to address the change.

In Figure 5, it was observed that in the hypolimnion, the simulated daily variation of DO is larger than that of temperature because of more sink terms. This hypolimnetic DO is particularly important because it regulates the phosphorus release from sediments. High phosphorus release from lake sediments is frequently reported as an important mechanism delaying lake recovery after external loading of phosphorus has been reduced [45–47]. A long-term survey of 35 lakes in Europe and North America concluded that internal release of phosphorus typically continues for 10–15 years after the external loading reduction [48] but in some lakes, the internal release may last longer than 20 years [46]. In shallow lakes, it is common to observe negligible changes in phosphorus concentrations in lake water even after external load diversion [49]. For example, Lake Trummen in Sweden remained hypereutrophic even after 11 years of sewage (primary source of external loading) diversion. Finally, the removal of 1 m of high phosphorus sediment reduced the internal loading dramatically [49]. Because of the numerous physical and biogeochemical processes involved, the development of a lake water quality model that enables estimating DO responses to the external/internal environment for short intervals is essential for understanding the dynamics of hypoxia. An hourly model will be useful in scientific research of hypoxia conditions and for planning and forecasting site-specific responses to different management scenarios. They are needed to provide advice to policy-makers about the probable effectiveness of various remedial actions at affordable costs.

4.4. Future Studies

In this study, the simulated hourly water temperatures were compared with the hourly observed water temperatures near the surface for five lakes, but the hourly DO data were not available. Moreover, due to scarcity of observed profile data, all data were used for model calibration purpose. In the future, the hourly DO data should be collected and compared with hourly simulated DO to advance the hourly DO model and understand the complex DO diurnal dynamics. With long-term profile data available, the observed data should be divided into two parts: one part to be used for model calibration and the other for model validation purpose. One drawback of these model simulations was to not consider inflow to the lake which can be important for some shallow eutrophic lakes. The daily inflow/outflow model needs to be modified/improved for hourly simulation. Only five shallow lakes in Minnesota were simulated in this study. In the future, the model can be applied to more lakes of different characteristics and different climate/geographic areas and should be improved for more general use.

5. Conclusions

Both water temperature and DO in lakes exhibit noticeable diurnal changes due to changing weather conditions, and solar-radiation-dependent phytoplankton and benthic activities. The one-dimensional hourly lake water quality model MINLAKE2018 was developed from the daily MINLAKE2012 model to evaluate diurnal variations in Minnesota lakes. The simulated hourly water temperatures and DO concentrations were compared with available observed hourly near-surface water temperatures and measured temperature and DO profiles at specific times in 36–87 days (Table 4) over several years. Simulation results from the hourly MINLAKE2018 model provide the following conclusions:

1. MINLAKE2018 was calibrated against measured profiles in five shallow Minnesota lakes (Table 4) with an average standard error of 1.48 °C for temperature and 2.02 mg/L

for DO. With the help of available surface water temperature hourly data, the average RMSE of long-term water temperature simulation was 1.50 °C with a standard deviation of 0.32 °C. For Pearl Lake, the average RMSE for water temperature simulation at six different depths is 1.30 °C with a standard deviation of 0.15 °C.

2. When compared with the daily MINLAKE2012 model, for Pearl Lake (H_{max} = 5.6 m), the hourly model calculated 12% and 13% more temperature stratification for ice cover period and open water season, respectively (Table 6). Similarly, for DO, stratification increases were 14% and 20% for ice cover period and open water season, respectively. For other lakes, hourly model simulation also resulted in increased stratification percentages for water temperature and DO. The hourly model can capture diurnal changes and mixing events that lasted a few hours within a day, which the daily model ignores. Moreover, it was observed that the daily model could not predict most of the weak stratifications of shallow lakes in the fall season (Figure 12). As a result, to ensure desired water quality for aquatic organisms and fish habitat, the hourly model is suitable for shallow lakes all year round.

3. The hourly model MINLAKE2018 performs better than the daily model MINLAKE2012 in water temperature and DO profile simulation (Figure 2). The RMSEs of temperature and DO from MINLAKE2018 decreased by 17.3% and 18.2%, respectively, and Nash-Sutcliffe efficiency increased by 10.3% and 66.7%, respectively, in comparison to MINLAKE2012.

4. Sediment heating subroutine was modified to include direct heating of sediment from solar radiation for all sediment layers. After modification, the sediment heat flux pattern became coincident with the solar radiation pattern eliminating the lag time between the change in solar radiation and the change in heat flux to appear. The magnitude of sediment heat flux was reduced for both cases (heat flux going from water to sediment or sediment to water) after the sediment subroutine was modified.

Author Contributions: B.T. developed the final version of the MINLAKE2018 program, conducted the simulations, analyzed the result, and prepared the manuscript draft and revisions. J.A.J. developed the first version of the program and prepared input data for lakes. X.F. supervised model development, simulation runs, data analysis, and revised the manuscript. Y.Z. and J.S.H. supervised the writing and revised the manuscript. All authors made contributions to the study and writing the manuscript. All authors have read and agreed to the published version of the manuscript.

Funding: The study is partially supported by funding from the OUC-AU Joint Center for Aquaculture and Environmental Science for the project *"Forecasting the Ecological Health of Coastal Waters in Alabama and China."* Alan E. Wilson is PI, Xing Fang and Joel Hayworth are Co-PIs at Auburn University; Sun Dajiang is PI, Yangen Zhou, Kai You, and Hongwei Shan are Co-PIs from OUC (Ocean University of China).

Institutional Review Board Statement: Not applicable.

Informed Consent Statement: Not applicable.

Data Availability Statement: Some or all data, models, or code generated or used during the study are proprietary or confidential and may only be provided with restrictions. The model input and output data are specifically designed for a research numerical model. They are available upon request but are not useful for the general public.

Conflicts of Interest: The authors declare no conflict of interest.

References

1. Coutant, C.C. Striped bass, temperature, and dissolved oxygen: A speculative hypothesis for environmental risk. *Trans. Am. Fish. Soc.* **1985**, *14*, 31–61. [CrossRef]
2. USACE. *CE-QUAL-R1: A Numerical One-Dimensional Model of Reservoir Water Quality*; User's Manual; U.S. Army Corps of Engineers Waterways Experiment Station: Vicksburg, MS, USA, 1995.
3. Chapra, S.C. *Surface Water-Quality Modeling*; Waveland Press: Salem, WI, USA, 2008.
4. Fang, X.; Jiang, L.; Jacobson, P.C.; Fang, N.Z. Simulation and validation of cisco habitat in Minnesota lakes using the lethal-niche-boundary curve. *Br. J. Environ. Clim. Chang.* **2014**, *4*, 444–470. [CrossRef]

5. Riley, M.J.; Stefan, H.G. MINLAKE: A dynamic lake water quality simulation model. *Ecol. Model.* **1988**, *43*, 155–182. [CrossRef]
6. Fang, X.; Alam, S.R.; Stefan, H.G.; Jiang, L.; Jacobson, P.C.; Pereira, D.L. Simulations of water quality and oxythermal cisco habitat in Minnesota lakes under past and future climate scenarios. *Water Qual. Res. J. Can.* **2012**, *47*, 375–388. [CrossRef]
7. Riley, M.J.; Stefan, H.G. *Dynamic Lake Water Quality Simulation Model "MINLAKE"*; St. Anthony Falls Hydraulic Laboratory, University of Minnesota: Minneapolis, MN, USA, 1987; 140p.
8. Fang, X.; Stefan, H.G. *Modeling of Dissolved Oxygen Stratification Dynamics in Minnesota Lakes under Different Climate Scenarios*; St. Anthony Falls Hydraulic Laboratory, University of Minnesota: Minneapolis, MN, USA, 1994; 260p.
9. Hondzo, M.; Stefan, H.G. Regional water temperature characteristics of lakes subjected to climate change. *Clim. Chang.* **1993**, *24*, 187–211. [CrossRef]
10. Stefan, H.G.; Fang, X. Dissolved oxygen model for regional lake analysis. *Ecol. Model.* **1994**, *71*, 37–68. [CrossRef]
11. Fang, X.; Stefan, H.G. *Temperature and Dissolved Oxygen Simulations for a Lake with Ice Cover*; Project Report 356; St. Anthony Falls Hydraulic Laboratory, University of Minnesota: Minneapolis, MN, USA, 1994.
12. Fang, X.; Stefan, H.G. Dynamics of heat exchange between sediment and water in a lake. *Water Resour. Res.* **1996**, *32*, 1719–1727. [CrossRef]
13. Fang, X.; Ellis, C.R.; Stefan, H.G. Simulation and observation of ice formation (freeze-over) in a lake. *Cold Reg. Sci. Technol.* **1996**, *24*, 129–145. [CrossRef]
14. Fang, X.; Alam, S.R.; Jacobson, P.; Pereira, D.; Stefan, H.G. *Simulations of Water Quality in Cisco Lakes in Minnesota*; St. Anthony Falls Laboratory: Minneapolis, MN, USA, 2010.
15. Xu, Z.; Xu, Y.J. A deterministic model for predicting hourly dissolved oxygen change: Development and application to a shallow eutrophic lake. *Water* **2016**, *8*, 41. [CrossRef]
16. Martin, J.L. *Hydro-Environmental Analysis: Freshwater Environments*; CRC Press: Boca Raton, FL, USA, 2013.
17. Heddam, S. Modeling hourly dissolved oxygen concentration (DO) using two different adaptive neuro-fuzzy inference systems (ANFIS): A comparative study. *Environ. Monit. Assess.* **2013**, *186*, 597–619. [CrossRef]
18. Kisi, O.; Alizamir, M.; Gorgij, A.D. Dissolved oxygen prediction using a new ensemble method. *Environ. Sci. Pollut. Res.* **2020**, *27*, 9589–9603. [CrossRef] [PubMed]
19. Granger, R.J.; Hedstrom, N. Modelling hourly rates of evaporation from small lakes. *Hydrol. Earth Syst. Sci.* **2011**, *15*, 267–277. [CrossRef]
20. Hondzo, M.; Stefan, H.G. Lake water temperature simulation model. *J. Hydraul. Eng.* **1993**, *119*, 1251–1273. [CrossRef]
21. Jamily, J.A. *Developing an Hourly Water Quality Model to Simulate Diurnal Water Temperature and Dissovled Oxygen Variations in Shallow Lakes*; Auburn University: Auburn, AL, USA, 2018.
22. Gu, R.; Stefan, H.G. Year-round temperature simulation of cold climate lakes. *Cold Reg. Sci. Technol.* **1990**, *18*, 147–160. [CrossRef]
23. Ji, Z.-G. *Hydrodynamics and Water Quality: Modeling Rivers, Lakes, and Estuaries*; John Wiley & Sons: Hoboken, NJ, USA, 2008.
24. Megard, R.O.; Tonkyn, D.W.; Senft, W.H. Kinetics of oxygenic photosynthesis in planktonic algae. *J. Plankton Res.* **1984**, *6*, 325–337. [CrossRef]
25. Holley, E. Oxygen transfer at the air-water interface. *Transp. Process. Lakes Ocean.* **1977**, *7*, 117.
26. Thomann, R.V.; Mueller, J.A. *Principles of Surface Water Quality Modeling and Control*; Harper Collins Publishers Inc.: New York, NY, USA, 1987; p. xii.
27. Utley, B.C.; Vellidis, G.; Lowrance, R.; Smith, M.C. Factors affecting sediment oxygen demand dynamics in blackwater streams of Georgia's coastal plain. *J. Am. Water Resour. Assoc.* **2008**, *44*, 742–753. [CrossRef]
28. Truax, D.D.; Shindala, A.; Sartain, H. Comparison of two sediment oxygen demand measurement techniques. *J. Environ. Eng.* **1995**, *121*, 619–624. [CrossRef]
29. Fang, X.; Stefan, H.G. Long-term lake water temperature and ice cover simulations/measurements. *Cold Reg. Sci. Technol.* **1996**, *24*, 289–304. [CrossRef]
30. Fang, X.; Stefan, H.G. Temperature variability in the lake sediments. *Water Resour. Res.* **1998**, *34*, 717–729. [CrossRef]
31. Fang, X.; Stefan, H.G.; Davis, M.B. *Status of Climate Change Effect Simulations for Mirror Lake, New Hampshire*; St. Anthony Falls Hydraulic Laboratory, University of Minnesota: Minneapolis, MN, USA, 1993; 52p.
32. Walters, R.A.; Carey, G.F.; Winter, D.F. Temperature computation for temperate lakes. *Appl. Math. Mod.* **1978**, *2*, 41–48. [CrossRef]
33. Wang, B.; Ma, Y.; Ma, W.; Su, Z. Physical control on half-hourly, daily and monthly turbulent flux and energy budget over a high-altitude small lake on the Tibetan Plateau. *J. Geophys. Res. Atmos.* **2017**, *122*, 2289–2303. [CrossRef]
34. NAS; NAE. *Water Quality Criteria 1972—A Report of the Committee on Water Quality Criteria*; Environmental Protection Agency: Washington, DC, USA, 1973.
35. Fang, X.; Stefan, H.G. Chapter 16 Impacts of Climatic Changes on Water Quality and Fish Habitat in Aquatic Systems. In *Handbook of Climate Change Mitigation*; Chen, W.-Y., Seiner, J.M., Suzuki, T., Lackner, M., Eds.; Springer: Berlin/Heidelberg, Germany, 2012; Volume 1, pp. 531–570.
36. Nash, J.E.; Sutcliffe, J.V. River flow forecasting through conceptual models part I—A discussion of principles. *J. Hydrol.* **1970**, *10*, 282–290. [CrossRef]
37. Buzzelli, C.P.; Powers, S.P.; Luettich, R.A., Jr.; McNinch, J.E.; Peterson, C.H.; Pinckey, J.L.; Paerl, H.W. Estimating the spatial extent of bottom water hypoxia and benthic fishery habitat degradation in the Neuse River Estuary, NC. *Mar. Ecol. Prog. Ser.* **2002**, *230*, 103–112. [CrossRef]

38. Waldhauer, R.; Draxler, A.F.J.; McMillan, D.G.; Zetlin, C.A.; Leftwich, S.; Matte, A.; O'Reilly, J.E. Biological, physical and chemical dynamics along a New York Bight transect and their relation to hypoxia. *Estuaries* **1985**, *8*, 129.
39. Yin, K.; Lin, Z.; Ke, Z. Temporal and spatial distribution of dissolved oxygen in the Pearl River Estuary and adjacent coastal waters. *Cont. Shelf Res.* **2004**, *24*, 1935–1948. [CrossRef]
40. Pena, M.A.; Katsev, S.; Oguz, T.; Gilbert, D. Modeling dissolved oxygen dynamics and hypoxia. *Biogeosciences* **2010**, *7*, 933–957. [CrossRef]
41. Levin, L.A.; Ekau, W.; Gooday, A.J.; Jorissen, F.; Middelburg, J.; Naqvi, J.; Neira, S.W.A.; Rabalais, N.N.; Zhang, F. Effects of natural and human-induced hypoxia on coastal benthos. *Biogeosciences* **2009**, *6*, 2063–2098. [CrossRef]
42. Ekau, W.; Auel, H.; Portner, H.O.; Gilbert, D. Impacts of hypoxia on the structure and processes in the pelagic community (zooplankton, macro-invertebrates and fish). *Biogeosciences* **2009**, *6*, 5073–5144.
43. D' Avanzo, C.; Kremer, J.N. Diel oxygen dynamics and anoxic events in an eutrophic estuary of Waquoit Bay, Massachusetts. *Estuaries* **1994**, *171*, 131–139. [CrossRef]
44. Shen, J.; Wang, T.; Herman, J.; Masson, P.; Arnold, G.L. Hypoxia in a coastal embayment of the Chesapeake Bay: A model diagnostic study of oxygen dynamics. *Estuaries Coasts* **2008**, *31*, 652–663. [CrossRef]
45. Marsden, M.W. Lake restoration by reducing external phosphorus loading; the influence of sediment phosphorus release. *Freshw. Biol.* **1989**, *21*, 139–162. [CrossRef]
46. Sondergaard, M.; Jeppesen, E.; Jensen, J.P.; Amsinck, S.L. Water framework directive: Ecological classification of Danish lakes. *J. Appl. Ecol.* **2005**, *42*, 616–629. [CrossRef]
47. Philips, G.; Kelly, A.; Pitt, J.A.; Sanderson, R.; Taylor, E. The recovery of a very shallow eutrophic lake, 20 years after the control of effluent derived phosphorus. *Freshw. Biol.* **2005**, *50*, 1628–1638. [CrossRef]
48. Jeppesen, E.; Sondergaard, M.; Jensen, J.P.; Havens, K.E.; Anneville, O.; Carvalho, L.; Coveney, M.F.; Deneke, R.; Dokulil, M.T.; Foy, B.; et al. Lake responses to reduced nutrient loading: An analysis of contemporary long-term data from 35 case studies. *Freshw. Biol.* **2005**, *50*, 1747–1771. [CrossRef]
49. Welch, E.B.; Cooke, G.D. Internal phosphorus loading in shallow lakes: Importance and control. *Lake Reserv. Manag.* **2009**, *21*, 209–217. [CrossRef]

Article

Gas Pressure Dynamics in Small and Mid-Size Lakes

Bertram Boehrer [1,*], Sylvia Jordan [2], Peifang Leng [1,3,4], Carolin Waldemer [1], Cornelis Schwenk [1,5], Michael Hupfer [2] and Martin Schultze [1]

[1] Helmholtz-Centre for Environmental Research—UFZ, 39114 Magdeburg, Germany; peifang.leng@ufz.de (P.L.); carolin.waldemer@ufz.de (C.W.); cornelis.schwenk@ufz.de (C.S.); martin.schultze@ufz.de (M.S.)
[2] Leibniz-Institute of Freshwater Ecology and Inland Fisheries—IGB, 12587 Berlin, Germany; jordan@igb-berlin.de (S.J.); hupfer@igb-berlin.de (M.H.)
[3] Key Laboratory of Ecosystem Network Observation and Modeling, Institute of Geographic Sciences and Natural Resources Research, Chinese Academy of Sciences, Beijing 100101, China
[4] College of Resources and Environment, University of Chinese Academy of Sciences, Beijing 100049, China
[5] Institute for Environmental Physics, University Heidelberg, 69120 Heidelberg, Germany
* Correspondence: bertram.boehrer@ufz.de

Abstract: Dissolved gases produce a gas pressure. This gas pressure is the appropriate physical quantity for judging the possibility of bubble formation and hence it is central for understanding exchange of climate-relevant gases between (limnic) water and the atmosphere. The contribution of ebullition has widely been neglected in numerical simulations. We present measurements from six lacustrine waterbodies in Central Germany: including a natural lake, a drinking water reservoir, a mine pit lake, a sand excavation lake, a flooded quarry, and a small flooded lignite opencast, which has been heavily polluted. Seasonal changes of oxygen and temperature are complemented by numerical simulations of nitrogen and calculations of vapor pressure to quantify the contributions and their dynamics in lacustrine waters. In addition, accumulation of gases in monimolimnetic waters is demonstrated. We sum the partial pressures of the gases to yield a quantitative value for total gas pressure to reason which processes can force ebullition at which locations. In conclusion, only a small number of gases contribute decisively to gas pressure and hence can be crucial for bubble formation.

Keywords: dissolved gas; Henry law; total gas pressure; ebullition; greenhouse gases; lacustrine waters

1. Introduction

Dissolved gases in aquatic systems have moved into the focus of limnological studies recently because of their central role in the carbon cycle and hence their relevance for the climate [1,2]. Lakes are known for the burial of organic material but also as sources of methane (CH_4) and carbon dioxide (CO_2). Lakes contribute decisively to fluxes of CH_4 and CO_2 into the atmosphere by both diffusive processes and ebullition [1]. CH_4 is a highly potent greenhouse gas, i.e., a multiple of CO_2 at equal concentrations [3]. The concentrations of both gases keep rising in the atmosphere. This fact emphasizes the need for elucidating the involvement of lakes and rivers in global carbon fluxes [4]. As a consequence, many recent studies have aimed at quantifying the fluxes from limnic waters into the atmosphere. In particular, reservoirs are known for releasing methane—especially in shallow or dry-falling areas. This fact may put the reputation of hydropower as green energy at stake at least in some cases [2,5].

Beyond their recognition as being climate-relevant, gases are central players in the ecology of limnic waters, especially oxygen (O_2) for all breathing organisms and carbon dioxide (CO_2) for photosynthetic organisms. Furthermore, dissolved gases that are conceived as less reactive such as nitrogen become relevant for nitrogen fixation when supply with inorganic nitrogen runs short (e.g., [6,7]).

Any dissolved gas produces a gas pressure. The contributions of all gases add up to the total gas pressure. Though not in wide use in limnology, total gas pressure is the proper physical quantity to judge proximity to spontaneous bubble formation and ebullition [8]. The ratios between partial pressures determine the composition of forming bubbles (e.g., [9,10]) and the exchange with the surrounding water while ascending through the waterbody to the surface [11]. In conclusion, putatively irrelevant gases have a decisive impact on the removal of ecologically relevant gases. Hence, gas pressure is central for quantifying gas fluxes to the atmosphere and for understanding ebullition.

Despite its relevance, gas pressure is not widely referred to in the limnological literature and appears nearly exclusively in connection with large-scale ebullition events—so-called limnic eruptions (e.g., [12]). Catastrophic events of spontaneous gas ebullition from deep waters (Lake Nyos and Lake Monoun—both in Cameroon, Africa) have cost the lives of many humans in single events [12–15]. Since then, a number of other lakes with gas pressures of concern have been reported in the literature (e.g., Lake Kivu: [16]) and assessed for the danger of limnic eruptions (Lake Kivu: [17], Guadiana pit Lake: [18]).

Distribution of gases in the water column and chemical reactions—most of them biologically mediated—change gas concentrations and hence affect gas pressure. However, a good overview of processes increasing gas pressure to the level of spontaneous ebullition is missing in the limnological literature as well as the physical limnology literature. The same accounts for the localization of these processes where gas pressure may be raised sufficiently. What limits the gas pressure and, if ebullition sets in, what controls the bubble composition and hence the removed or stripped gas? In conclusion, a closer competent view on the gas pressure in lakes with appropriate depictions is urgently needed to effectively impart the knowledge to the wider limnological community.

With this paper, we attempt to fill this gap. We present new data from six lakes in the German state of Saxony-Anhalt, including natural and artificial lakes reflecting the broad variety of limnic waters. Observations of extreme gas pressures in lakes are referred to in the discussion. Solubilities of the most relevant gases are listed in comparison. We demonstrate the contributions of the most relevant gases to gas pressure and complement the gas measurements with profiles from numerical model simulations to finally depict them together with their contributions to gas pressure. This gas pressure can be affected by chemical reactions (produced or removed) or temperature change. We demonstrate under which conditions total gas pressure can be raised to absolute pressure to finally result in bubble formation and ebullition.

2. Environmental Gases and Methods

2.1. Solubility of Gases

When a water surface gets into contact with the atmosphere, atmospheric gas flux goes into the water until an equilibrium concentration c_i is reached, which is described by the Henry law:

$$c_i = k_{H,i} \cdot p_i \quad (1)$$

where p_i represents partial pressure in the gas phase (e.g., the atmosphere) and i is the marker for the different gases. Henry coefficients k_H are specific for gases and depend on the temperature (and much weaker on other dissolved substances and pressure) (see Table 1 or [19]). The temperature effect is remarkable and the value roughly drops to half from 0 °C to 30 °C for many gases. A simplified quantitative description (e.g., Sander 2015) is

$$k_H = k_H(25\ °C) \cdot \exp\left(T_E \cdot \left(\frac{1}{T} - \frac{1}{298.15\ K}\right)\right) \quad (2)$$

where $T_E = -(\Delta_{sol}H)/R$ has the dimension of (absolute) temperature and is generally determined empirically (also listed in Table 1; $\Delta_{sol}H$—dissolution enthalpy, general gas constant $R = N_A \cdot k$ is the product of Avogadro number and Boltzmann constant).

Table 1. Henry coefficients (from [19] Sander, 2015); volumetric portion in dry atmosphere (N_2, O_2, Ar from [20] Roedel 1992, CO_2 from [21] Worch 2015, CH_4 by [22] Saunois et al. 2020); Ostwald coefficient calculated from values in [19] and coefficients for Equation (4) for N_2 and O_2 ([23] Weiss 1970), Ar ([24] Jenkins et al. 2019), CH_4 ([25] Wiesenburg and Guinasso 1979) and CO_2 ([26] Weiss 1974); changes due to the introduction of the new temperature standard ITS-90 in 1990 are too small to show up on our scale and temperature range.

	Atmosphere	k_H (25 °C)	k_H (25 °C)	T_E	k_H (25 °C)	A_1	A_2	A_3	u
	[%]	[mol/m³/Pa]	[mol/L/bar]	[K]	[-]	\multicolumn{3}{c}{Coefficients for Equation (4)}	
N_2	78.09	6.4×10^{-6}	6.4×10^{-4}	1300	0.016	−59.6274	85.7661	24.3696	986.9/ 22391
O_2	20.95	1.3×10^{-5}	1.3×10^{-3}	1500	0.032	−58.3877	85.8079	23.8439	
Ar	0.93	1.4×10^{-5}	1.4×10^{-3}	1400	0.035	−55.6578	82.0262	22.5929	
CH_4	0.00019	1.4×10^{-5}	1.4×10^{-3}	1600	0.035	−68.8862	101.4956	28.7314	
CO_2	~0.039	3.3×10^{-4}	3.3×10^{-2}	2400	0.82	−58.0931	90.5069	22.2940	1/ 1.01325

If concentration is given in mol/L and partial pressure in bar, the Henry coefficient has the unit of (mol/L)/bar. However, both the concentration in the liquid phase as well as the partial pressure can be given in various units. Hence Henry coefficients can have differing units with accordingly differing values. A particularly interesting version of Henry coefficients results from replacing partial pressure p_i with the concentration in the gas space $c_{g,i}$, by applying the ideal gas law

$$c_{g,i} = \frac{p_i}{R \cdot T} \qquad (3)$$

As a consequence of having concentration on either side of Equation (1), the Henry coefficient does not possess a unit, and is commonly referred to as Bunsen coefficient (at 25 °C) or Ostwald coefficient (temperature dependent). This version of Henry coefficient relates concentration in the water directly to concentrations in the gas phase. Most common gases have a Bunsen coefficient in the range of 0.01 to 0.03 with the important exception of carbon dioxide, which has a Bunsen coefficient of the order of 1. In conclusion, most gases have concentrations of a factor 50 lower in solution than in the adjacent gas phase. Of the listed gases, only carbon dioxide is present at nearly the same concentration in equilibrated water as in air.

The conversion between units for concentrations is straightforward between mols and grams by multiplication by molar mass (in g/mol). However the conversion between molar units and other concentration units (e.g., molal units (mol/(kgH_2O) or permille (or practical salinity units (psu) or g/(kgSample))) or partial pressures can be complex for a mixture of solutes [19,21,27].

A more accurate temperature fit compared to Equation (2) is achieved, when the Clausius-Clapeyron equation is solved and the result developed into a Taylor series, of which the exponent of the first three terms has the following form fitted with coefficients A_i (see also Supplementary Materials):

$$k_H = \exp\left(A_1 + A_2 \frac{100}{T} + A_3 \ln\left(\frac{T}{100}\right)\right) \cdot u \qquad (4)$$

where we added a unit conversion factor u, as solubilities have traditionally been presented in a variation of different units.

Intuitively the Henry law is understood as a limited water volume in contact with an infinite atmosphere. However, the law also applies for a closed system with a limited air (or gas) space. Moreover in this case, partial pressures and concentrations are coupled: conditions inside the gas space are set by the concentrations in the water as used in headspace extractions for measurements of gas concentrations.

2.2. Gas Pressure, Saturation, Total Gas Pressure

By solving Henry's law (Equation (1)) for pressure, we find that each dissolved volatile substance is connected to a gas pressure of its own, which is proportional to its concentration. Hence we can use the specific Henry coefficients to evaluate the gas pressures from concentration profiles of each gas. The temperature dependence of the Henry coefficient results in a temperature dependence of the gas pressure at given gas concentration (Equation (2), Table 1). However, in equilibrium between air and (e.g., surface) water, the dissolved gas in the water produces the same gas pressure as the partial pressure in the gas space. For a water surface at sea level and normal pressure, this amounts to 21% of 1013.25 mbar for oxygen: accordingly less for higher altitude, low air pressure and moist air.

Gas concentrations in the water can change due to sources and sinks. As a consequence, also gas pressures are affected. In addition, heating can raise gas pressures. The ratio between gas pressure and partial pressure in the adjacent air volume is defined as the saturation and usually given as a percentage with 100% representing equilibrium with dissolved gas and the adjacent air space. The conditions of partial pressures at the lake surface, which depend on air pressure and humidity at the time of measurement, function as the conventional reference, as instruments are usually calibrated on site. This reference, however, is variable in the range of few percent over the year due to changing weather conditions.

In the usual range of gas concentrations and pressures, gas pressures of all gases can be added to a total gas pressure: non-linearities and mutual interaction only play a role at extreme conditions: e.g., Lake Kivu [28]. A (hypothetical) bubble in the water column is subject to the gas pressure of the ambient water and it will eventually collapse, if local pressure p_{abs} lies above total gas pressure p_{tdg} [8,11]. As a consequence, the bubble formation limits the increase of total gas pressure to absolute pressure in natural waters (mainly hydrostatic and atmospheric pressure), which is a function of water depth. At greater depth, higher amounts of gases are soluble as a consequence.

$$p_{tdg} \leq p_{abs} \quad (5)$$

2.3. Relevant Gases

Clearly, the number of detectable gases in natural water bodies goes far beyond those listed in Table 1. Unlisted gases may be of central ecological relevance and others are used for tracing water bodies, but only in very extreme cases, they may contribute considerably to the total gas pressure (e.g., [6,29–31]). We use the simple approximation of an exponential temperature dependence as this facilitates an easy intercomparison of solubility and temperature dependence. For many purposes, these approximations are sufficiently accurate, but calculations of high accuracy must use more sophisticated numerical approximations: for N_2 and O_2 [23], Ar [24], CH_4 [25], and CO_2 [26]. The deviation of the exponential fit and the more sophisticated approach lies within about 3% for temperature 10 to 35 °C; at lower temperatures deviation are even larger (see Supplementary Materials and Figure S1). The effect of dissolved solids on the Henry coefficient is small for freshwater (<3 g/L of dissolved solids) and quantifications are not available for salt compositions in inland waters (e.g., [28,32]) and hence has not been included in our evaluation.

In addition to the dissolved gases, water itself develops a vapor pressure which contributes to the total gas pressure as described by the Magnus equation. We follow the recommendation of Alduchov and Eskridge [33] and propose the simple formula for vapor pressure:

$$E(\theta) = 6.1094 \exp\left(\frac{17.625\,\theta}{\theta + 243.04}\right) \quad (6)$$

with temperature θ in °C.

This equation approximates the curve [34], which is recommended by the International Association of Properties of Water and Steam—IAPWS by better than 0.385% in the range

0 to 40 °C (see Supplementary Materials and Figure S2). Another good approximation (better than 0.006% even for temperatures beyond 40 °C) is proposed by Huang [35] (see Supplementary Materials).

Gas concentration in natural waters tend to equilibrate with the atmosphere while their surfaces are exposed to the atmosphere. However source and sink terms modify the concentrations. This can involve inflows, but also geochemical processes with many of them controlled by organisms. We included supporting information for a brief overview of the most important sources and sinks of the most important gases. Beyond this, we refer to textbooks on geochemistry of natural waters (such as [21,31]).

2.4. Simulations

To complement measurements for gases that have not been in field survey programmes, and to produce vertical profiles of good resolution, we implemented a simple one-dimensional model for a conservative gas, i.e., without sources and sinks in the water. We divided the water column into 48 equally spaced layers of d = 1 m thickness each. The simulations were run in MATLAB. Diffusion was implemented by exchanging half of each layer in steps of t = 30 days with neighboring layers to implement a turbulent diffusivity of the order $k = d^2/t = 4 \times 10^{-7}$ m^2/s. Equilibrium conditions were implemented for the gas for the entire epilimnion according to the measured water temperatures in spring and summer (as justified below in the results with measurements of oxygen in the epilimnion). As a result, we gained continuous and vertically coherent profiles for conservative gases.

3. Measurements
3.1. Investigated Lakes

The investigated lakes are located in Saxony-Anhalt, one federal state in Central Germany. This state has only few natural lakes as a consequence of its locations outside the area of the last glaciation. We present six lacustrine water bodies, including (1) a natural lake, (2) a drinking water reservoir, (3) one gravel pit lake, (4) one salt-affected mine pit lake and (5) one flooded quarry, and finally (6) one mine pit lake temporarily used as a dumping side (Figure 1). These water bodies are representative of the range of lacustrine waters in this area, which has been densely populated since the Middle Ages and hence intensively used for agriculture, forestry, settling, and exploitation of ore and salt deposits. Since the industrial revolution around 1870, it has also been heavily affected by traffic, lignite mining, and industrial production. Precipitation is generally low (around 550–600 mm per annum, except for the catchment of Rappbode Reservoir, where precipitation can reach up to 1700 mm per annum in the highest areas of the Harz mountains).

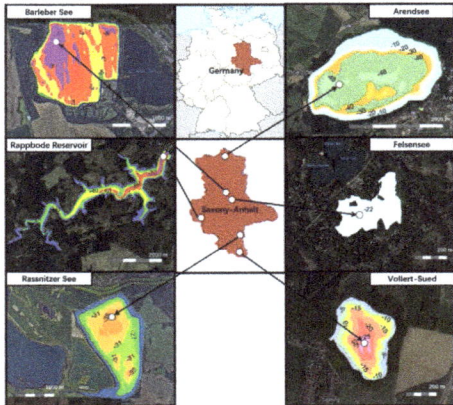

Figure 1. Location of investigated lakes; depth contour maps Arendsee, Rappbode Reservoir, Rassnitzer See, Barleber See, Vollert-Sued, Felsensee.

The lakes in detail:

1. Arendsee is a natural lake and originates from subsidence caused by dissolution of a salt deposit deeper in the ground [36]. The eutrophic lake has no river inflow and is mainly fed by groundwater [37,38].
2. Rappbode Reservoir was built in the 1950s for flood protection and drinking water supply and is in full operation since 1959 [39–41].
3. Rassnitzer See formed in the abandoned lignite mine Merseburg Ost 1b in the 1990s. Fresh and salty groundwater filled the void, resulting in a salinity-stratified water body, which does not overturn completely in winter. The final water level was reached by introducing freshwater from the nearby river Weisse Elster in 2002 [42–45].
4. Barleber See is the residual of a gravel pit (gravel excavations took place at the beginning of the 1930s). As the local open air swimming facility of the city of Magdeburg, it is intensively used for recreation. Increasing nutrient concentrations led to heavy algal and cyanobacteria blooms and a restoration by alum treatment in 1986 [46,47]. A further use as recreational area required a second chemical treatment of the waters with poly-aluminum chloride from 9th July to 15th October 2019. Inflow and outflow exclusively happen through exchange with groundwater [48].
5. Felsensee is a small lake, which formed in a former quarry. After stone production ceased, the quarry filled with groundwater. The water level has reached about 22 m. Higher conductivity groundwater inflows have turned the lake meromictic [49].
6. Lake Vollert-Sued is a flooded opencast lignite mine. The pit was (until 1969) used to dispose of wastewater from lignite processing. There is no surficial inflow or outflow but exchange with groundwater balancing the evaporation deficit and causing groundwater contamination in the near vicinity of the lake [50,51]. Hence it is heavily affected through its history as a dumping site. The water has been treated in 1999 to reduce the unpleasant smell and the impact on animals in the area [52,53]. The lake has since been meromictic.

The first three lakes have been selected to demonstrate the oxygen dynamics under usual conditions, while the latter three were selected to demonstrate special features of gas production and accumulation in lakes.

3.2. Equipment

From most lakes, we could retrieve profiles of temperature and electrical conductivity as indicators of density stratification, as well as oxygen profiles at three times of the year: one profile in early spring before stratification set in, one in summer when surface temperatures were high, and one in autumn, when the cooling surface forced a deeper recirculation of the lake water. In the case of Felsensee, we only show measurements of one sampling date, as well as for Lake Vollert-Sued, where we include data from Horn et al. [53].

Following pieces of equipment were used:

1. Arendsee: CTD profiles 2017: YSI 6600 V2, 2019: EXO2 from YSI, USA; optical oxygen sensor; CO_2 and gas pressure measurement in a gas volume behind a permeable membrane; CO_2 detection by IR spectrometry) CONTROS HydroC® CO_2 from Kongsberg Maritime, Germany;
2. Rappbode Reservoir: CTM90 from Sea & Sun Technology, Germany; optical oxygen sensor;
3. Rassnitzer See: CTM90 from Sea & Sun Technology, Germany; optical oxygen sensor;
4. Barleber See: CTM90 from Sea & Sun Technology, Germany; optical oxygen sensor;
5. Felsensee: Ocean Seven 316 from Idronaut, Italy; amperometric oxygen sensor;
6. Vollert Sued: CTD + O_2: Ocean Seven 316 from Idronaut, Italy; amperometric oxygen sensor; gas pressure: (TDG-sensor pressure measurement in a gas filled permeable silicon tube) Hydrolab, USA; CH_4, CO_2 and N_2: samples in GC thermal conductivity detector (see [53]).

4. Results

4.1. General Picture of Circulation and Atmospheric Recharge

Profiles with a multiparameter probe documented the stratification in the lakes Arendsee, Rappbode Reservoir, and (mine pit lake) Rassnitzer See (see Figure 1 and Table 2). During deep recirculation (profiles in March), the oxygen concentration was homogenized over the entire circulated water body (Figure 2, middle panel); the deep recirculation comprised the entire water body in holomictic Arendsee and Rappbode Reservoir. However, due to its meromictic character, the bottom 7 m of Rassnitzer See were not included in the deep recirculation. Oxygen levels remained at zero in the bottom waters. Profiles of oxygen saturation indicated that the entire circulated water body showed close to 100% saturation and hence was equilibrated with the atmosphere. Moreover, during summer and later in autumn, surface waters showed values close to 100% saturation.

Table 2. Properties of the investigated lakes (residence time was calculated as volume by outflow); the origin of the data is supplied in the text of Section 3.1 for each lake in separate.

Lake Name	Surface Area [km^2]	Volume [10^6 m^3]	Max. Depth [m]	Inflow [10^6 m^3/y]	Outflow [10^6 m^3/y]	Residence Time [y]	Age in 2020 [y]	Origin
Arendsee	5.1	149	49	6.03	2.65	56	>10,000	Dissolution of salt dome and subsidence
Rappbode Reservoir	3.95	113	89	109.8	89.6	0.942	61	Artificial dam
Rassnitzer See	3.1	68	38	2.07	0.4	170	18	Lignite surface mine
Barleber See	1.03	6.9	9.8	1.18	0.53	13	88	Gravel and sand excavations
Felsensee	0.085	Unknown (~1)	22	little GW flow	little GW flow	Unknown	>55	Stone quarry
Vollert-Sued	0.09	2	27	0.055	0.005	400	51	Polluted opencast lignite mine

During the stratification period, vertical exchange was largely reduced; hence local production and local depletion of oxygen could be observed in the water column. Both in Arendsee and Rappbode Reservoir, a reduction of oxygen concentrations could be measured: both lakes formed a metalimnetic oxygen minimum. This could possibly be attributed to the decomposition of organic material below the epilimnion, while the metalimnetic and hypolimnetic water remained disconnected from the supply with new oxygen from the atmosphere (see also [54,55] for Rappbode Reservoir; [56,57] for Arendsee).

On the contrary in Rassnitzer See, we saw a clear rise of oxygen saturation beyond 100% at depths of the thermocline and below. A small part of this could be attributed to primary production—assuming that sufficient light could enter deep enough to allow for photosynthesis: however, concentration profiles indicated that not much oxygen was added to the loading from spring deep recirculation. As a consequence, most of the rising saturation values had to be attributed to rising temperatures due to solar irradiation and turbulent diffusive heat transport from above beyond possible consumption and diffusive loses over the stratification period.

In general, we expected a very similar recharge and equilibration behavior of other gases, e.g., nitrogen and argon. Detailed documentations were not available as those concentrations are not particularly relevant for the ecology of a lake (see Section 2). However, contrary to mixing and recharge, we anticipated no relevant concentration changes over a stratification period due to geochemical reaction (as in the case of oxygen).

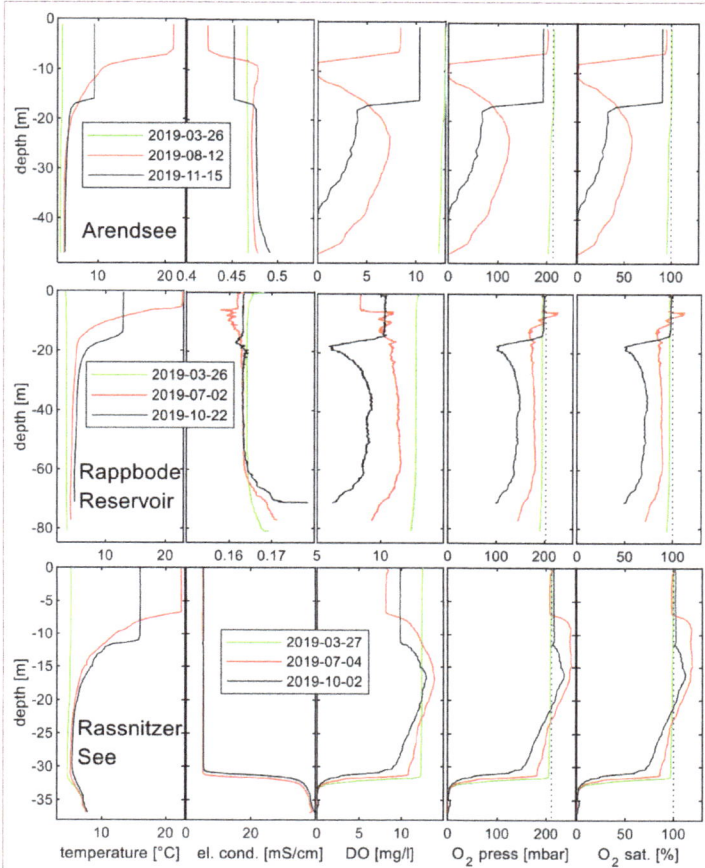

Figure 2. Profiles measured in Arendsee, Rappbode Reservoir, Rassnitzer See (all located in Saxony-Anhalt, Germany) on three sampling dates (green: spring, red; summer, black: autumn) in 2019: temperature, electrical conductivity, dissolved oxygen concentration, oxygen gas pressure (dotted line: atmospheric gas pressure at lake surface), oxygen saturation (dotted line: 100% saturation).

4.2. Complementing Gas Concentrations for Gas Pressure

We selected the 16th of August 2017 to produce a full set of profiles of relevant gases in Arendsee. Oxygen was measured with an optical sensor. It showed a minimum in the metalimnion (around 10 m depth), high values in the epilimnion, and lower values in the hypolimnion tending to zero toward the lake bed (Figure 3, left panel).

Carbon dioxide was low in the epilimnion due to the direct coupling to (the low) atmospheric concentrations. However, the deeper waters showed considerably higher concentrations. These concentrations corresponded to the missing O_2 in the water column quite well, but they were not equal as the amount of produced CO_2 from degrading biomass was not strictly tied to a stoichiometric value of 1, and some of the produced CO_2 could be forwarded into bicarbonate (HCO_3^-) as a result of the carbonate equilibrium. Due to the strong depletion of oxygen in Arendsee, CO_2 reached the same order of magnitude as O_2.

Nitrogen (N_2) is not as closely documented in lakes as oxygen. It is much less reactive and hence less relevant for ecological processes (see also discussion). Although, N_2 is part of the nitrogen cycle, the supply probably never runs short. Usually, lake waters show N_2 concentrations close to the atmospheric equilibrium even in meromictic lakes (e.g., [53]). As a consequence, N_2 is often considered conservative, if no better information is available (e.g., [12]). Since also in our field programme nitrogen (N_2) concentrations were not measured, we included simulated profiles from our simple 1D lake model assuming equilibration of the epilimnion with atmospheric concentrations of nitrogen and a turbulent diffusive exchange between layers (see above Section 2.4). As the deep water was recharged with N_2 at a lower temperature (higher Henry coefficient), the hypolimnion showed higher concentrations than the epilimnion, which was equilibrated at summery temperatures. The transition through the metalimnion was smoothened by implementing diffusive transport. At all depths, N_2 was obviously the gas with the highest concentration.

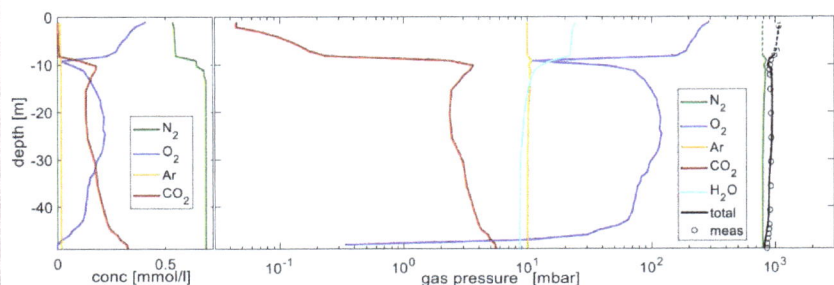

Figure 3. Left panel: concentration of gases in Arendsee on 16 August 2017: right panel: Gas pressures. Oxygen and carbon dioxide gas pressures were measured, while nitrogen and argon were modelled (see text), whereas vapor pressure was calculated from a temperature profile; total gas pressure was calculated from adding the partial pressures of displayed volatile substances and drawn with directly measured total gas pressure from the CO_2 probe (symbols). Broken lines (for epilimnion values of N_2 and total gas pressure) do not represent the real situation.

We also used the model for argon, where the conservative assumptions (no sources nor sinks) were satisfied even better. The profile looks nearly the same though concentrations were considerably lower due to the lower concentration in air compared to nitrogen; the numerical simulation profiles that demonstrate the vertical structure of a gas pressure profile.

Hydrogen sulphide has not been reported at noticeable concentrations in the open waters of Arendsee. Similarly, methane is removed during deep recirculation from this holomictic lake and cannot start to accumulate before oxygen is depleted. Measurements in the year 2019 [58] reconfirmed concentrations in the range of 100 nmol/L. Hence both gases could be neglected in the total gas pressure.

From the displayed concentration profiles, we calculated gas pressures by implementing a temperature-dependent Henry coefficient (Equation (1) solved for p_i); temperatures were used from a CTD-probe profile. The calculation yielded the (by far) leading contribution from nitrogen N_2; O_2 provided a smaller contributions, while CO_2—due to the high Henry coefficient—and argon (similar shape to N_2, though at lower values)—due to low concentration—contributed only a much subordinate gas pressure. We also added vapor pressure of H_2O, which was calculated as a function of temperature (Equation (6)) and hence was the third biggest gas pressure contribution in the epilimnion.

The gas pressures of all gases could be added to total gas pressure (Figure 3, right panel). We saw a local minimum in the metalimnion where oxygen had been depleted. Throughout the hypolimnion, gas pressures fell toward the lake bed as a consequence of reduced oxygen concentrations. Direct measurements of total gas pressure with the CO_2 probe confirmed the structure of the total gas pressure profile well. Smaller deviations were attributed to the fact that the sensor was not optimized for total gas pressure but for CO_2. The long response time might also have contributed to some additional error.

4.3. Elevated Gas Pressure Observations

Measurements in (the small gravel pit lake) Barleber See showed clearly elevated oxygen concentrations in the epilimnion during summer reaching a saturation of 140%, i.e., a gas pressure of 70 mbar above atmospheric (Figure 4, upper row). If N_2 gas pressure was present at atmospheric gas pressure, then total gas pressure would have surmounted local pressure in the upper 70 cm of the water column and bubbles would be formed. Most probably bubbling had happened before and nitrogen had been stripped until total gas pressure lay below local pressure (observations of N_2 to confirm this were not done).

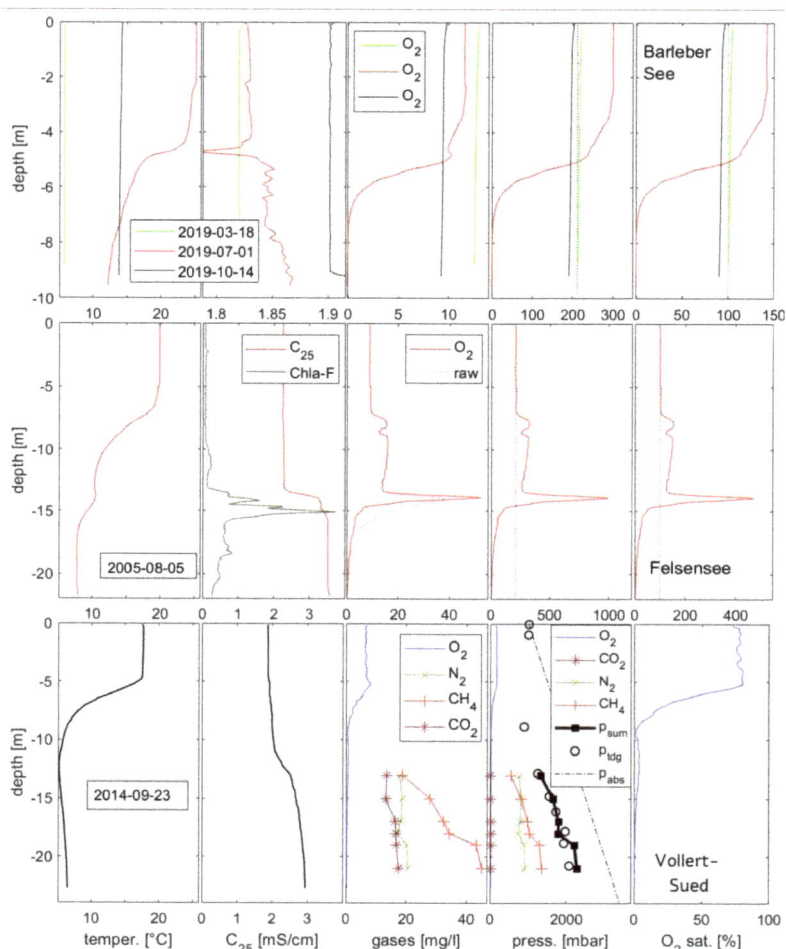

Figure 4. Profiles from the lakes: Barleber See, Felsensee, Vollert-Sued showing measured profiles of temperature, el. conductivity, gas concentrations, gas pressures (dotted line: atmospheric O_2 partial pressure above lake surface) and oxygen saturation (dotted line: 100% saturation). Felsensee O_2 data have been compensated for 6.5 s response time of the amperometric sensor from raw data. Vollert-Sued: p_{sum}: sum of calculated gas pressures; p_{tdg}: directly measured total gas pressure; p_{abs}: absolute pressure calculated as atmospheric plus hydrostatic.

In (the small quarry lake) Felsensee, a deep oxygen maximum was documented at 15 m depth in early August. The oxygen peak was accompanied by high values of chlorophyll-a fluorescence (Figure 4, middle row). Hence oxygen levels were attributed to photosynthetic activity of floating organisms at the upper edge of the monimolimnion. Obviously the organisms could profit from the chemical setting (nutrients or CO_2) at this depth. Moreover, clearly below the epilimnion at depths 7–13 m, oxygen concentrations and gas pressures lay clearly above atmospheric oxygen pressure.

Finally gas concentrations in (the small polluted pit lake) Vollert-Sued were measured (data from Horn et al. [53]). In the monimolimnion, methane had accumulated and raised the gas pressure by more than 1 bar. Together with nitrogen, total gas pressure in the monimolimnion came into the range of the absolute pressure. Methane had been created by degradation of organic material in the sediment. Methane either diffused out of the sediment or formed bubbles which released part of their methane into the ambient (monimolimnetic) water while ascending through the water column. Bubbles could always be observed at the water surface.

5. Discussion

In most lacustrine waters, gas pressure is clearly dominated by nitrogen N_2 and oxygen O_2. While nitrogen is quite conservative and changes in N_2 concentrations happen at a small rate, O_2 is produced and used by aquatic organisms and hence O_2 often shows a highly dynamic behavior in limnic systems. As a consequence, the oxygen contribution is the leading <u>variable</u> component of gas pressure in limnic waters. This is particularly visible in productive small lakes (e.g., Barleber See, Figure 4), as longer periods of weak winds allow a clear decoupling from the atmosphere. However, nitrogen always contributes a large portion to the total gas pressure and hence forms a large portion of the gas in bubbles.

In addition, gas pressure (and hence oxygen saturation) is affected by heating (e.g., solar irradiation into a density stratified layer, see upper hypolimnion of Rassnitzer See in Figure 2). Although concentrations may not change, the temperature dependence of the Henry coefficient will increase gas pressure when temperatures rise. Henry coefficients of N_2, O_2, Ar, and CH_4 drop by a factor of about 2 over the limnologically interesting range from 0 °C to 30 °C. The temperature dependence of Henry coefficients of CO_2 (and e.g., helium or neon) is noticeably different (see also noble gas thermometer [59]).

N_2 concentrations can be retrieved from samples (as done for Lake Vollert-Sued ([53] Figure 4). However, such data sets have limited vertical resolution and limited vertical comparability as a consequence of possible error of chemical analysis between samples. Hence we decided to create continuous N_2 profiles with a model approach from our understanding and assumption of conservative behavior. The resulting profile was realistic and included the typical features of the N_2 profile (Figure 4). An increased N_2 gas pressure in the metalimnion was the consequence of faster transport of heat (diffusive and by solar irradiation) than the N_2 molecules.

Denitrification (forming N_2 from inorganic nitrogen) and N_2 fixation (forming organic nitrogen compounds from N_2) happen at low rates in general. However a closer look at denitrification rates (e.g., [6]) reveals that even holomictic lakes may experience a production of N_2 that may affect the N_2 budget significantly over a stratification period (putatively in the range up to 10%). In monimolimnia, where more time would be available for accumulation (see below), the replenishment of nitrate is limited. On the other hand, nitrogen fixation probably can reduce the N_2 budget and hence total gas pressure only in very special configurations. As a consequence, we encourage measurements of N_2 when close investigation of total gas pressure in lakes are envisaged and accurate methods are available.

Furthermore in the range of 10 mbars, dissolved argon and vapor pressure contribute to gas pressure, which in general is in the order of 1 percent of total gas pressure. Hence, both gases are often neglected: in the case of argon, it is conveniently included in the nitrogen part (as gas chromatograph columns often do not separate Ar from O_2 and N_2, e.g., [9]). In the cold hypolimnetic water, the vapor pressure does not play an important role, but it does in warmer water: at 25 °C: water vapor pressure amounts to about 30 mbar and is the biggest contribution after N_2 and O_2 in the epilimnion (see Figure 3). As a consequence, moist samples in a head space contain about 3% of H_2O, which can be found missing in the recovery rate of an accurate gas chromatography, if water vapor is not detected in separate (e.g., [32]).

Though the production and the removal of carbon dioxide is directly connected with the oxygen dynamics, its contribution to gas pressure is not relevant in holomictic lakes, as the Henry coefficient of CO_2 is nearly two orders of magnitude larger. In addition, the involvement of the carbonate system extenuates the variability. For an accumulation that becomes relevant for gas pressure, long time periods must be provided (as in meromictic lakes) or a very strong source, e.g., from volcanic vents, has to supply gas [60,61].

Methane is usually not produced fast enough in the water column to contribute to gas pressure considerably; however conditions in the sediment are much more favorable [62]. Limnic sediments can provide a nearly inexhaustible amount of degradable organic material and anoxic conditions prevent the further oxidation of methane. Methane can diffuse out of sediment pores into the open water or alternatively form bubbles (see below), from which part of the methane is released into the water column while ascending to the lake surface.

5.1. Ebullition

The release of gases from the water body by bubbles is called ebullition. Bubbles can be formed and sustained, when gas pressure reaches absolute pressure (Figure 5; [8,63,64]). Such bubbles migrate toward the water surface due to their buoyancy. Mainly two gases are produced in natural waters by organisms to raise total gas pressure enough to force bubble formation: oxygen and methane.

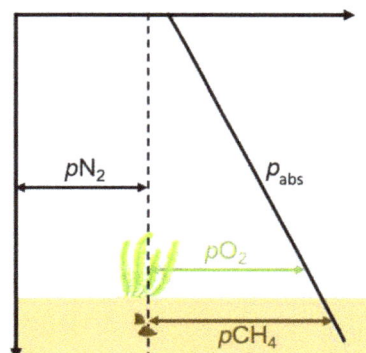

Figure 5. Gas pressure against depth in an idealized lake, gas pressure of argon is included in the N_2 value, gas pressures of other gases and water vapor pressure are neglected.

In sunny conditions, primary production can form oxygen. Close to the surface, not much raising of the oxygen gas pressure is required to reach absolute pressure (e.g., Barleber See in Figure 4). However at greater depth, the gas pressure needs to be raised considerably before bubbles are form (e.g., Felsensee in Figure 4). As a consequence, elevated levels of oxygen can be seen before bubbles form. Deep chlorophyll maxima sustain primary production when they are exposed to favorable light conditions [6]. At a depth of 14 m in Felsensee (Figure 4) for example, an absolute pressure of 2.4 bar (1 bar atmospheric pressure plus 1.4 bar hydrostatic pressure) needs to be overcome. If gas pressure of N_2 (and Ar) is at atmospheric equilibrium (i.e., 0.78 bar and 0.01 bar, respectively), the remainder to absolute pressure has to be accomplished by oxygen production until an O_2 gas pressure of 2.4–0.79 bar = 1.61 bar is reached. 1.61 bar of oxygen gas pressure are commensurate with 750% of oxygen saturation (as 100% saturation mean 0.21 bar partial pressure of O_2). In Felsensee, we detected close to 950 mbar (a saturation of nearly 450%) on the day of measurement. Whether this lake has ever reached the limit for ebullition has not been documented.

In general, the gas pressure of oxygen needs to be raised to

$$p_{O2} + p_{rest} = p_{abs} \tag{7}$$

where the nitrogen gas pressure is the leading part in p_{rest}, and also gas pressures of argon and vapor are part of it (Figure 5).

Ebullition from macrophytes and algal mats on the lake bed has been documented much better: bubbles remain attached to the plants before they have grown enough to detach and ascend through the water column [65–67]. As long as a bubble remains attached to a plant, it is subject to gas exchange with the surrounding water [11]. Gases diffuse in and out, and—provided that enough time has been available—the gas composition (partial pressures) will reflect the gas pressure inside the water. The data presented from two small lakes [9] are compliant with this prediction.

The situation is very similar for methane bubbles. Methane is formed from biodegradation (see Supporting Information). This largely happens in the upper zone of the sediment. Methane can be accumulated in the pore space and finally form bubbles when total gas pressure reaches the absolute pressure (Figure 5). When the buoyancy is sufficient to escape from the sediment, methane bubbles enter the bottom water of the lake and start ascending through the water column [10,68,69].

Like oxygen bubbles, methane bubbles do not consist of pure methane, but contain mainly nitrogen and traces of other gases. As the bubbles also remove N_2 from the pore water, this may be observable and even indicative of how much methane has been produced [10,70,71].

In conclusion, we could identify three possible zones of bubble release: (1) Close to the surface by oxygen release through primary production close to the water surface at roughly atmospheric composition Barleber See (Figure 4); (2) release of bubbles by accumulating oxygen in deep photosynthetically active areas (of phytoplankton of macrophytes: oxygen content is higher than atmospheric content ([9], see also Felsensee Figure 4); (3) Bubbles formed by methane production in the sediment mainly consist of methane and nitrogen ([10,53], see also Vollert-Sued Figure 4).

5.2. Other Mechanisms for Raising Gas Pressure and Releasing Gas Bubbles

At the surface, gas pressure could be raised by surface warming during the day. Warming releases bubbles of oxygen:nitrogen of 21:78. In Barleber See, oxygen saturation has clearly risen beyond 100% (see Figure 4, top row) and hence 0.21 bar. As the heating of the metalimnion is faster than the diffusive removal of nitrogen, a zone of high nitrogen pressure forms in the metalimnion (see Arendsee in Figure 3). This could contribute to forming bubbles in lakes where also oxygen is produced by photosynthesis at the same time. In addition, currently the most feared trigger mechanism for a catastrophic release of gas from Lake Kivu is submerged volcanic activity, bringing hot lava in contact with highly gas-charged deep water layers. Higher temperatures would correspond to lower Henry coefficients and hence to higher gas pressures. A chain reaction known as "limnic eruption" had been feared as a possible result.

Maeck et al. [72] showed that the release of methane from the river Saar/Mosel was triggered by surface waves originating from opening ship locks. The arriving wave trough lowered the pressure at the river bed, which resulted in bubbling. A similar connection exists with air pressure, wind, or water level changes [73–75]. When air pressure rises, Horn et al. [53] for example showed that ebullition decelerated while a falling air pressure increased the ebullition flux.

Water motions can impact the release of gases. Obviously if water parcels are moved vertically, they experience a lower hydrostatic pressure and release of gases can be triggered. In Lake Nyos, such an event following a land slide is the most commonly accepted trigger mechanism for the limnic eruption in 1986 [13].

Finally we want to mention that technical measures can raise gas pressure to values that even endanger aquatic life. To oppose oxygen depletion in hypolimnetic waters, aeration—i.e., introduction of air bubbles in the deep water can be considered. However the dissolution of bubbles under high pressure facilitates to raise gas pressure beyond atmospheric pressure, which can result in problems in particular for organisms that move vertically in the water column, such as fish (e.g., [76–78]). Excessively high gas pressures can be avoided by introducing only oxygen or—at least in part—by forcing the bubble plume to reach the water surface at the cost of money or changing the ecosystem. In addition, elevated gas pressures have been reported from reservoirs belonging to pumped storage plants.

5.3. Accumulation of Gases in Monimolimnia

Permanently stratified water bodies like meromictic lakes provide the preconditions for an accumulation of methane over long time scales at concentrations that eventually become relevant or even dominant for total gas pressure (e.g., [1,79,80]). Vollert-Sued is one example (Figure 3, bottom row), where degradable sediments have provided the material for methane formation. Methane and nitrogen form the gas pressure in the deep water. Famous other examples for a methane dominated gas pressure in monimolimnia are Lake Kivu, East Africa [60,81], and Lake Monticchio Piccolo, Italy [82].

Monimolimnia also accumulate carbon dioxide as an end product of biogeochemical degradation paths for organic matter. However, due to the good solubility of carbon dioxide, it takes a long time to form a CO_2 gas pressure of concern only from organic degradation. The famous examples of extreme CO_2 gas pressure: Lake Nyos, Lake Monoun (both in Cameroon), Kabuno Bay of Lake Kivu (in D.R. Congo), and main basin of Lake Kivu (in Rwanda and D.R. Congo) have a volcanic origin for the supply of CO_2 [13,14,83]. In the special case of Guadiana Pit Lake (in Spain), acid rock drainage dissolved carbonate deposits in the underground and stored the formed CO_2 in the monimolimnion [18,84].

The gas either originates from volcanic sources (e.g., Lake Nyos, Lake Monoun, Lake Kivu, Lake Monticchio Piccolo) or from geochemical interaction (Guadiana Pit Lake: [85]) or from biodegradation (Vollert-Sued pit lake). In all cases, either methane (Lake Kivu, Vollert-Sued, Monticchio Poccolo) or carbon dioxide (Nyos; Monoun; Guadiana Pit Lake: [18,86]) provide the leading contribution beyond the N_2 background. The most prominent representative is Lake Kivu with its huge methane storage for commercial interest [16,28,32,83]. Especially high gas pressures of carbon dioxide are dangerous, as a sudden release can liberate an immense volume of gas to the atmosphere and threaten the lives of humans in the vicinity. Disastrous degassing happened at Lake Nyos and Lake Monoun in Cameroon in the 1980s [13–15].

Monimolimnia of meromictic lakes are shielded from direct exchange with the atmosphere. Water properties are renewed at a very slow rate. This yields ages of monimolimnia up to 800 years (Lake Kivu, Africa—Schmid et al. 2005) and more than 6000 years in the case of Salsvatn, Norway [87] or even 11,000 years in the case of Powell Lake, Canada [88,89]. Hence meromixis can provide a long time scale for the accumulation of solutes such as gases (see [79,90,91]).

It is clear that we may not have knowledge of all lakes with extreme gas pressure. Often H_2S is reported as dangerous gas dissolved at high concentrations (k_H = 0.1 mol/L/bar, T_E = 2100 K [19]). Very high concentrations of sulphide (2.5 mmol/L) are found in Alatsee (Bavaria, Germany [92]). Depending on the pH, only part is present as dissolved H_2S ($K_s = 9.77 \times 10^{-6}$) and hence contributes to gas pressure. At most, i.e., in acidic conditions (pH not reported), this corresponds to a gas pressure of 0.015 bar gas pressure at reported monimolimnion temperatures of 6 °C. Trapped ocean water could potentially produce about ten times as much sulphide from its dissolved sulphate, if losses by diffusion and precipitation would be small over the time period required for reducing the sulphate. We measured our highest sulphide concentrations (10 and 12 mmol/L) in the monimolimnion of Hufeisensee (Saxony-Anhalt, Germany, see also [93]), which corresponded to about 0.05 bar partial pressure of H_2S (at measured pH = 6.8 and T = 6 °C). In conclusion, only in extreme cases, sulphide will provide a considerable contribution to the total gas pressure.

5.4. Final Remarks

From literature review and our own results, we confirm that in limnic waters only very few gases play an important role for gas pressure:

1. Nitrogen always contributes to gas pressure decisively and must be considered;
2. Oxygen from the atmosphere and from photosynthesis can contribute decisively to the gas pressure;

As far as documented, only under meromictic conditions in the deep anoxic waters (monimolimnion),

3. Methane (mainly from biodegradation or volcanic sources) or
4. Carbon dioxide (from external sources such as volcanic vents and geochemical reactions) can become an important or even the leading contribution to gas pressure.

In holomictic lakes, the gas pressure contribution of methane and carbon dioxide is small (usually even smaller than vapor and argon). At a smaller scale (gas pressure of tens of millibars), we can detect gas pressure of

5. Vapor pressure from water;
6. Argon from atmospheric sources.

All other gases play a much subordinate role. Other noble gases, chlorofluorocarbons, sulfurhexafluoride SF_6 (e.g., [94,95]), and other gases may be used for tracing waters and dating the last intensive exchange with the atmosphere. It can be relevant to know their partial pressure to quantify the concentrations. However for total gas pressure, they do not play a role.

Bubbles in the open lake water can be created by photosynthetic oxygen production. This may be accomplished by submerged macrophytes and algal mats or planktonic algae closer to the surface. Additional heating may help forming bubbles as solubility of gases is temperature dependent. On the contrary, when bubbles form in the upper layer of the sediment, they originate from the decomposition of organic material where the produced methane is the leading component for raising the gas pressure. Biodegradation also produces carbon dioxide, but this contribution to gas pressure and hence for the release of bubbles is much lower due to the good solubility of carbon dioxide.

Nitrogen (N_2) is produced by biogeochemical reactions, but rates usually are small in comparison to the N_2 background. Hence nitrogen is rarely made responsible for forming bubbles. However, due to the high background of nitrogen gas pressure, it is very important for total gas pressure and hence contributes to a high portion of gas in the bubbles. The composition of the bubbles can be quantitatively calculated from partial pressures. In general, the required gas pressure of the produced gas (oxygen in open water or methane in the sediment) increases with depth (Equation (7)) and so does its portion in the bubble. Ascending bubbles are subject to exchange with the surrounding water (stripping).

Supplementary Materials: The following are available online at https://www.mdpi.com/article/10.3390/w13131824/s1, Figure S1: solubilities of the most relevant gases for gas pressure in lakes: left column: solubilities in mol/l/bar: simple exponential fits from Sander (2015) [19] in comparison to accurate parametrizations for N_2 and O_2 (Weiss 1970) [23], Ar (Jenkins et al. 2019) [24], CH_4 (Wiesenburg and Guinasso 1979) [25] and CO_2 (Weiss 1974) [26]. Right column: difference between approaches. Figure S2: Vapour pressure against temperature. Left panel: Huang (2018) [35] compared to the IAPWS recommended curve (Wagner and Pruss, 1993) [34]; right panel: Magnus equation compared to the recommendation of IAPWS (Wagner and Pruss, 1993) [34]. References [96–129] are cited in Supplementary Materials.

Author Contributions: Conceptualization, B.B. and C.W.; data curation, B.B., S.J., M.H. and M.S.; formal analysis, B.B., P.L., C.S. and M.S.; funding acquisition, M.H.; investigation, B.B., S.J., C.S. and M.S.; methodology, C.W.; resources, B.B., M.H. and M.S.; validation, B.B.; visualization, P.L. and C.S.; writing—original draft, B.B., S.J., P.L., C.W., C.S., M.H. and M.S.; writing—review and editing, B.B., C.W., C.S., M.H. and M.S. All authors have read and agreed to the published version of the manuscript.

Funding: The authors thank the city of Magdeburg for funding measurements on Barleber See, Deutsche Forschungsgemeinschaft—DFG for funding the project "newMOM" (ID: RI2040/4-1) and Talsperrenbetrieb Sachsen-Anhalt for supporting measurements on Rappbode Reservoir.

Institutional Review Board Statement: Not applicable.

Informed Consent Statement: Not applicable.

Data Availability Statement: Not applicable.

Acknowledgments: The authors would like to thank Karsten Rahn for measurements in Rassnitzer See, Rappbode Reservoir, Barleber See, and Vollert-Sued; Burkhard Kuehn for measurements in Barleber See and Thomas Hintze for the technical support at Arendsee. Many thanks to Matthias Koschorreck for comments on the manuscript.

Conflicts of Interest: The authors declare no conflict of interest.

References

1. Bastviken, D.; Tranvik, L.J.; Downing, J.A.; Crill, P.; Enrich-Prast, A. Freshwater Methane Emissions Offset the Continental Carbon Sink. *Science* **2011**, *331*, 50. [CrossRef] [PubMed]
2. Cole, J.J.; Prairie, Y.T.; Caraco, N.F.; McDowell, W.H.; Tranvik, L.J.; Striegl, R.G.; Duarte, C.M.; Kortelainen, P.; Downing, J.A.; Middelburg, J.J.; et al. Plumbing the global carbon cycle: Integrating inland waters into the terrestrial carbon budget. *Ecosystems* **2007**, *10*, 171–184.
3. Myhre, G.; Shindell, D.; Bréon, F.; Collins, W.; Fuglestvedt, J.; Huang, J. Anthropogenic and natural radiative forcing. In *Climate Change the Physical Science Basis. Contribution of Working Group I to the Fifth Assessment Report of the Intergovernmental Panel on Climate Change*; Stocker, T.F., Qin., D., Plattner, G., Tignor, M., Allen, S.K., Boschung, J., Nauels, A., Xia, Y., Bex, V., Midgley, P.M., Eds.; Cambridge University Press: Cambridge, UK, 2013.
4. Turner, A.J.; Frankenberg, C.; Wennberg, P.O.; Jacob, D.J. Ambiguity in the causes for decadal trends in atmospheric methane and hydroxyl. *Proc. Nat. Acad. Sci. USA* **2017**, *114*, 5367–5372. [CrossRef] [PubMed]
5. Keller, P.S.; Catalán, N.; Von Schiller, D.; Grossart, H.-P.; Koschorreck, M.; Obrador, B.; Frassl, M.A.; Karakaya, N.; Barros, N.; Howitt, J.A.; et al. Global CO_2 emissions from dry inland waters share common drivers across ecosystems. *Nat. Commun.* **2020**, *11*, 1–8. [CrossRef]
6. Wetzel, R.G. *Limnology: Lake and River Ecosystems*, 3rd ed.; Academic Press: Cambridge, MA, USA, 2001.
7. Higgins, S.N.; Paterson, M.J.; Hecky, R.E.; Schindler, D.W.; Venkiteswaran, J.J.; Findlay, D.L. Biological Nitrogen Fixation Prevents the Response of a Eutrophic Lake to Reduced Loading of Nitrogen: Evidence from a 46-Year Whole-Lake Experiment. *Ecosystems* **2018**, *21*, 1088–1100. [CrossRef]
8. Miyake, Y. The Possibility and the Allowable Limit of Formation of Air Bubbles in the Sea. *Pap. Meteorol. Geophys.* **1951**, *2*, 95–101. [CrossRef]
9. Koschorreck, M.; Hentschel, I.; Boehrer, B. Oxygen Ebullition from Lakes. *Geophys. Res. Lett.* **2017**, *44*, 9372–9378. [CrossRef]
10. Langenegger, T.; Vachon, D.; Donis, D.; McGinnis, D.F. What the bubble knows: Lake methane dynamics revealed by sediment gas bubble composition. *Limnol. Oceanogr.* **2019**, *64*, 1526–1544. [CrossRef]
11. McGinnis, D.F.; Greinert, J.; Artemov, Y.; Beaubien, S.E.; Wüest, A. Fate of rising methane bubbles in stratified waters: How much methane reaches the atmosphere? *J. Geophys. Res. Space Phys.* **2006**, *111*. [CrossRef]
12. Halbwachs, M.; Sabroux, J.-C.; Grangeon, J.; Kayser, G.; Tochon-Danguy, J.-C.; Felix, A.; Beard, J.-C.; Villevieille, A.; Vitter, G.; Richon, P.; et al. Degassing the "killer lakes" Nyos and Monoun, Cameroon. *Eos Transacti. Am. Geophys.* **2004**, *85*, 281–285. [CrossRef]
13. Sigurdsson, H.; Devine, J.; Tchua, F.; Presser, F.; Pringle, M.; Evans, W. Origin of the lethal gas burst from Lake Monoun, Cameroun. *J. Volcanol. Geotherm. Res.* **1987**, *31*, 1–16. [CrossRef]
14. Kling, G.; Clark, M.A.; Wagner, G.N.; Compton, H.R.; Humphrey, A.M.; Devine, J.D.; Evans, W.C.; Lockwood, J.; Tuttle, M.L.; Koenigsberg, E.J. The 1986 Lake Nyos Gas Disaster in Cameroon, West Africa. *Science* **1987**, *236*, 169–175. [CrossRef]
15. Freeth, S.J.; Kling, G.; Kusakabe, M.; Maley, J.; Tchoua, F.M.; Tietze, K.; Freeth, G.W.K.S.J. Conclusions from Lake Nyos disaster. *Nature* **1990**, *348*, 201. [CrossRef]
16. Wüest, A.; Jarc, L.; Bürgmann, H.; Pasche, N.; Schmid, M. Methane Formation and Future Extraction in Lake Kivu. In *Lake Kivu*; Springer: Berlin/Heidelberg, Germany, 2012; pp. 165–180.
17. Lorke, A.; Tietze, K.; Halbwachs, M.; Wüest, A. Response of Lake Kivu stratification to lava inflow and climate warming. *Limnol. Oceanogr.* **2004**, *49*, 778–783. [CrossRef]
18. Boehrer, B.; Yusta, I.; Magin, K.; Sanchez-España, J. Quantifying, assessing and removing the extreme gas load from meromictic Guadiana pit lake, Southwest Spain. *Sci. Total Environ.* **2016**, *563–564*, 468–477. [CrossRef]
19. Sander, R. Compilation of Henry's law constants (version 4.0) for water as solvent. *Atmosph. Chem. Phys. Discuss.* **2015**, *15*, 4399–4981. [CrossRef]
20. Roedel, W. *Physik Unserer Umwelt: Die Atmosphäre*; Springer: Heidelberg, Germany, 1992.
21. Worch, E. *Hydrochemistry. Basic Concepts and Exercises*; De Gruyter: Berlin, Germany, 2015; (De Gruyter Graduate).

22. Saunois, M.; Stavert, A.R.; Poulter, B.; Bousquet, P.; Canadell, J.G.; Jackson, R.B.; Raymond, P.A.; Dlugokencky, E.J.; Houweling, S.; Patra, P.K.; et al. The Global Methane Budget 2000–2017. *Earth Syst. Sci. Data* **2020**, *12*, 1561–1623. [CrossRef]
23. Weiss, R. The solubility of nitrogen, oxygen and argon in water and seawater. *Deep Sea Res. Oceanogr. Abstr.* **1970**, *17*, 721–735. [CrossRef]
24. Jenkins, W.; Lott, D.; Cahill, K. A determination of atmospheric helium, neon, argon, krypton, and xenon solubility concentrations in water and seawater. *Mar. Chem.* **2019**, *211*, 94–107. [CrossRef]
25. Wiesenburg, D.A.; Guinasso, N.L. Equilibrium solubilities of methane, carbon monoxide and hydrogen in water and seawater. *J. Chem. Eng. Data* **1979**, *24*, 357–360. [CrossRef]
26. Weiss, R. Carbon dioxide in water and seawater: The solubility of a non-ideal gas. *Mar. Chem.* **1974**, *2*, 203–215. [CrossRef]
27. Dietz, S.; Lessmann, D.; Boehrer, B. Contribution of Solutes to Density Stratification in a Meromictic Lake (Waldsee/Germany). *Mine Water Environ.* **2012**, *31*, 129–137. [CrossRef]
28. Bärenbold, F.; Boehrer, B.; Grilli, R.; Mugisha, A.; von Tümpling, W.; Umutoni, A.; Schmid, M. Updated dissolved gas concentrations in Lake Kivu from an intercomparison project. *PLoS ONE* **2020**, *15*, e0237836.
29. Kipfer, R.; Aeschbach-Hertig, W.; Peeters, F.; Stute, M. Noble Gases in Lakes and Ground Waters. *Rev. Miner. Geochem.* **2002**, *47*, 615–700. [CrossRef]
30. Christenson, B.; Tassi, F. Gases in Volcanic Lake Environments. In *Advances in Volcanology*; Springer: Berlin/Heidelberg, Germany, 2015; pp. 125–153.
31. Worch, E. *Wasser und Wasserinhaltsstoffe. Eine Einführung in die Hydrochemie.* Wiesbaden, s.l.; Vieweg+Teubner Verlag: Berlin, Germany, 1997.
32. Boehrer, B.; von Tuempling, W.; Mugisha, A.; Rogemont, C.; Umutoni, A. Reliable reference for the methane concentrations in Lake Kivu at the beginning of industrial exploitation. *Hydrol. Earth Syst. Sci.* **2019**, *23*, 4707–4716. [CrossRef]
33. Alduchov, O.A.; Eskridge, R.E. Improved Magnus Form Approximation of Saturation Vapor Pressure. *J. Appl. Meteorolog.* **1996**, *4*, 601–609. [CrossRef]
34. Wagner, W.; Pruss, A. International Equations for the Saturation Properties of Ordinary Water Substance. Revised According to the International Temperature Scale of Addendum to 1990. *J. Phys. Chem. Ref. Data* **1993**, *16*, 783–787. [CrossRef]
35. Huang, J. A Simple Accurate Formula for Calculating Saturation Vapor Pressure of Water and Ice. *J. Appl. Meteorol. Clim.* **2018**, *57*, 1265–1272. [CrossRef]
36. Hartmann, O.; Schönberg, G. Geologische Entwicklungsgeschichte und Untersuchungsergebnisse am Arendsee. *Nachr. Arb. Unterwasserarchäologie* **2009**, *15*, 58–64.
37. Meinikmann, K.; Lewandowski, J.; Nützmann, G. Lacustrine groundwater discharge: Combined determination of volumes and spatial patterns. *J. Hydrol.* **2013**, *502*, 202–211. [CrossRef]
38. Hannappel, S.; Köpp, C.; Rejman-Rasinska, E. Aufklärung der Ursachen zur Phosphorbelastung des oberflächennahen Grundwassers im hydraulischen Zustrom zum Arendsee in der Altmark. *Hydrol. Wasserbewirtsch.* **2020**, *62*, 25–38.
39. Rinke, K.; Kuehn, B.; Bocaniov, S.; Wendt-Potthoff, K.; Büttner, O.; Tittel, J.; Schultze, M.; Herzsprung, P.; Rönicke, H.; Rink, K.; et al. Reservoirs as sentinels of catchments: The Rappbode Reservoir Observatory (Harz Mountains, Germany). *Environ. Earth Sci.* **2013**, *69*, 523–536. [CrossRef]
40. Pöhlein, F.; Schultze, M.; Donner, J.; Dietze, M.; Wilhayn, S. Reaktionen eines zweistufigen Talsperrensystems auf Veränderungen des Stoffeintrags am Beispiel des Bodewerks. *Forum Hydrol. Wasserbewirtsch.* **2004**, *34*, 137–144.
41. Wentzky, V.C.; Tittel, J.; Jäger, C.G.; Rinke, K. Mechanisms preventing a decrease in phytoplankton biomass after phosphorus reductions in a German drinking water reservoir-results from more than 50 years of observation. *Freshw. Biol.* **2018**, *63*, 1063–1076. [CrossRef]
42. Boehrer, B.; Heidenreich, H.; Schimmele, M.; Schultze, M. Numerical prognosis for salinity profiles of future lakes in the opencast mine Merseburg-Ost. *Int. J. Salt Lake Res.* **1998**, *7*, 235–260. [CrossRef]
43. Heidenreich, H.; Boehrer, B.; Kater, R.; Hennig, G. Gekoppelte Modellierung geohydraulischer und limnophysikalischer Vorgänge in Tagebaurestseen und ihrer Umgebung. *Grundwasser* **1999**, *4*, 49–54. [CrossRef]
44. Trettin, R.; Gläßer, W.; Lerche, I.; Seelig, U.; Treutler, H.-C. Flooding of lignite mines: Isotope variations and processes in a system influenced by saline groundwater. *Isot. Environ. Health Stud.* **2006**, *42*, 159–179. [CrossRef]
45. Boehrer, B.; Kiwel, U.; Rahn, K.; Schultze, M. Chemocline erosion and its conservation by freshwater introduction to meromictic salt lakes. *Limnology* **2014**, *44*, 81–89. [CrossRef]
46. Kong, X.; Seewald, M.; Dadi, T.; Friese, K.; Mi, C.; Boehrer, B.; Schultze, M.; Rinke, K.; Shatwell, T. Unravelling winter diatom blooms in temperate lakes using high frequency data and ecological modeling. *Water Res.* **2021**, *190*, 116681. [CrossRef]
47. Rönicke, H.; Frassl, M.A.; Rinke, K.; Tittel, J.; Beyer, M.; Kormann, B.; Gohr, F.; Schultze, M. Suppression of bloom-forming colonial cyanobacteria by phosphate precipitation: A 30 years case study in Lake Barleber (Germany). *Ecol. Eng.* **2021**, *162*, 106171. [CrossRef]
48. Hannappel, S.; Strom, A. Methode zur Ermittlung des Phosphoreintrags über das Grundwasser in den Barleber See bei Magdeburg. *Korresp. Wasserwirtsch.* **2020**, *13*, 24–30.
49. Spott, D. Zum Chemismus der künstlichen Seen des Steinbruchgebietes zwischen Plötzky, Gommern und Pretzien. *Nat. Nat. Heim. Bez. Halle Magdebg.* **1967**, *4*, 43–53.
50. Eccarius, B. Groundwater Monitoring and Isotope Investigation of Contaminated Wastewater from an Open Pit Mining Lake. *Environ. Geosci.* **1998**, *5*, 156–161. [CrossRef]

51. Eccarius, B.; Christoph, G.; Ebhardt, G.; Glaser, W. Grundwassermodellierung zur Gefährdungsabschätzung eines phenolverseuchten Tagebaurestsees. *Grundwasser* **2001**, *6*, 61–70. [CrossRef]
52. Stottmeister, U.; Kuschk, P.; Wiessner, A. Full-scale bioremediation and long-term monitoring of a phenolic wastewater disposal lake. *Pure Appl. Chem.* **2010**, *82*, 161–172. [CrossRef]
53. Horn, C.; Metzler, P.; Ullrich, K.; Koschorreck, M.; Boehrer, B. Methane storage and ebullition in monimolimnetic waters of polluted mine pit lake Vollert-Sued, Germany. *Sci. Total Environ.* **2017**, *584–585*, 1–10. [CrossRef]
54. Wentzky, V.C.; Frassl, M.A.; Rinke, K.; Boehrer, B. Metalimnetic oxygen minimum and the presence of Planktothrix rubescens in a low-nutrient drinking water reservoir. *Water Res.* **2019**, *148*, 208–218. [CrossRef]
55. Mi, C.; Shatwell, T.; Ma, J.; Wentzky, V.C.; Boehrer, B.; Xu, Y.; Rinke, K. The formation of a metalimnetic oxygen minimum exemplifies how ecosystem dynamics shape biogeochemical processes: A modelling study. *Water Res.* **2020**, *175*, 115701. [CrossRef]
56. Boehrer, B.; Schultze, M. Stratification of lakes. *Rev. Geophys.* **2008**, *46*. [CrossRef]
57. Kreling, J.; Bravidor, J.; Engelhardt, C.; Hupfer, M.; Koschorreck, M.; Lorke, A. The importance of physical transport and oxygen consumption for the development of a metalimnetic oxygen minimum in a lake. *Limnol. Oceanogr.* **2016**, *62*, 348–363. [CrossRef]
58. Xiao, S.; Liu, L.; Wang, W.; Lorke, A.; Woodhouse, J.; Grossart, H.-P. A Fast-Response Automated Gas Equilibrator (FaRAGE) for continuous in situ measurement of CH_4 and CO_2 dissolved in water. *Hydrol. Earth Syst. Sci.* **2002**, *24*, 3871–3880. [CrossRef]
59. Aeschbach-Hertig, W.; Peeters, F.; Beyerle, U.; Kipfer, R. Interpretation of dissolved atmospheric noble gases in natural waters. *Water Resour. Res.* **1999**, *35*, 2779–2792. [CrossRef]
60. Schmid, M.; Tietze, K.; Halbwachs, M.; Lorke, A.; McGinnis, D.F.; Wüest, A. How hazardous is the gas accumulation in Lake Kivu? Arguments for a risk assesment in light of the Nyiragongo volcano eruption of 2002. *Acta Vulcanol.* **2004**, *14–15*, 115–122.
61. Avagyan, A.; Sahakyan, L.; Meliksetian, K.; Karakhanyan, A.; Lavrushin, V.; Atalyan, T.; Hovakimyan, H.; Avagyan, S.; Tozalakyan, P.; Shalaeva, E.; et al. New evidences of Holocene tectonic and volcanic activity of the western part of Lake Sevan (Armenia). *Geol. Q.* **2020**, *64*, 288–303. [CrossRef]
62. Bastviken, D. Methan. In *Encyclopedia of Inland Waters*; Elsevier: Oxford, UK, 2009; pp. 783–805.
63. Ramsey, W.L. Bubble growth from dissolved oxygen near the sea surface. *Limnol. Oceanogr.* **1962**, *7*, 1–7. [CrossRef]
64. D'Aoust, B.G. Technical Note: Total Dissolved Gas Pressure (TDGP) Sensing in the Laboratory. *Dissolut. Technol.* **2007**, *14*, 38–41. [CrossRef]
65. Long, M.H.; Sutherland, K.; Wankel, S.D.; Burdige, D.J.; Zimmerman, R.C. Ebullition of oxygen from seagrasses under supersaturated conditions. *Limnol. Oceanogr.* **2020**, *65*, 314–324. [CrossRef]
66. Pedersen, O.; Colmer, T.D.; Esand-Jensen, K. Underwater Photosynthesis of Submerged Plants—Recent Advances and Methods. *Front. Plant. Sci.* **2013**, *4*, 140. [CrossRef]
67. Mendoza-Lera, C.; Federlein, L.L.; Knie, M.; Mutz, M. The algal lift: Buoyancy-mediated sediment transport. *Water Resourc. Res.* **2016**, *52*, 108–118. [CrossRef]
68. Boudreau, B.P. The physics of bubbles in surficial, soft, cohesive sediments. *Mar. Pet. Geol.* **2012**, *38*, 1–18. [CrossRef]
69. Schmid, M.; Ostrovsky, I.; McGinnis, D.F. Role of gas ebullition in the methane budget of a deep subtropical lake: What can we learn from process-based modeling? *Limnol. Oceanogr.* **2017**, *62*, 2674–2698. [CrossRef]
70. Reeburgh, W.S. Observations of gases in chesapeake bay sediments. *Limnol. Oceanogr.* **1969**, *14*, 368–375. [CrossRef]
71. Brennwald, M.; Kipfer, R.; Imboden, D. Release of gas bubbles from lake sediment traced by noble gas isotopes in the sediment pore water. *Earth Planet. Sci. Lett.* **2005**, *235*, 31–44. [CrossRef]
72. Maeck, A.; Hofmann, H.; Lorke, A. Pumping methane out of aquatic sediments—Ebullition forcing mechanisms in an impounded river. *Biogeosciences* **2014**, *11*, 2925–2938. [CrossRef]
73. Joyce, J.; Jewell, P.W. Physical controls on methane ebullition from reservoirs and lakes. *Environ. Eng. Geosci.* **2003**, *9*, 167–178. [CrossRef]
74. Chanton, J.P.; Martens, C.S.; Kelley, C.A. Gas transport from methane-saturated, tidal freshwater and wetland sediments. *Limnol. Oceanogr.* **1989**, *34*, 807–819. [CrossRef]
75. Varadharajan, C.; Hemond, H.F. Time-series analysis of high-resolution ebullition fluxes from a stratified, freshwater lake. *J. Geophys. Res. Space Phys.* **2012**, *117*. [CrossRef]
76. Gächter, R.; Wehrli, B. Ten years of artificial mixing and oxygenation: No effect on the internal phosphorus loading of two eutrophic lakes. *Environ. Sci. Technol.* **1998**, *32*, 3659–3665. [CrossRef]
77. Bürgi, H.; Stadelmann, P. Change of phytoplankton composition and biodiversity in Lake Sempach before and during restoration. *Hydrobiologia* **2002**, *469*, 33–48. [CrossRef]
78. Cooke, G.D.; Welch, E.B.; Peterson, S.A.; Nicols, S.L. *Restoration and Management of Lakes and Reservoirs*, 3rd ed.; CRC Press: Boca Raton, FL, USA, 2005.
79. Boehrer, B.; Von Rohden, C.; Schultze, M. Physical Features of Meromictic Lakes: Stratification and Circulation. In *Mediterranean-Type Ecosystems*; Springer: Berlin/Heidelberg, Germany, 2017; pp. 15–34.
80. Schultze, M.; Boehrer, B.; Wendt-Potthoff, K.; Katsev, S.; Brown, E.T. Chemical Setting and Biogeochemical Reactions in Meromictic Lakes. In *Mediterranean-Type Ecosystems*; Springer: Berlin/Heidelberg, Germany, 2017; pp. 35–59.
81. Schmid, M.; Halbwachs, M.; Wehrli, B.; Wüest, A. Weak mixing in Lake Kivu: New insights indicate increasing risk of uncontrolled gas eruption. *Geochem. Geophys. Geosyst.* **2005**, *6*, Q07009. [CrossRef]
82. Caracausi, A.; Nuccio, P.M.; Favara, R.; Nicolosi, M.; Paternoster, M. Gas hazard assessment at the Monticchio crater lakes of Mt. Vulture, a volcano in Southern Italy. *Terra Nova* **2009**, *21*, 83–87. [CrossRef]

83. Tietze, K.; Geyh, M.; Müller, H.; Schröder, L.; Stahl, W.; Wehner, H. The genesis of the methane in Lake Kivu (Central Africa). *Acta Diabetol.* **1980**, *69*, 452–472. [CrossRef]
84. Sánchez-España, J.; Boehrer, B.; Yusta, I. Extreme carbon dioxide concentrations in acidic pit lakes provoked by water/rock interaction. *Environ. Sci. Technol.* **2014**, *48*, 4273–4281. [CrossRef]
85. Sánchez-España, J.; Falagán, C.; Ayala, D.; Wendt-Potthoff, K. Adaptation of *Coccomyxa* sp. to extremely low light conditions causes deep chlorophyll and oxygen maxima in acidic pit lakes. *Microorganisms* **2020**, *8*, 1218. [CrossRef] [PubMed]
86. Sánchez-España, J.; Yusta, I.; Boehrer, B. Degassing Pit Lakes: Technical Issues and Lessons Learnt from the HERCO$_2$ Project in the Guadiana Open Pit (Herrerías Mine, SW Spain). *Mine Water Environ.* **2020**, *39*, 517–534. [CrossRef]
87. Strøm, K.; Str, K. A Lake with Trapped Sea-Water? *Nat. Cell Biol.* **1957**, *180*, 982–983. [CrossRef]
88. Williams, P.M.; Mathews, W.H.; Pickard, G.L. A Lake in British Columbia containing Old Sea-Water. *Nat. Cell Biol.* **1961**, *191*, 830–832. [CrossRef]
89. Sanderson, B.; Perry, K.; Pedersen, T. Vertical diffusion in meromictic Powell Lake, British Columbia. *J. Geophys. Res. Space Phys.* **1986**, *91*, 7647. [CrossRef]
90. Gulati, R.D.; Zadereev, E.S.; Degermendzhi, A.G. Ecology of meromictic lakes. In *Ecological Studies 228*; Springer: Berlin/Heidelberg, Germany, 2017.
91. Zadereev, E.S.; Gulati, R.D.; Camacho, A. Biological and Ecological Features, Trophic Structure and Energy Flow in Meromictic Lakes. In *Mediterranean-Type Ecosystems*; Springer: Berlin/Heidelberg, Germany, 2017; pp. 61–86.
92. Oikomonou, A.; Filker, S.; Breiner, H.-W.; Stoeck, T. Protistan diversity in a permanently stratified meromictic lake (Lake Alatsee, SW Germany). *Environ. Microbiol.* **2015**, *17*, 2144–2157. [CrossRef]
93. Nitzsche, H.M. Chemical and Isotope Investigations in Dissolved Gases of a Meromictic Residual Lake. *Isotop. Environ. Health Stud.* **1999**, *35*, 63–73. [CrossRef]
94. Maiss, M.; Ilmberger, J.; Münnich, K.O. Vertical mixing in Überlingersee (Lake Constance) traced by SF 6 and heat. *Aquat. Sci.* **1994**, *56*, 329–346. [CrossRef]
95. Von Rohden, C.; Ilmberger, J.; Boehrer, B. Assessing groundwater coupling and vertical exchange in a meromictic mining lake with an SF6-tracer experiment. *J. Hydrol.* **2009**, *372*, 102–108. [CrossRef]
96. Aeschbach-Hertig, W.; Kipfer, R.; Hofer, M.; Imboden, D.; Wieler, R.; Signer, P. Quantification of gas fluxes from the subcontinental mantle: The example of Laacher See, a maar lake in Germany. *Geochim. Cosmochim. Acta* **1996**, *60*, 31–41. [CrossRef]
97. Bärenbold, F.; Schmid, M.; Brennwald, M.S.; Kipfer, R. Missing atmospheric noble gases in a large, tropical lake: The case of Lake Kivu, East-Africa. *Chem. Geol.* **2020**, *532*, 119374. [CrossRef]
98. Baulch, H.M.; Dillon, P.J.; Maranger, R.; Schiff, S.L. Diffusive and ebullitive transport of methane and nitrous oxide from streams: Are bubble-mediated fluxes important? *J. Geophys. Res. Space Phys.* **2011**, *116*. [CrossRef]
99. Bräuer, K.; Geissler, W.H.; Kämpf, H.; Niedermann, S.; Rman, N. Helium and carbon isotope signatures of gas exhala-tions in the westernmost part of the Pannonian Basin (SE Austria/NE Slovenia): Evidence for active lithospheric mantle de-gassing. *Chem. Geol.* **2016**, *422*, 60–70. [CrossRef]
100. Burton, M.R.; Sawyer, G.M.; Granieri, D. Deep Carbon Emissions from Volcanoes. *Rev. Miner. Geochem.* **2013**, *75*, 323–354. [CrossRef]
101. Chambers, L.; Gooddy, D.; Binley, A. Use and application of CFC-11, CFC-12, CFC-113 and SF6 as environmental tracers of groundwater residence time: A review. *Geosci. Front.* **2019**, *10*, 1643–1652. [CrossRef]
102. Emerson, S.; Quay, P.D.; Stump, C.; Wilbur, D.; Schudlich, R. Chemical tracers of productivity and respiration in the subtropical Pacific Ocean. *J. Geophys. Res. Space Phys.* **1995**, *100*, 15873. [CrossRef]
103. Etiope, G.; Klusman, R. Microseepage in drylands: Flux and implications in the global atmospheric source/sink budget of methane. *Glob. Planet. Chang.* **2010**, *72*, 265–274. [CrossRef]
104. Etiope, G.; Lollar, B.S. Abiotic methane on Earth. *Rev. Geophys.* **2013**, *51*, 276–299. [CrossRef]
105. Etiope, G.; Ciotoli, G.; Schwietzke, S.; Schoell, M. Gridded maps of geological methane emissions and their isotopic signature. *Earth Syst. Sci. Data* **2019**, *11*, 1–22. [CrossRef]
106. Frondini, F.; Cardellini, C.; Caliro, S.; Beddini, G.; Rosiello, A.; Chiodini, G. Measuring and interpreting CO_2 fluxes at regional scale: The case of the Appennines, Italy. *J. Geol. Soc.* **2019**, *176*, 408–416. [CrossRef]
107. Giggenbach, W.E.; Sano, Y.; Schmincke, H.U. CO_2-rich gases from Lakes Nyos and Monoun, Cameroon; Laacher See, Germany; Dieng, Indonesia, and Mt. Gambier, Australia—Variations on a common theme. *J. Volcanol. Geotherm. Res.* **1991**, *45*, 311–323. [CrossRef]
108. Guidotti, T.L. Hydrogen Sulphide. *Occup. Med.* **1996**, *46*, 367–371. [CrossRef] [PubMed]
109. Hamersley, M.R.; Turk, K.; Leinweber, A.; Gruber, N.; Zehr, J.; Gunderson, T.; Capone, D. Nitrogen fixation within the water column associated with two hypoxic basins in the Southern California Bight. *Aquat. Microb. Ecol.* **2011**, *63*, 193–205. [CrossRef]
110. Hamersley, M.R.; Woebken, D.; Boehrer, B.; Schultze, M.; Lavik, G.; Kuypers, M.M. Water column anammox and denitrification in a temperate permanently stratified lake (Lake Rassnitzer, Germany). *Syst. Appl. Microbiol.* **2009**, *32*, 571–582. [CrossRef] [PubMed]
111. Hunt, J.A.; Zafu, A.; Mather, T.A.; Pyle, D.M.; Barry, P. Spatially Variable CO_2 Degassing in the Main Ethiopian Rift: Implications for Magma Storage, Volatile Transport, and Rift-Related Emissions. *Geochem. Geophys. Geosystems* **2017**, *18*, 3714–3737. [CrossRef]
112. Kämpf, H.; Bräuer, K.; Schumann, J.; Hahne, K.; Strauch, G. CO_2 discharge in an active, non-volcanic continental rift area (Czech Republic): Characterisation (δ13C, 3He/4He) and quantification of diffuse and vent CO_2 emissions. *Chem. Geol.* **2013**, *339*, 71–83. [CrossRef]

113. Kelemen, P.B.; Manning, C.E. Reevaluating carbon fluxes in subduction zones, what goes down, mostly comes up. *Proc. Natl. Acad. Sci. USA* **2015**, *112*, E3997–E4006. [CrossRef]
114. Kerrick, D.M. Present and past nonanthropogenic CO_2 degassing from the solid earth. *Rev. Geophys.* **2001**, *39*, 565–585. [CrossRef]
115. Koutsoyiannis, D. Clausius–Clapeyron equation and saturation vapour pressure: Simple theory reconciled with practice. *Eur. J. Phys.* **2012**, *33*, 295–305. [CrossRef]
116. Lee, S.; Kang, N.; Park, M.; Hwang, J.Y.; Yun, S.H.; Jeong, H.Y. A review on volcanic gas compositions related to volcanic activities and non-volcanological effects. *Geosci. J.* **2018**, *22*, 183–197. [CrossRef]
117. Lewis, A.E. Review of metal sulphide precipitation. *Hydrometall.* **2010**, *104*, 222–234. [CrossRef]
118. Madigan, M.T.; Martinko, J.M.; Stahl, D.A.; Clark, D.P. *Brock Mikrobiologie*, 13th ed.; Pearson: Munich, Germany, 2013.
119. Macpherson, G. CO_2 distribution in groundwater and the impact of groundwater extraction on the global C cycle. *Chem. Geol.* **2009**, *264*, 328–336. [CrossRef]
120. Moreira, S.; Schultze, M.; Rahn, K.; Boehrer, B. A practical approach to lake water density from electrical conductivity and temperature. *Hydrol. Earth Syst. Sci.* **2016**, *20*, 2975–2986. [CrossRef]
121. Van de Graaf, A.A.; Mulder, A.; de Bruijn, P.; Jetten, M.S.; Robertson, L.A.; Kuenen, J.G. Anaerobic ammonium oxidation discovered in a denitri-fying fluidized bed reactor. *FEMS Microbiol. Ecol.* **1995**, *16*, 177–183. [CrossRef]
122. Poissant, L.; Constant, P.; Pilote, M.; Canário, J.; O'Driscoll, N.; Ridal, J.; Lean, D. The ebullition of hydrogen, carbon monoxide, methane, carbon dioxide and total gaseous mercury from the Cornwall Area of Concern. *Sci. Total Environ.* **2007**, *381*, 256–262. [CrossRef]
123. Press, F.; Siever, R.; Jordan, T.H.; Grontzinger, J. *Understanding Earth*, 4th ed.; W. H. Freeman & Co.: New York, NY, USA, 2003.
124. Raymond, P.A.; Hartmann, J.; Lauerwald, R.; Sobek, S.; McDonald, C.; Hoover, M.; Butman, D.; Striegl, R.; Mayorga, E.; Humborg, C.; et al. Global carbon dioxide emissions from inland waters. *Nature* **2013**, *503*, 355–359. [CrossRef]
125. Ross, K.A.; Gashugi, E.; Gafasi, A.; Wüest, A.; Schmid, M. Characterisation of the Subaquatic Groundwater Discharge That Maintains the Permanent Stratification within Lake Kivu; East Africa. *PLoS ONE* **2015**, *10*, e0121217. [CrossRef]
126. Schink, B. Energetics of syntrophic cooperation in methanogenic degradation. *Microbiol. Mol. Biol. Rev.* **1997**, *61*, 262–280. [CrossRef] [PubMed]
127. Schink, B. Microbially driven redox reactions in anoxic environments: Pathways, energetics, and biochemical conse-quences. *Eng. Life Sci.* **2006**, *6*, 228–233. [CrossRef]
128. Sobolewski, A. Metal species indicate the potential of constructed wetlands for long-term treatment of metal mine drainage. *Ecol. Eng.* **1996**, *6*, 259–271. [CrossRef]
129. Stumm, W.; Morgan, J.J. *Aquatic Chemistry*, 3rd ed.; John Wiley and Sons: New York, NY, USA, 1996.

Article

Relative Performance of 1-D Versus 3-D Hydrodynamic, Water-Quality Models for Predicting Water Temperature and Oxygen in a Shallow, Eutrophic, Managed Reservoir

Xiamei Man [1], Chengwang Lei [1,*], Cayelan C. Carey [2] and John C. Little [3]

1. Centre for Wind, Waves and Water, School of Civil Engineering, The University of Sydney, Sydney, NSW 2006, Australia; xman2403@uni.sydney.edu.au
2. Department of Biological Sciences, Virginia Polytechnic Institute and State University, 2025 Derring Hall, Virginia Tech, Blacksburg, VA 24061, USA; cayelan@vt.edu
3. Department of Civil and Environmental Engineering, Virginia Polytechnic Institute and State University, 401 Durham Hall, Virginia Tech, Blacksburg, VA 24061, USA; jcl@vt.edu
* Correspondence: chengwang.lei@sydney.edu.au; Tel.: +61-2-9351-2457

Abstract: Many researchers use one-dimensional (1-D) and three-dimensional (3-D) coupled hydrodynamic and water-quality models to simulate water quality dynamics, but direct comparison of their relative performance is rare. Such comparisons may quantify their relative advantages, which can inform best practices. In this study, we compare two 1-year simulations in a shallow, eutrophic, managed reservoir using a community-developed 1-D model and a 3-D model coupled with the same water-quality model library based on multiple evaluation criteria. In addition, a verified bubble plume model is coupled with the 1-D and 3-D models to simulate the water temperature in four epilimnion mixing periods to further quantify the relative performance of the 1-D and 3-D models. Based on the present investigation, adopting a 1-D water-quality model to calibrate a 3-D model is time-efficient and can produce reasonable results; 3-D models are recommended for simulating thermal stratification and management interventions, whereas 1-D models may be more appropriate for simpler model setups, especially if field data needed for 3-D modeling are lacking.

Keywords: hydrodynamic model; bubble plume; artificial mixing; GLM; Si3D

1. Introduction

One-dimensional (1-D) coupled hydrodynamic models, which simulate water balance and thermal stratification dynamics in lake and reservoir ecosystems [1,2], are popular due to their low computational requirements. Linked with biogeochemical and ecological modeling libraries, their computational efficiency allows 1-D models to quickly simulate vertical stratification in lake dynamics [3,4] including oxygen and long-term nutrient cycles [5,6]. Verified against field data using metrics such as root mean square error (RMSE) and normalized mean absolute error (NMAE) [7], 1-D models have been adopted to simulate water-quality variables such as dissolved oxygen (DO) and nutrient concentrations with adequate accuracy in many water bodies [8–10].

In contrast, three-dimensional (3-D) coupled hydrodynamic models are necessary to simulate spatially-resolved hydrodynamic and water-quality variables including oxygen [11,12] and plankton [13], especially in large aquatic ecosystems and the ocean [14–16]. Further, 3-D models may be particularly useful in waterbodies with complex bathymetry or that experience dynamic conditions [17]. For example, 3-D coupled hydrodynamic and water-quality models are useful for lake and reservoir management when engineered systems, including side steam supersaturation (SSS) and epilimnion mixing (EM), are installed to improve water quality [18,19]. One common perception [20] is that 3-D models are better in simulating engineered water bodies.

A direct comparison of the modeling results of 1-D and 3-D models may reveal the relative advantages of the different models in a quantitative manner. Fleischmann et al. [21] studied trade-offs between 1-D and 2-D regional river hydrodynamic models, but a similar study is absent in existing lake and reservoir research for 1-D and 3-D models. In addition, in certain situations, concurrent adoption of both 1-D and 3-D models may be needed to achieve simulation goals because of their complementary benefits.

One example of concurrent adoption of both 1-D and 3-D models is Romero et al. [22], who used a 3-D coupled hydrodynamic and aquatic ecological model to simulate a flood underflow through Lake Burragorang, a deep lake near Sydney, Australia. Due to the flood underflow, biogeochemical distributions varied spatially and temporally, which could not be captured by the 1-D model. At the same time, their team used a 1-D coupled hydrodynamic and aquatic ecological model to simulate Lake Burragorang for over two years due to the low computational needs for long-term calibration and validation of 1-D models. In this case, quantifying the relative advantages of the 1-D and 3-D models are crucial for obtaining reliable simulation results in a timely manner.

Romero et al. [22] also found that one set of biogeochemical parameters can be adopted in both 1-D and 3-D models to adequately simulate nutrient and plankton dynamics in the lake, which provides another reason for concurrent adoption of both 1-D and 3-D models. Parameter identification and sensitivity analysis are critical for performance evaluation when using ecological modeling libraries linked with hydrodynamic models [23], usually requiring thousands of repeated model runs. This is feasible with 1-D models due to their high computational efficiency. However, it is extremely time-consuming to do the same with 3-D models, for which a single one-year simulation may take up to a few days of real time even on high-performance computer clusters. Since most ecological parameters in 3-D models are chosen based on literature values or manual tuning, using 1-D models as a test-bed may be an efficient solution for parameter identification and sensitivity analysis for 3-D models.

Similar to the work carried out by Romero et al. [22], this study adopts a 1-D model as a test-bed environment to speed up the calibration of a 3-D hydrodynamic and water-quality model. Further, this study employs well-calibrated biogeochemical parameters of a 1-D coupled hydrodynamic and water-quality model in a 3-D model of a shallow eutrophic reservoir.

When performing numerical simulations, both the spatial and temporal resolutions may affect the numerical accuracy. Reducing spatial and temporal resolutions shortens computing time. However, a lower temporal resolution may lead to errors in energy balance [24], and a lower spatial resolution may fail to resolve the bathymetry of water bodies [25]. Therefore, it is necessary to compromise between the numerical accuracy and the computational efficiency when determining the spatial and temporal resolutions. Numerical tests are performed to determine a suitable cell size for calibration and simulation.

Temperature and DO field data collected over two one-year periods are used to verify the simulation results. The aim of this study is to quantify the relative advantages of 1-D and 3-D coupled hydrodynamic and water-quality models and determine the most time-efficient modeling approach with combined 1-D and 3-D calibration and modeling. In addition, this study is the first attempt to implement 1-D EM modeling, with two bubble plume model variants enabled in the 1-D and 3-D coupled models. Additional simulations of four EM periods over two years are carried out to compare the two bubble plume models and the 1-D and 3-D EM modeling approaches.

2. Study Methods

2.1. Study Sites and Artificial Mixing Systems

The study site is Falling Creek Reservoir (FCR) (Figure 1) in Virginia, USA (37°18'20" N, 79°50'19" W), a eutrophic, shallow, drinking-water reservoir managed by Western Virginia Water Authority (WVWA). Persistent hypolimnetic hypoxia and occasional algae blooms have caused water-quality problems in FCR during the stratified period [26,27], which shorten filter run times and increase soluble iron, manganese and phosphorus released from the sediment.

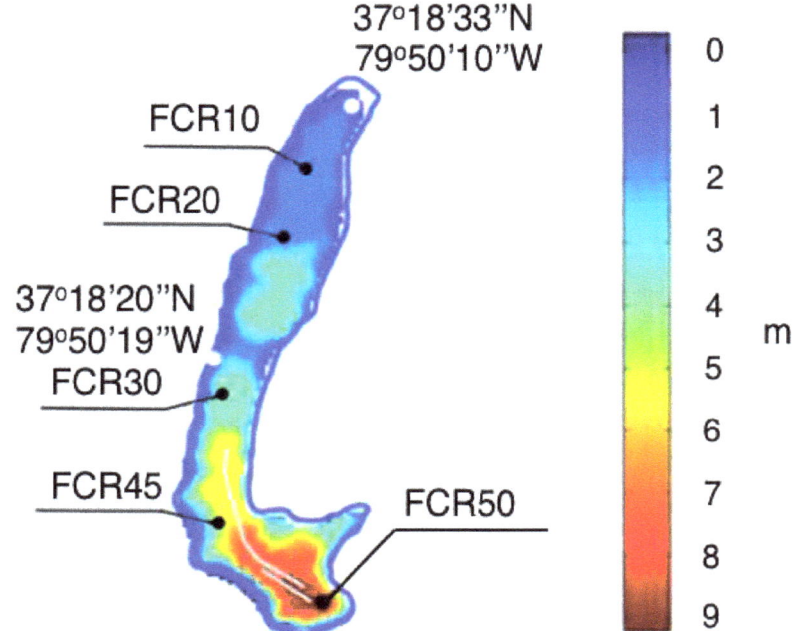

Figure 1. Bathymetry contours of Falling Creek Reservoir (FCR) and field data collection sites. The locations of the side stream supersaturation (SSS) system (short white line) and the epilimnion mixing (EM) system (long white line) are also shown.

FCR is equipped with a side-stream supersaturation system and an epilimnion mixing system to address these concerns. The SSS system consists of a submersible pump, inlet piping, oxygen source, oxygen contact chamber, outlet piping and distribution header with nozzles [28]. The EM system consists of a compressed air system on land and a distribution header submerged in the water [29]. The SSS system aims to add DO to the hypolimnion, while the EM system is designed to inject air to mix and deepen the surface mixed layer above the hypolimnion, thereby hindering the growth of algae [29,30]. The SSS distribution header is positioned 1 m above the sediment in the hypolimnion, while the EM diffuser is located approximately 5 m below the water surface.

Field data was collected in FCR continuously over 5 years from 2013 to 2018. Figure 1 presents the locations of the SSS distribution header, the EM diffuser and the five locations (FCR10, FCR20, FCR30, FCR45 and FCR50) where data were collected. FCR50 is at the deepest point near the intake of the water treatment plant.

The collected field data included water temperature, DO and meteorological data. An SBE 19 plus high-resolution (4 Hz sampling rate) conductivity, temperature, and depth (CTD) profiler customized with an SBE 43 DO probe (Seabird Electronics, Bellevue, WA, USA) and a ProODO meter (YSI Inc., Yellow Springs, OH, USA) was used to collect depth profiles of the temperature and DO. The CTD can collect data at 0.1 m increments in the water column with a response time of 1.4 s at 20 °C, and the ProODO meter was used to check the quality of the temperature and DO data collected by the CTD. The hourly meteorological data required for the numerical modeling were obtained from North American Land Assimilation System-2 (NLDAS-2) project (https://ldas.gsfc.nasa.gov/nldas/v2/models).

Table 1 shows the details of the field campaign, simulation periods and oxygenation settings for 2014 and 2015, and Table 2 presents the simulation dates of Si3D-AED2 and GLM-AED2, the EM operation periods and the corresponding flow rates in the years 2015 and 2016.

Table 1. Information about the operation of the oxygenation system in the years 2014 and 2015 and the corresponding simulation periods.

Year	Field Campaign DoY	Oxygenation Settings			Simulation Period (DoY)
		DoY	Oxygen Flow Rate (kg/day)	Water Flow Rate (L/min)	
2014	121–310	126–154	20	208	121–273
		180–210	20		
		230–273	25		
2015	90–331	125–152	15		90–331

Table 2. The Si3D-AED2 and GLM-AED2 simulation dates, the EM operation periods and the corresponding flow rates in the years 2015 and 2016.

Year	Name	Simulation DoY	EM Period DoY	Time	Flow Rate (L/min)
2015	EM15	146–154	151	12:00–15:00	708
			153	12:00–15:00	
2016	EM16–1	147–153	150	12:00–18:00	708
	EM16–2	172–183	178	12:00–19:00	425
				19:00–24:00	283
			179	0:00–12:00	
	EM16–3	202–213	206	12:00–17:00	227
				17:00–24:00	708
			207	0:00–12:00	
				12:00–24:00	340
			208	0:00–12:00	

2.2. The 1-D Model

To simulate freshwater systems at a global scale [31], the General Lake Model (GLM) was developed as a community 1-D hydrodynamic model for enclosed aquatic systems [32]. The open source model has been applied to natural and managed systems from wetlands and ponds to deep lakes with diverse climate conditions [33,34]. With continuous development of the GLM software and a growing modeling community, an increasing number of data visualization and processing tools are available for GLM [35,36].

GLM adopts a 1-D approach that resolves a series of horizontal layers [32], with core layer and mixing algorithms similar to the dynamic reservoir simulation model (DYRESM) [1] and the dynamic lake model (DLM) [37]. This approach defines each layer as a 'control volume' that can contract or expand in response to inflows/outflows and mixing with adjacent layers. The model solves the water balance in the lake domain, where the layers amalgamate, expand, contract or split due to water density changes caused by surface heating, vertical mixing and inflows/outflows. Regarding the energy balance approach, GLM estimates the amount of turbulent kinetic energy available for the surface mixed layer (surface mixing) and layers below the thermocline (deep mixing) respectively to simulate mixing dynamics. For surface mixing, the deepening rate of the surface mixed layer is calculated based on the balance between the available turbulent kinetic energy and the energy required for mixing to occur. The deep mixing is modeled using a characteristic vertical diffusivity, which is either approximated as a constant or using an equation introduced by Weinstock [38]. The vertical diffusivity is used to estimate the diffusion of scalars (including temperature and water-quality variables) between two neighboring layers.

Each layer also contains heat and other constituents, generally referred to as scalars. The scalars are conserved when layers change thickness or merge/split. The heat budget of the surface layer is determined by a balance of shortwave and longwave radiation fluxes as well as sensible and latent heat fluxes. In addition, a sediment heat module is available in GLM for modeling the heat exchange between the sediment and the water column. When the module is turned on, each water layer changes its temperature at a rate depending on the area in contact with the sediment and the local temperature gradient, as described by Hipsey et al. [32].

As already mentioned, coupled models are used to simulate the operation of the SSS and EM systems. GLM can include submerged inflows at any user-specified depth, which is used to simulate the SSS in FCR in the present investigation. The oxygenated water is first extracted then added to FCR at 8 m depth as a new layer.

A bubble plume module is enabled in GLM [39] to simulate epilimnion mixing. In this study, two variants of the bubble plume model are tested for GLM: the first bubble plume model detrains the entrained water at the depth of neutral buoyancy (GLM_DNB), and the second model detrains the entrained water at the depth of maximum plume rise (GLM_DMPR). The results of these two models are compared to determine which detrainment option simulates bubble plume mixing better in GLM, as described in Section 3 and Section S4 of Supplementary Materials.

GLM is pre-linked with the aquatic ecodynamic model (AED2) library to resolve the vertical profiles of water-quality variables of interest (such as oxygen, methane, etc.) [32]. GLM-AED2 was calibrated with field data collected over the period of 2013–2018 using manual adjustments and auto-calibrating scripts. The hydrodynamic model (GLM) was calibrated using Markov-chain Monte Carlo (MCMC) [40] tools, and the water-quality model (AED2) was calibrated by covariance matrix adaptation evolution strategy (CMA-ES) [41]. The version of GLM used in this study is 3.0.0 beta 10, and the version of AED2 is 1.3.4.

2.3. The 3-D Model

Si3D is a 3-D hydrodynamic model, which adopts a finite-difference method to obtain numerical solutions of the Navier–Stokes equations [42]. Si3D and its coupled models for simulating artificial mixing have been validated in previous studies of oxygenated reservoirs [12,43]. A coupled water-jet model with Si3D has been verified with field temperature data in FCR during oxygenation period [20]. The water-jet model accounts for both the momentum induced by the jet discharge and the ambient flow entrained by the expanding jet to resolve the small-scale jet flow within larger grid cells.

Details of the coupled bubble plume model with Si3D can be found in the literature [12,39], which has been verified for Spring Hollow Reservoir, VA, USA by Singleton et al. [12] and for FCR by Chen et al. [20]. Similar to the bubble plume models coupled with GLM, the bubble plume model in Si3D includes entrainment of ambient water and detrainment when the bubble-plume system is in operation. In this study, two variants of bubble plume models are tested for Si3D and the following acronyms are used to differentiate these two bubble plume models: Si3D_DNB for the plume model detraining the entrained water at the depth of neutral buoyancy; and Si3D_DMPR for the plume model detraining the entrained water at the depth of maximum plume rise (DMPR).

Si3D was coupled with AED2 through the Framework of Aquatic Biogeochemistry Model Library (FABM) [44]. FABM acts as an interface between the hydrodynamic host model (Si3D) and the coupled biogeochemical model (AED2). At each time-step, FABM first refers physical data (e.g., temperature, pressure) and tracer states (e.g., oxygen, phytoplankton) from the hydrodynamic host model to the coupled biogeochemical model, then the ecological variables (e.g., oxygen, phytoplankton) modelled by the biogeochemical model are returned to the hydrodynamic host model as biogeochemical source-sink terms [29]. With this coupling, the modules in AED2 can interactively simulate a range of water-quality variables of interest, including DO and sediment flux. The version of AED2 library coupled with Si3D is 1.3.4.

3. Numerical Tests

To determine appropriate spatial and temporal resolutions for the 1-D and 3-D models, GLM-AED2 time-step dependence test and Si3D-AED2 grid and time-step dependence tests are carried out.

In the existing studies using 1-D models including GLM, hourly time-steps are usually adopted [32]. These studies have demonstrated the capability of the 1-D models for correctly simulating water body temperature using hourly time-steps, confirming the high computational efficiency of the 1-D models. However, no GLM-AED2 time-step dependence test has been reported to reveal how model performance varies with the time-step. The purpose of the time-step dependence test is to examine the impact of the time-step on the simulated lake thermal structure. A number of time-steps ranging from one hour to one day are tested. Time-steps shorter than one hour are not tested here since the time resolution of meteorological data is one hour. These test runs with different time-steps simulate the temperature variation in FCR from 31 March (day of year—DoY 90) to 31 July (DoY 212) in 2015 without inflow/outflow and oxygenation. The details of the tested time-steps and the corresponding test results are presented in Section S1 of Supplementary Materials. Based on the time-step dependence test, the 3600 s time-step is adopted for the subsequent GLM-AED2 model runs due to its good compromise between the computational efficiency and numerical accuracy.

The sensitivity of the 3-D modeling results to the grid resolution and time-step is also examined to determine a suitable grid resolution and time-step setting for subsequent calculations. Starting with a relatively coarse grid of 10 m $\times$ 40 m $\times$ 0.6 m and a relatively large time-step of 10 s, the grid is refined twice by halving the cell sizes and in the meantime, the time-step is also halved twice to keep the CFL (Courant–Freidrich–Lewy) number [45] the same among the three sets of calculations. The tests run with the different cell sizes and time-steps calculate the temperature and DO in FCR using Si3D-AED2 over the period of 31 March (DoY 90) to 27 November (DoY 331) in 2015 without inflow/outflow and oxygenation. Further details of the test settings and the corresponding results are presented in Section S2 of Supplementary Materials. Based on these test results, the 5 m $\times$ 20 m $\times$ 0.3 m cell size is adopted for subsequent Si3D-AED2 simulations to ensure good accuracy in both stratified and mixed periods. Here, the time period between the spring and fall turnover is referred to as the stratified period, and the rest of the year is referred to as the mixed period. Spring and fall turnover is defined as the days when the temperature at 1 m equals the temperature at 8 m using observations made every 15 min throughout the monitoring period by two optical INW DO2 DO sondes (Seametrics, WA, USA) [46].

In addition to the dependence on the spatial and temporal resolutions, different model configurations may also affect the numerical results. The sediment heat module is an optional module available for GLM-AED2, which accounts for the heat transfer between the sediment and the water body [32]. For a shallow reservoir like FCR, the impact of the water/sediment heat exchange may be significant. Accordingly, GLM-AED2 is tested with and without the sediment heat module. The test results are given in Section S3 of Supplementary Materials. Based on this test, the configuration with the sediment heat module turned on will be adopted for comparison with Si3D-AED2, which assumes zero heat flux at the water-sediment interface.

Further, both GLM-AED2 and Si3D-AED2 have two detrainment options for the coupled bubble plume models, denoted by GLM_DNB, GLM_DMPR, Si3D_DNB and Si3D_DMPR respectively (refer to Section S4 of Supplementary Materials for further information). GLM_DNB and Si3D_DNB detrain the entrained plume water at the depth of neutral buoyancy, while GLM_DMPR and Si3D_DMPR detrain the entrained plume water at the depth of maximum plume rise. The two detrainment options are tested for the

1-D and 3-D models to identify the better options for GLM-AED2 and Si3D-AED2. The comparisons are given in Section S4. Based on this test, GLM_DNB and Si3D_DNB are adopted for comparison in the following section.

4. Results and Discussion

4.1. Year 2014 and 2015 Simulations

Information about the simulation periods, the oxygenation durations and the corresponding flow rates in the years 2014 and 2015 has been presented in Section 2.1, Table 1. The 2014 simulation covers 152 days with the oxygenation system turned on and off three times over the period, while the 2015 simulation covers 241 days with the oxygenation system turned on for most of the period. The numerically obtained temperature and DO data for these two years is analyzed to study the performance of the 1-D and 3-D models under both dynamic and steady-state oxygenation conditions and for both the stratified and mixed periods. To evaluate the performance of the models, the RMSEs and NMAEs of the temperature and DO are calculated using Equations (1) and (2) respectively:

$$RMSE = \sqrt[2]{\frac{\sum_1^n \left(v_{field} - v_{sim}\right)^2}{n}} \quad (1)$$

$$NMAE = \frac{\sum_1^n \left|v_{field} - v_{sim}\right|}{n\overline{v}_{field}} \quad (2)$$

where v_{field} and v_{sim} are respectively the field measurement and simulated data of the variable of interest, and n is the total number of field data points.

In what follows, we examine the RMSEs of the temperature and DO at different depths and the whole-lake NMAEs based on the 2014 and 2015 simulations. For 2015, the RMSEs of the temperature and DO is calculated separately for the stratified and mixed periods.

4.2. Qualitative Comparison of Simulated Temperature and DO Profiles of 2014 and 2015

The temperature and DO concentration profiles obtained from the 2014 and 2015 simulations are shown in Figure 2 (temperature) and Figure 3 (DO) respectively. Although the same initial and surface/flow boundary conditions are applied to the GLM-AED2 and Si3D-AED2 models, it is clear in Figures 2 and 3 that there are major variations between the 1-D and 3-D model results, and they are also different from the field data to certain extent.

It can be observed in Figure 2 that GLM-AED2 simulates hypolimnion temperature better, whereas Si3D-AED2 simulates surface mixed layer temperature better. In terms of DO plots of the two models (refer to Figure 3), Si3D-AED2 and GLM-AED2 yield similar results. Both models manage to simulate anoxia in hypolimnion during stratified period (blue color in the field DO plot) and the hypolimnion oxygenation due to SSS operation (green color in the hypolimnion in the field DO plot). From visual observation, Si3D-AED2 simulates the field DO slightly better than GLM-AED2 for relatively deep water below the epilimnion.

While the different heat boundary conditions at the sediment/water interface (refer to Section 3 and Supplementary Materials) contribute to the variation of the simulation results between GLM-AED2 and Si3D-AED2, the main cause of the variation is considered to be related to the different solution methods of the 1-D and 3-D models: the 1-D model adopts an energy balance approach, with which the mixing dynamics are based on the estimation of the available amount of turbulent kinetic energy [32]; and the 3-D model solves the full momentum and energy equations. In addition, GLM-AED2 adopts a Lagragian method, with water layers that move vertically, expand or contract [4], whereas Si3D-AED2 adopts an Eulerian method, with water layers of fixed thicknesses and volumes [42]. The difference in these two approaches helps to explain the different results that GLM-AED2 and Si3D-AED2 have produced in the 2014 and 2015 simulations.

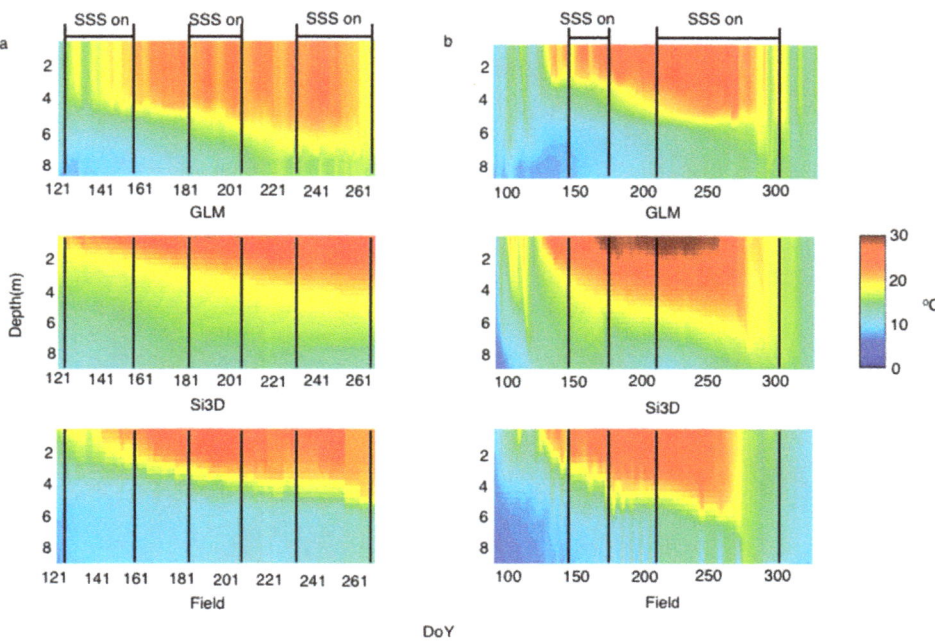

Figure 2. Comparisons of the one-dimensional (1-D, sediment module on, top row) and three dimensional (3-D) model temperature results (middle row) with field data (bottom row). (**a**) Temperature plot of 2014. (**b**) Temperature plot of 2015.

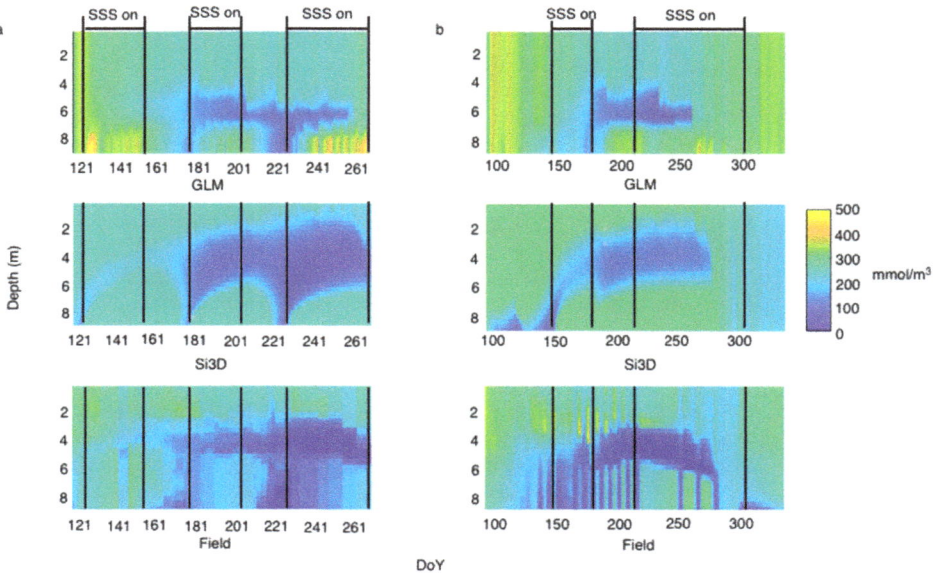

Figure 3. Comparisons of the 1-D (sediment module on, top row) and 3-D model DO results (middle row) with field data (bottom row). (**a**) Dissolved oxygen (DO) plot of 2014. (**b**) DO plot of 2015.

To quantitatively evaluate the relative performance of the 1-D and 3-D models, the RMSEs and NMAEs of the predicted temperature and DO by GLM-AED2 and Si3D-AED2 for the 2014 and 2015 simulation periods are presented in the following section.

4.3. Quantitative Comparison of Simulated Temperature and DO Profiles of Year 2014 and 2015

To quantitatively compare the performances of the 1-D and 3-D models, temperature and DO RMSEs of 2014 and 2015 simulations at the depths of 0.1 m, 3 m, 6 m and 9 m and for the whole-lake are presented in Tables 3 and 4. Further, to investigate the performance of the 1-D and 3-D models at all depths between 0.1–9 m, profiles of the temperature and DO RMSEs of both models for 2014 and 2015 mixed periods are shown in Figure 4.

Referring to Tables 3 and 4, the GLM-AED2 temperature results have relatively lower whole-lake RMSEs for 2014 and 2015 than those of Si3D-AED2 (over both the stratified and mixed periods). In 2014, and the mixed period of 2015, GLM-AED2 produces lower temperature RMSEs at almost all depths from 0.1 m to 9.0 m (Figure 4) and more than 50% lower whole-lake temperature RMSEs in the mixed period of 2015. The surprisingly good GLM-AED2 temperature prediction indicates the capability of the 1-D model to simulate temperature equally well or even better than the 3-D model in certain circumstances. A similar observation was made by Ladwig et al. [47], who found the 1-D model favorable for a long simulation period after studying summer anoxia dynamics in a eutrophic lake for a 37-year period. They suggested that 1-D model was adequate due to its lower computational needs and limited field data although the spatial extent of summer anoxia is fundamentally 3-D [48].

Table 3. The root mean square errors (RMSEs) of the predicted temperatures (in °C) at different depths for 2014 and 2015. The lower RMSEs between the 1-D and 3-D model results for both stratified and mixed conditions are shown in bold.

	1-D			3-D		
	2014 *	2015		2014 *	2015	
		Stratified	Mixed		Stratified	Mixed
0.1 m	**1.05**	**1.18**	0.92	2.64	3.40	4.07
3.0 m	**1.68**	**2.39**	1.06	1.64	1.64	3.25
6.0 m	**2.54**	**3.12**	0.92	3.17	4.54	3.03
9.0 m	2.45	**1.42**	1.36	**1.20**	3.84	3.21
Whole lake **	**1.99**	**1.98**	1.17	2.28	3.14	3.25

* FCR is stratified during the 2014 simulation period. ** The whole-lake RMSEs calculate the RMSEs of the variable of interest over 0.1–9 m depths with 1 m resolution.

Table 4. The RMSEs of the predicted DO (in mmol/m^3) at different depths for 2014 and 2015. Lower RMSEs between the 1-D and 3-D model results are shown in bold.

	1-D			3-D		
	2014 *	2015		2014 *	2015	
		Stratified	Mixed		Stratified	Mixed
0.1 m	**15.87**	**36.29**	73.63	43.36	51.63	56.24
3.0 m	78.71	130.04	**48.34**	**38.51**	**70.42**	57.91
6.0 m	86.88	**59.80**	**42.02**	**59.23**	88.37	95.42
9.0 m	103.25	**143.63**	**64.93**	**95.95**	153.58	163.59
Whole lake **	72.84	**96.50**	**56.31**	**56.59**	89.44	98.79

* FCR is stratified during the 2014 simulation period. ** The whole-lake RMSEs calculate the RMSEs of the variable of interest over 0.1–9 m depths with 1 m resolution.

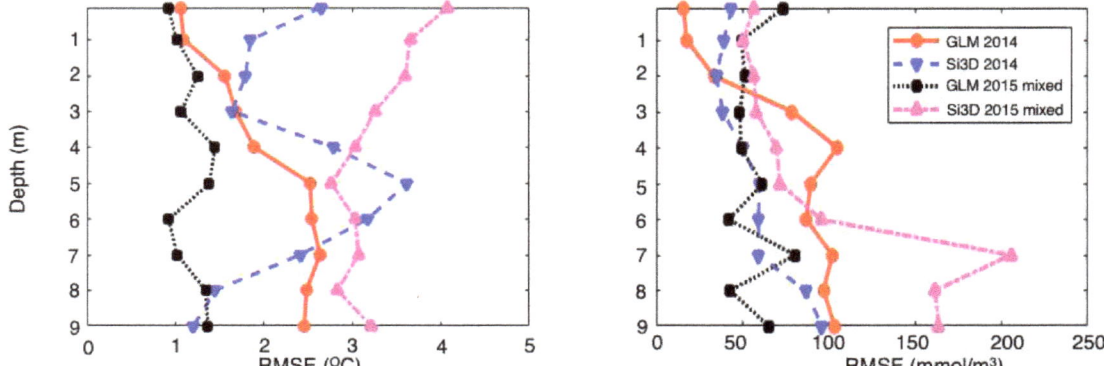

Figure 4. Profiles of the RMSEs of the temperature (left) and DO (right) RMSEs of the 1-D and 3-D models for 2014 and the mixed period of 2015.

Si3D-AED2 produces less accurate results for the temperature in 2015. It should be noted that SSS was operating during most of the stratified period in 2015. Therefore, 2015 represents a relatively stable engineering scenario without changing from an anoxic hypolimnion to an oxic one during the simulation period. In this scenario, the temperature and DO are not likely to vary temporally to a great extent, which is suitable for the 1-D model. Regarding the higher DO RMSEs yielded by GLM-AED2, it should be noted that, although the temperature does not vary much spatially within a small reservoir like FCR [28], DO is spatially sensitive within the shallow and eutrophic reservoir due to spatially varying water column depths and sediment composition [46]. This explains why a better temperature result, but not DO result, is obtained with GLM-AED2 compared to that with Si3D-AED2.

Regarding the better DO result obtained by Si3D-AED2, Si3D-AED2 produces lower whole-lake RMSEs of the DO than GLM-AED2 for both 2014 and the stratified period of 2015. It is also seen in Figure 4 that Si3D-AED2 yields lower RMSEs of the DO at all depths below 3.0 m than GLM-AED2 in 2014. However, for the mixed period of 2015, the comparison is entirely the opposite. GLM-AED2 produces lower RMSEs of the DO at all depths below 3.0 m compared to Si3D-AED2 (Figure 4). Combined with the temperature comparison above, it indicates that GLM-AED2 has a better performance than Si3D-AED2 in the mixed period. Given that water temperature and DO in the water column do not vary much with depth in the mixed period, this provides further support for recommending the 1-D model for relatively stable engineering scenarios.

In addition to the RMSEs of the simulated temperature and DO for the whole-lake and various depths in 2014 and 2015 by both Si3D-AED2 and GLM-AED2, the whole-lake NMAEs are also calculated (Table 5). It is seen in Table 5 that the simulated temperature and DO in 2014 and 2015 by both Si3D-AED2 and GLM-AED2 are within the reported ranges in the literature [8,9,49]. This result indicates the validity of adopting calibrated GLM-AED2 parameters for Si3D-AED2 to produce satisfying water-quality results. The present finding is supported by the study of McDonald et al. [50], who developed a 1-D model as an efficient test-bed environment, in which model parameters were estimated by Markov chain Monte Carlo approach for a 3-D model for Lake Superior. The calibrated 3-D model reproduced major features of the observed concentration profiles of nutrients, dissolved organic carbon and chlorophyll at the calibration location.

Table 5. The whole-lake normalized mean absolute error (NMAEs) of the predicted temperature and DO for 2014 and 2015. The literature range [8,9,49] is shown in the last row. The relatively lower NMAEs between the 1-D and 3-D model results are shown in bold.

	1-D			3-D			Literature Range
	2014 *	2015		2014 *	2015		
		Stratified	Mixed		Stratified	Mixed	
Temperature	**0.072**	**0.081**	0.11	0.084	0.12	0.12	0.037–0.12
DO	0.29	0.29	**0.18**	**0.24**	**0.23**	0.32	0.054–0.33

* FCR is stratified during the Year 2014 simulation period.

4.4. Quantitative Analysis of the Predicted Thermal Structures

To further assess the performance of Si3D-AED2 and GLM-AED2, the RMSEs of the predicted thermocline depth and metalimnion bottom depth for 2015 are calculated, and the results are presented in Table 6. Thermocline is the layer with the largest density gradient, and the metalimnion is the layer with the steepest thermal gradient in a stratified water body [36]. It is seen in Table 6 that the simulated thermocline depths and metalimnion bottom depths of Si3D-AED2 have similar RMSEs as those of GLM-AED2 in 2014 and have lower RMSEs than those of GLM-AED2 in 2015. The lower RMSEs in 2015 indicate that Si3D-AED2 simulates the reservoir heat structure in the stratification period better. It supports the finding in Section 4.3 that Si3D-AED2 results are more accurate than GLM-AED2 in relatively sophisticated engineering scenarios (referring to the stratified period in the previous section). To further evaluate the validity of this finding, GLM-AED2 and Si3D-AED2 simulations are compared with artificial mixing as another example of sophisticated engineering scenarios in the following section (Section 4.5).

Table 6. The RMSEs of the predicted thermocline depths and the metalimnion bottom depths for the entire stratified period of 2014 and 2015 from Si3D and GLM-AED2 simulations. The relatively lower RMSEs between the 1-D and 3-D model results are shown in bold.

	Si3D-AED2		GLM-AED2	
Year	2014	2015	2014	2015
RMSE of the thermocline depth (m)	1.7	**1.2**	1.7	3.0
RMSE of the metalimnion bottom depth (m)	2.4	**0.92**	2.3	2.2

4.5. Comparison of GLM-AED2 and Si3D-AED2 with EM

Table 2 presents the simulation dates of Si3D-AED2 and GLM-AED2, the EM operation periods and the corresponding flow rates in 2015 and 2016. The model results are compared with the field data in Figures 5 and 6.

Figures 5 and 6 respectively show the field and simulated temperature contours of EM15 and EM16–3. For both EM15 and EM16–3, the bubble plume model coupled with GLM-AED2 and Si3D-AED2 is able to simulate the EM mixing effect [20], reasonably representing the deepening of the surface mixed layer. For EM15, the Si3D-AED2 temperature contours agree well with the field temperature structure and are evidently better than the GLM-AED2 temperature contours since GLM-AED2 underestimates the surface mixed layer temperature. For EM16–3, the situation is different. After EM is turned on (after DoY 206), Si3D-AED2 over-predicts mixing in epilimnion, while the thermal structure predicted by GLM-AED2 is similar to the field thermal structure. One possible explanation for the conflicting predictions of EM15 and EM16–3 is that the time resolutions for the GLM-AED2 and Si3D-AED2 plots are both 1 h, but the field data time resolutions are much coarser (see the black triangles in Figures 5 and 6). With such a coarse temporal resolution, the field temperature contours may not reflect the characteristics of the transient thermal structures in the field during the EM periods. Quantitative analysis will be carried out below to compare between GLM-AED2 and Si3D-AED2 (Figure 7 and Table 7).

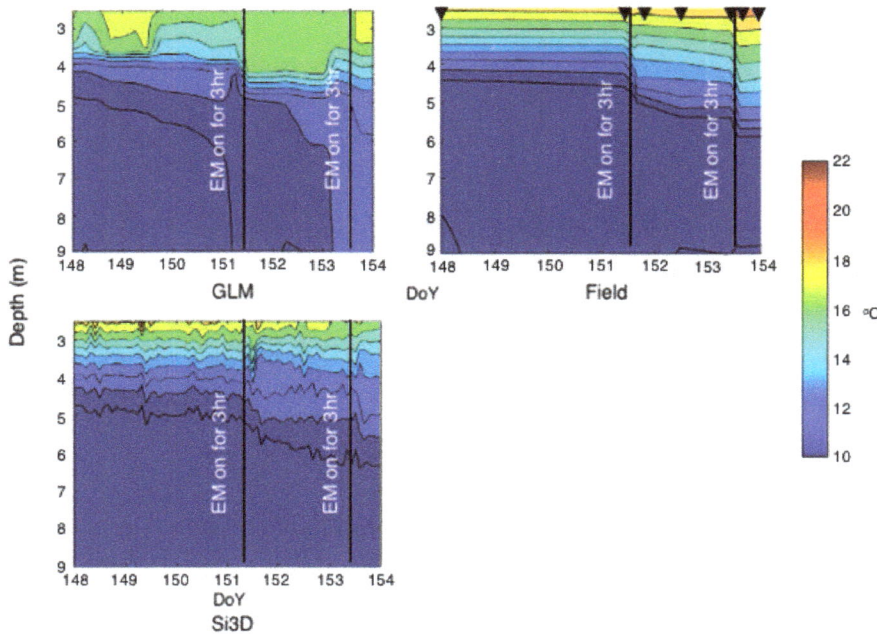

Figure 5. Comparison of the simulated temperature and field data for EM15 at FCR50 (year 2015). The time of the field data collection is indicated by black inverse triangles on the plot of the field data.

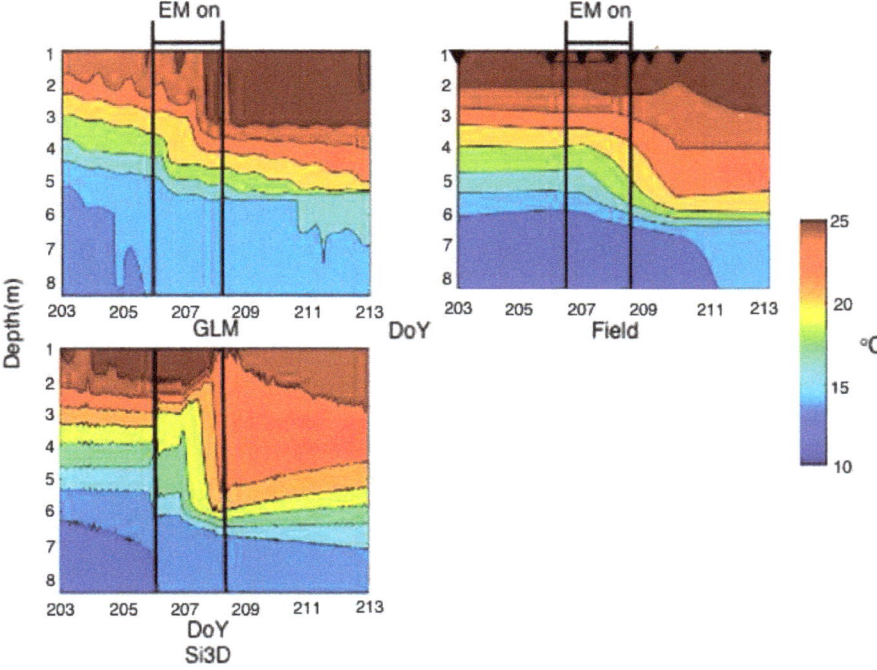

Figure 6. Comparison of the simulated temperature and field data for EM16–3 at FCR50. The time of the field data collection is indicated by black inverse triangles on the plot of the field data.

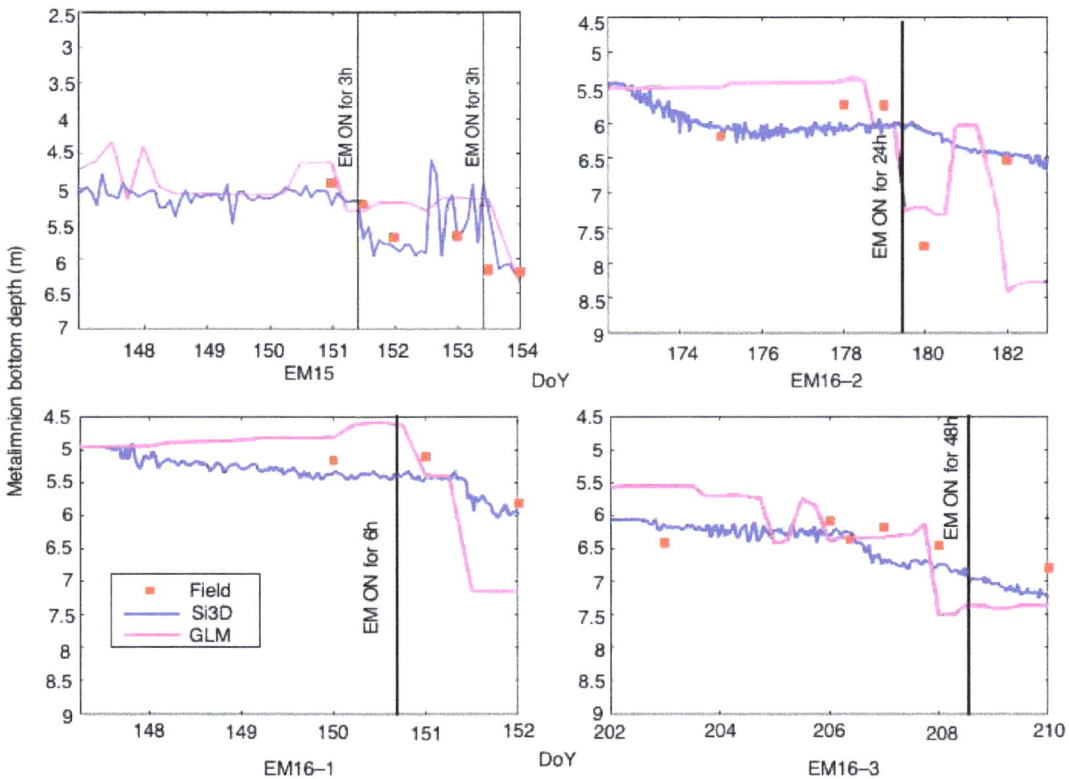

Figure 7. The field and simulated metalimnion bottom depths during the EM periods of 2015 and 2016.

Table 7. RMSEs (in m) of simulated metalimnion bottom depths during the EM periods of 2015 and 2016.

	EM15	EM16–1	EM16–2	EM16–3	Weighted Average
GLM	0.6	0.6	1	0.7	0.7
Si3D	0.5	0.2	0.8	0.3	0.5

To further evaluate the performance of the 1-D and 3-D models with EM, the metalimnion bottom depths are calculated for the field and simulated data, which are plotted in Figure 7. While both GLM-AED2 and Si3D-AED2 reasonably simulate the increase of metalimnion bottom depth during EM operation, it is clear in Figure 7 that Si3D-AED2 predicts the plume temperature structure much better than GLM-AED2 for all the EM periods. Quantitative analysis presented in Table 7 also proves this with lower RMSEs of the simulated metalimnion bottom depth in all four simulations by Si3D compared to those by GLM. This observation indicates that Si3D-AED2 instead of GLM-AED2 should be adopted if artificial mixing is in operation. A similar conclusion was drawn by Chen et al. [20], who studied the 3-D effect of artificial mixing on thermal structures in FCR and found that EM leads to 3-D heterogeneity in the thermal structures, which is impossible to capture by 1-D models. Accordingly, they concluded that 3-D models are more suitable than 1-D models for the design and operation of engineered systems although artificial mixing in lakes and reservoirs has mostly been studied with 2-D models [12,51]. The better performance of 3-D models may be due to their capability to simulate the horizontal momentum added by the injected bubble, while 1-D models do not include horizontal mixing.

4.6. Performance of Si3D-AED2

Table 8 presents the NMAEs of the predicted temperature and DO concentration at FCR20, FCR30 and FCR45 for the 2014 simulation using the 3-D model. The average NMAE is 0.080 for temperature and 0.18 for DO, which compare with the NMAE ranges of 0.037–0.12 for temperature and 0.054–0.33 for DO reported in the literature [8,9,49]. It shows that Si3D-AED2 simulates 3-D temperature and DO well even in a long simulation period.

Table 8. The NMAEs of the temperature and DO of the 3-D model at FCR20, FCR30 and FCR45 for the 2014 simulation.

	FCR20	FCR30	FCR45	Average
Temperature	0.077	0.077	0.085	0.080
DO	0.17	0.16	0.20	0.18

5. Conclusions

Choosing between 1-D and/or 3-D coupled hydrodynamic and water-quality models for efficient and accurate simulations of natural water bodies is important. In certain circumstances, for example, where the water body is managed to control water quality, it is reasonable to adopt both 1-D and 3-D models for one water body but for different simulation periods. To determine an ideal simulation setup for different engineering scenarios, a community 1-D model and a 3-D hydrodynamic model are coupled with artificial mixing models and the same water-quality model library. The bubble plume model, which is coupled with the community 1-D model for the first time in this study, adequately simulates the bubble plume mixing effect. It is found that the 3-D coupled model adopting the calibrated water-quality parameters from the 1-D model produces satisfactory water-quality results. Based on the comparison between two one-year simulation results using the 1-D and 3-D models for a shallow, eutrophic, managed reservoir, the relative performance of these models is evaluated quantitatively. The 1-D model is recommended when stratification and artificial mixing do not substantially vary during the simulation period, while the 3-D model is recommended to simulate stratification, artificial mixing and spatially sensitive water-quality variables during dynamic periods.

Supplementary Materials: The following are available online at https://www.mdpi.com/2073-4441/13/1/88/s1. The details and results of the numerical tests described in Section 3 are presented in Supplementary Materials.

Author Contributions: The following statements should be used Conceptualization, X.M., C.L., C.C.C., J.C.L.; methodology, X.M., C.L., C.C.C., J.C.L.; software, X.M., C.C.C.; validation, X.M., C.C.C.; formal analysis, X.M., C.C.C.; data curation, X.M, C.C.C.; writing—original draft preparation, X.M.; writing—review and editing, X.M., C.L., C.C.C., J.C.L.; visualization, X.M.; supervision, C.L., C.C.C., J.C.L.; project administration, C.L., C.C.C., J.C.L. All authors have read and agreed to the published version of the manuscript.

Funding: This research was supported by U.S. National Science Foundation grants 1737424 and 1753639.

Institutional Review Board Statement: Not applicable.

Informed Consent Statement: Not applicable.

Data Availability Statement: Software and data used for this study is available through the GitHub repositories. https://github.com/xman2403/1D3D_codes.

Acknowledgments: The authors thank the Reservoir Group at Virginia Tech and FCR Carbon team for field data collection and feedback.

Conflicts of Interest: The authors declare no conflict of interest.

References

1. Patterson, J.C.; Imberger, J. Simulation of bubble plume destratification systems in reservoirs. *Aquat. Sci.* **1989**, *51*, 3–18. [CrossRef]
2. Stepanenko, V.M.; Martynov, A.; Joehnk, K.D.; Subin, Z.M.; Perroud, M.; Fang, X.; Beyrich, F.; Mironov, D.; Goyette, S. A one-dimensional model intercomparison study of thermal regime of a shallow, turbid midlatitude lake. *Geosci. Model Dev.* **2013**, *6*, 1337–1352. [CrossRef]
3. Gal, G.; Imberger, J.; Zohary, T.; Antenucci, J.; Anis, A.; Rosenberg, T. Simulating the thermal dynamics of Lake Kinneret. *Ecol. Model.* **2003**, *162*, 69–86. [CrossRef]
4. Hamilton, D.P.; Schladow, S. Prediction of water quality in lakes and reservoirs. Part I—Model description. *Ecol. Model.* **1997**, *96*, 91–110. [CrossRef]
5. Hu, F.; Bolding, K.; Bruggeman, J.; Jeppesen, E.; Flindt, M.R.; Van Gerven, L.P.A.; Janse, J.H.; Janssen, A.B.G.; Kuiper, J.J.; Mooij, W.M.; et al. FABM-PCLake—Linking aquatic ecology with hydrodynamics. *Geosci. Model Dev.* **2016**, *9*, 2271–2278. [CrossRef]
6. Peeters, F.; Traile, D.I.S.; Orke, A.L.; Ivingstone, D.M.L. Earlier onset of the spring phytoplankton bloom in lakes of the temperate zone in a warmer climate. *Glob. Chang. Biol.* **2007**, *13*, 1898–1909. [CrossRef]
7. Perroud, M.; Goyette, S.; Martynov, A.; Beniston, M.; Annevillec, O. Simulation of multiannual thermal profiles in deep Lake Geneva: A comparison of one-dimensional lake models. *Limnol. Oceanogr.* **2009**, *54*, 1574–1594. [CrossRef]
8. Farrell, K.J.; Ward, N.K.; Krinos, A.I.; Hanson, P.C.; Daneshmand, V.; Figueiredo, R.J.; Carey, C.C. Ecosystem-scale nutrient cycling responses to increasing air temperatures vary with lake trophic state. *Ecol. Model.* **2020**, *430*, 109134. [CrossRef]
9. Snortheim, C.A.; Hanson, P.C.; McMahon, K.D.; Read, J.S.; Carey, C.C.; Dugan, H.A. Meteorological drivers of hypo-limnetic anoxia in a eutrophic, north temperate lake. *Ecol. Model.* **2017**, *343*, 39–53. [CrossRef]
10. Ward, N.K.; Steele, B.G.; Weathers, K.C.; Cottingham, K.L.; Ewing, H.A.; Hanson, P.C.; Carey, C.C. Differential Responses of Maximum Versus Median Chlorophyll- a to Air Temperature and Nutrient Loads in an Oligotrophic Lake Over 31 Years. *Water Resour. Res.* **2020**, *56*. [CrossRef]
11. McKinley, G.; Follows, M.J.; Marshall, J. Mechanisms of air-sea CO2flux variability in the equatorial Pacific and the North Atlantic. *Glob. Biogeochem. Cycles* **2004**, *18*, 2011. [CrossRef]
12. Singleton, V.L.; Gantzer, P.; Little, J.C. Linear bubble plume model for hypolimnetic oxygenation: Full-scale validation and sensitivity analysis. *Water Resour. Res.* **2007**, *43*. [CrossRef]
13. Six, K.D.; Maier-Reimer, E. Effects of plankton dynamics on seasonal carbon fluxes in an ocean general circulation model. *Glob. Biogeochem. Cycles* **1996**, *10*, 559–583. [CrossRef]
14. Cerco, C.F.; Cole, T. Three-Dimensional Eutrophication Model of Chesapeake Bay. *J. Environ. Eng.* **1993**, *119*, 1006–1025. [CrossRef]
15. Follows, M.J.; Dutkiewicz, S.; Grant, S.; Chisholm, S.W. Emergent Biogeography of Microbial Communities in a Model Ocean. *Science* **2007**, *315*, 1843–1846. [CrossRef]
16. Hoyer, A.B.; Wittmann, M.E.; Chandra, S.; Schladow, S.G.; Rueda, F.J. A 3D individual-based aquatic transport model for the assessment of the potential dispersal of planktonic larvae of an invasive bivalve. *J. Environ. Manag.* **2014**, *145*, 330–340. [CrossRef]
17. Acosta, M.; Anguita, M.; Fernández-Baldomero, F.J.; Ramón, C.L.; Schladow, S.G.; Rueda, F.J. Evaluation of a nested-grid implementation for 3D finite-difference semi-implicit hydrodynamic models. *Environ. Model. Softw.* **2015**, *64*, 241–262. [CrossRef]
18. Preece, E.P.; Moore, B.C.; Skinner, M.M.; Child, A.; Dent, S. A review of the biological and chemical effects of hypo-limnetic oxygenation. *Lake Reserv. Manag.* **2019**, *35*, 229–246. [CrossRef]
19. Singleton, V.L.; Little, J.C. Designing Hypolimnetic Aeration and Oxygenation Systems—A Review. *Environ. Sci. Technol.* **2006**, *40*, 7512–7520. [CrossRef]
20. Chen, S; Little, J.C.; Carey, C.C.; McClure, R.P.; Lofton, M.E.; Lei, C. Three-Dimensional Effects of Artificial Mixing in a Shallow Drinking-Water Reservoir. *Water Resour. Res.* **2018**, *54*, 425–441. [CrossRef]
21. Fleischmann, A.S.; Paiva, R.C.D.; Collischonn, W.; Siqueira, V.A.; Paris, A.; Moreira, D.M.; Papa, F.; Bitar, A.A.; Parrens, M.; Aires, F.; et al. Trade-Offs Between 1-D and 2-D Regional River Hydrodynamic Models. *Water Resour. Res.* **2020**, *56*, e2019WR026812. [CrossRef]
22. Romero, J.; Antenucci, J.; Imberger, J. One- and three-dimensional biogeochemical simulations of two differing reservoirs. *Ecol. Model.* **2004**, *174*, 143–160. [CrossRef]
23. Saloranta, T.M.; Andersen, T. MyLake—A multi-year lake simulation model code suitable for uncertainty and sensitivity analysis simulations. *Ecol. Model.* **2007**, *207*, 45–60. [CrossRef]
24. Alobaid, F.; Baraki, N.; Epple, B. Investigation into improving the efficiency and accuracy of CFD/DEM simulations. *Particuology* **2014**, *16*, 41–53. [CrossRef]
25. Andersson, A.G.; Hellström, J.G.I.; Andreasson, P.; Lundström, T.S. Effect of Spatial Resolution of Rough Surfaces on Numerically Computed Flow Fields with Application to Hydraulic Engineering. *Eng. Appl. Comput. Fluid Mech.* **2014**, *8*, 373–381. [CrossRef]
26. Gerling, A.B.; Munger, Z.W.; Doubek, J.P.; Hamre, K.D.; Gantzer, P.A.; Little, J.C.; Carey, C.C. Whole-Catchment Manipulations of Internal and External Loading Reveal the Sensitivity of a Century-Old Reservoir to Hypoxia. *Ecosystems* **2016**, *19*, 555–571. [CrossRef]
27. Munger, Z.W.; Carey, C.C.; Gerling, A.B.; Doubek, J.P.; Hamre, K.D.; McClure, R.P.; Schreiber, M.E. Oxygenation and hydrologic controls on iron and manganese mass budgets in a drinking-water reservoir. *Lake Reserv. Manag.* **2019**, *35*, 277–291. [CrossRef]

28. Gerling, A.B.; Browne, R.G.; Gantzer, P.A.; Mobley, M.H.; Little, J.C.; Carey, C.C. First report of the successful operation of a side stream supersaturation hypolimnetic oxygenation system in a eutrophic, shallow reservoir. *Water Res.* **2014**, *67*, 129–143. [CrossRef]
29. Chen, S.; Carey, C.C.; Little, J.C.; Lofton, M.E.; McClure, R.P.; Lei, C. Effectiveness of a bubble-plume mixing system for managing phytoplankton in lakes and reservoirs. *Ecol. Eng.* **2018**, *113*, 43–51. [CrossRef]
30. Hamre, K.D.; Lofton, M.E.; McClure, R.P.; Munger, Z.W.; Doubek, J.P.; Gerling, A.B.; Schreiber, M.E.; Carey, C.C. In situ fluorometry reveals a persistent, perennial hypolimnetic cyanobacterial bloom in a seasonally anoxic reservoir. *Freshw. Sci.* **2018**, *37*, 483–495. [CrossRef]
31. Hanson, P.C.; Weathers, K.C.; Kratz, T.K. Networked lake science: How the Global Lake Ecological Observatory Network (GLEON) works to understand, predict, and communicate lake ecosystem response to global change. *Inland Waters* **2016**, *6*, 543–554. [CrossRef]
32. Hipsey, M.R.; Bruce, L.C.; Boon, C.; Busch, B.; Carey, C.C.; Hamilton, D.P.; Hanson, P.C.; Read, J.S.; De Sousa, E.; Weber, M.; et al. A General Lake Model (GLM 3.0) for linking with high-frequency sensor data from the Global Lake Ecological Observatory Network (GLEON). *Geosci. Model Dev.* **2019**, *12*, 473–523. [CrossRef]
33. Bruce, L.C.; Frassl, M.A.; Arhonditsis, G.B.; Gal, G.; Hamilton, D.P.; Hanson, P.C.; Hetherington, A.L.; Melack, J.M.; Read, J.S.; Rinke, K.; et al. A multi-lake comparative analysis of the General Lake Model (GLM): Stress-testing across a global observatory network. *Environ. Model. Softw.* **2018**, *102*, 274–291. [CrossRef]
34. Read, J.S.; Winslow, L.A.; Hansen, G.J.; Hoek, J.V.D.; Hanson, P.C.; Bruce, L.C.; Markfort, C.D. Simulating 2368 temperate lakes reveals weak coherence in stratification phenology. *Ecol. Model.* **2014**, *291*, 142–150. [CrossRef]
35. Hamilton, D.P.; Carey, C.C.; Arvola, L.; Arzberger, P.; Brewer, C.; Cole, J.J.; Gaiser, E.; Hanson, P.C.; Ibelings, B.W.; Jennings, E.; et al. A Global Lake Ecological Observatory Network (GLEON) for synthesising high–frequency sensor data for validation of deterministic ecological models. *Inland Waters* **2015**, *5*, 49–56. [CrossRef]
36. Read, J.S.; Hamilton, D.P.; Jones, I.D.; Muraoka, K.; Winslow, L.A.; Kroiss, R.; Wu, C.H.; Gaiser, E. Derivation of lake mixing and stratification indices from high-resolution lake buoy data. *Environ. Model. Softw.* **2011**, *26*, 1325–1336. [CrossRef]
37. Chung, E.G.; Schladow, S.G.; Pérez-Losada, J.; Robertson, D.M. A linked hydrodynamic and water quality model for the Salton Sea. *Hydrobiologia* **2008**, *604*, 57–75. [CrossRef]
38. Weinstock, J. Vertical turbulence diffusivity for weak or strong stable stratification. *J. Geophys. Res. Space Phys.* **1981**, *86*, 9925. [CrossRef]
39. Wüest, A.; Brooks, N.H.; Imboden, D.M. Bubble plume modeling for lake restoration. *Water Resour. Res.* **1992**, *28*, 3235–3250. [CrossRef]
40. Haario, H.; Laine, M.; Mira, A.; Saksman, E. DRAM: Efficient adaptive MCMC. *Stat. Comput.* **2006**, *16*, 339–354. [CrossRef]
41. Hansen, N.; Müller, S.D.; Koumoutsakos, P. Reducing the Time Complexity of the Derandomized Evolution Strategy with Covariance Matrix Adaptation (CMA-ES). *Evol. Comput.* **2003**, *11*, 1–18. [CrossRef] [PubMed]
42. Smith, P.E. A Semi-Implicit, Three-Dimensional Model for Estuarine Circulation. In *Open-File Report*; U.S. Geological Survey: Reston, VA, USA, 2006.
43. Rueda, F.; Singleton, V.; Stewart, M.; Little, J.; Lawrence, G. Modeling the fate of oxygen artificially injected in the hypolimnion of a lake with multiple basins: Amisk Lake revisited, paper presented at Environmental Hydraulics. In Proceedings of the 6th International Symposium on Environmental Hydraulics, Athens, Greece, 23–25 June 2010; Volume 1, pp. 379–384. [CrossRef]
44. Bruggeman, J.; Bolding, K. A general framework for aquatic biogeochemical models. *Environ. Model. Softw.* **2014**, *61*, 249–265. [CrossRef]
45. Courant, R.; Friedrichs, K.; Lewy, H. On the Partial Difference Equations of Mathematical Physics. *IBM J. Res. Dev.* **1967**, *11*, 215–234. [CrossRef]
46. McClure, R.P.; Hamre, K.D.; Niederlehner, B.; Munger, Z.W.; Chen, S.; Lofton, M.E.; Schreiber, M.E.; Carey, C.C. Metalimnetic oxygen minima alter the vertical profiles of carbon dioxide and methane in a managed freshwater reservoir. *Sci. Total. Environ.* **2018**, *636*, 610–620. [CrossRef]
47. Ladwig, R.; Hanson, P.C.; Dugan, H.A.; Carey, C.C.; Zhang, Y.; Shu, L.; Duffy, C.J.; Cobourn, K.M. Lake thermal structure drives inter-annual variability in summer anoxia dynamics in a eutrophic lake over 37 years. *Hydrol. Earth Syst. Sci. Discuss* **2020**. in review. [CrossRef]
48. Biddanda, B.A.; Weinke, A.D.; Kendall, S.T.; Gereaux, L.C.; Holcomb, T.M.; Snider, M.J.; Dila, D.K.; Long, S.A.; Vandenberg, C.; Knapp, K.; et al. Chronicles of hypoxia: Time-series buoy observations reveal annually recurring seasonal basin-wide hypoxia in Muskegon Lake—A Great Lakes estuary. *J. Great Lakes Res.* **2018**, *44*, 219–229. [CrossRef]
49. Kara, E.L.; Hanson, P.C.; Hamilton, D.P.; Hipsey, M.R.; McMahon, K.D.; Read, J.S.; Winslow, L.A.; Dedrick, J.; Rose, K.; Carey, C.C.; et al. Time-scale dependence in numerical simulations: Assessment of physical, chemical, and biological predictions in a stratified lake at temporal scales of hours to months. *Environ. Model. Softw.* **2012**, *35*, 104–121. [CrossRef]
50. McDonald, C.P.; Bennington, V.; Urban, N.R.; McKinley, G.A. 1-D test-bed calibration of a 3-D Lake Superior biogeo-chemical model. *Ecol. Model.* **2012**, *225*, 115–126. [CrossRef]
51. Toffolon, M.; Serafini, M. Effects of artificial hypolimnetic oxygenation in a shallow lake. Part 2: Numerical modelling. *J. Environ. Manag.* **2013**, *114*, 530–539. [CrossRef]

MDPI
St. Alban-Anlage 66
4052 Basel
Switzerland
Tel. +41 61 683 77 34
Fax +41 61 302 89 18
www.mdpi.com

Water Editorial Office
E-mail: water@mdpi.com
www.mdpi.com/journal/water